Chemistry of
PETROCHEMICAL
PROCESSES

2nd Edition

Chemistry of
PETROCHEMICAL
PROCESSES

2nd Edition

Sami Matar
Lewis F. Hatch

G|P Gulf Professional Publishing
P|Y an imprint of Butterworth-Heinemann

Boston Oxford Johannesburg Melbourne New Delhi Singapore

This book is dedicated to the memory of Professor Lewis Hatch (1912–1991), a scholar, an educator, and a sincere friend.

Gulf Professional Publishing is an imprint of Butterworth–Heinemann.

Copyright © 2001 by Butterworth–Heinemann

 A member of the Reed Elsevier group

Recognizing the importance of preserving what has been written, Butterworth–Heinemann prints its books on acid-free paper whenever possible.

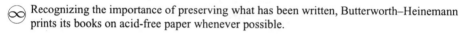 Butterworth–Heinemann supports the efforts of American Forests and the Global ReLeaf program in its campaign for the betterment of trees, forests, and our environment.

Library of Congress Cataloging-in-Publication Data

ISBN 0-88415-315-0

British Library Cataloguing-in-Publication Data
A catalogue record for this book is available from the British Library.

The publisher offers special discounts on bulk orders of this book.
For information, please contact:

Manager of Special Sales
Butterworth–Heinemann
225 Wildwood Avenue
Woburn, MA 01801-2041
Tel: 781-904-2500
Fax: 781-904-2620

For information on all Butterworth-Heinemann publications available, contact our World Wide Web home page at: http://www.bh.com

10 9 8 7 6 5 4 3 2 1

Printed in the United States of America

Contents

Preface to Second Edition

When the first edition of *Chemistry of Petrochemical Processes* was written, the intention was to introduce to the users a simplified approach to a diversified subject dealing with the chemistry and technology of various petroleum and petrochemical process. It reviewed the mechanisms of many reactions as well as the operational parameters (temperature, pressure, residence times, etc.) that directly effect products' yields and composition. To enable the readers to follow the flow of the reactants and products, the processes were illustrated with simplified flow diagrams.

Although the basic concept and the arrangement of the chapters is this second edition are the same as the first, this new edition includes many minor additions and updates related to the advances in processing and catalysis.

The petrochemical industry is a huge field that encompasses many commercial chemicals and polymers. As an example of the magnitude of the petrochemical market, the current global production of polyolefins alone is more than 80 billion tons per year and is expected to grow at a rate of 4–5% per year. Such growth necessitates much work be invested to improve processing technique and catalyst design and ensure good product qualities. This is primarily achieved by the search for new catalysts that are active and selective. The following are some of the important additions to the text:

- Because ethylene and propylene are the major building blocks for petrochemicals, alternative ways for their production have always been sought. The main route for producing ethylene and propylene is steam cracking, which is an energy extensive process. Fluid catalytic cracking (FCC) is also used to supplement the demand for these light olefins. A new process that produces a higher percentage of light olefins than FCC is deep catalytic cracking (DCC), and it is described in Chapter 3.

- The search for alternative ways to produce monomers and chemicals from sources other than oil, such as coal, has revived working using Fisher Tropseh technology, which produces in addition to fuels, light olefins, sulfur, phenols, etc. These could be used as feedstocks for petrochemicals as indicated in Chapter 4.
- Catalysts for many petroleum and petrochemical processes represent a substantial fraction of capital and operation costs. Heterogeneous catalysts are more commonly used due to the ease of separating the products. Homogeneous catalysts, on the other hand, are normally more selective and operate under milder conditions than heterogeneous types, but lack the simplicity and ease of product separation. This problem has successfully been solved for the oxo reaction by using rhodium modified with triphenylphosphine ligands that are water soluble. Thus, lyophilic products could be easily separated from the catalyst in the aqueous phase. A water soluble cobalt cluster can effectively hydroformylate higher olefins in a two-phase system using polyethylene glycol as the polar medium. This approach is described in Chapter 5.
- In the polymer filed, new-generation metallocenes, which are currently used in many polyethylene and polypropylene processes, can polymerize proplylene in two different modes: alternating blocks of rigid isotactic and flexible atactic. These new developments and other changes and approaches related to polymerization are noted in Chapters 11 and 12.

I hope the new additions that I felt necessary for updating this book are satisfactory to the readers.

Sami Matar, Ph.D.

Preface to First Edition

Petrochemicals in general are compounds and polymers derived directly or indirectly from petroleum and used in the chemical market. Among the major petrochemical products are plastics, synthetic fibers, synthetic rubber, detergents, and nitrogen fertilizers. Many other important chemical industries such as paints, adhesives, aerosols, insecticides, and pharmaceuticals may involve one or more petrochemical products within their manufacturing steps.

The primary raw materials for the production of petrochemicals are natural gas and crude oil. However, other carbonaceous substances such as coal, oil shale, and tar sand can be processed (expensively) to produce these chemicals.

The petrochemical industry is mainly based on three types of intermediates, which are derived from the primary raw materials. These are the C_2-C_4 olefins, the C_6-C_8 aromatic hydrocarbons, and synthesis gas (an H_2/CO_2 mixture).

In general, crude oils and natural gases are composed of a mixture of relatively unreactive hydrocarbons with variable amounts of nonhydrocarbon compounds. This mixture is essentially free from olefins. However, the C_2 and heavier hydrocarbons from these two sources (natural gas and crude oil) can be converted to light olefins suitable as starting materials for petrochemicals production.

The C_6-C_8 aromatic hydrocarbons—though present in crude oil—are generally so low in concentration that it is not technically or economically feasible to separate them. However, an aromatic-rich mixture can be obtained from catalytic reforming and cracking processes, which can be further extracted to obtain the required aromatics for petrochemical use. Liquefied petroleum gases (C_3-C_4) from natural gas and refinery gas streams can also be catalytically converted into a liquid hydrocarbon mixture rich in C6-C8 aromatics.

Synthesis gas, the third important intermediate for petrochemicals, is generated by steam reforming of either natural gas or crude oil fractions. Synthesis gas is the precursor of two big-volume chemicals, ammonia and methanol.

From these simple intermediates, many important chemicals and polymers are derived through different conversion reactions. The objective of this book is not merely to present the reactions involved in such conversions, but also to relate them to the different process variables and to the type of catalysts used to get a desired product. When plausible, discussions pertinent to mechanisms of important reactions are included. The book, however, is an attempt to offer a simplified treatise for diversified subjects dealing with chemistry, process technology, polymers, and catalysis.

As a starting point, the book reviews the general properties of the raw materials. This is followed by the different techniques used to convert these raw materials to the intermediates, which are further reacted to produce the petrochemicals. The first chapter deals with the composition and the treatment techniques of natural gas. It also reviews the properties, composition, and classification of various crude oils. Properties of some naturally occurring carbonaceous substances such as coal and tar sand are briefly noted at the end of the chapter. These materials are targeted as future energy and chemical sources when oil and natural gas are depleted. Chapter 2 summarizes the important properties of hydrocarbon intermediates and petroleum fractions obtained from natural gas and crude oils.

Crude oil processing is mainly aimed towards the production of fuels, so only a small fraction of the products is used for the synthesis of olefins and aromatics. In Chapter 3, the different crude oil processes are reviewed with special emphasis on those conversion techniques employed for the dual purpose of obtaining fuels as well as olefinic and aromatic base stocks. Included also in this chapter, are the steam cracking processes geared specially for producing olefins and diolefins.

In addition to being major sources of hydrocarbon-based petrochemicals, crude oils and natural gases are precursors of a special group of compounds or mixtures that are classified as nonhydrocarbon intermediates. Among these are the synthesis gas mixture, hydrogen, sulfur, and carbon black. These materials are of great economic importance and are discussed in Chapter 4.

Chapter 5 discusses chemicals derived directly or indirectly from methane. Because synthesis gas is the main intermediate from methane, it is again further discussed in this chapter in conjunction with the major chemicals based on it.

Higher paraffinic hydrocarbons than methane are not generally used for producing chemicals by direct reaction with chemical reagents due to their lower reactivities relative to olefins and aromatics. Nevertheless, a few derivatives can be obtained from these hydrocarbons through oxidation, nitration, and chlorination reactions. These are noted in Chapter 6.

The heart of the petrochemical industry lies with the C_2-C_4 olefins, butadiene, and C_6-C_8 aromatics. Chemicals and monomers derived from these intermediates are successively discussed in Chapters 7-10.

The use of light olefins, diolefins, and aromatic-based monomers for producing commercial polymers is dealt with in the last two chapters. Chapter 11 reviews the chemistry involved in the synthesis of polymers, their classification, and their general properties. This book does not discuss the kinetics of polymer reactions. More specialized polymer chemistry texts may be consulted for this purpose.

Chapter 12 discusses the use of the various monomers obtained from a petroleum origin for producing commercial polymers. Not only does it cover the chemical reactions involved in the synthesis of these polymers, but it also presents their chemical, physical and mechanical properties. These properties are well related to the applicability of a polymer as a plastic, an elastomer, or as a fiber.

As an additional aid to readers seeking further information of a specific subject, references are included at the end of each chapter. Throughout the text, different units are used interchangeably as they are in the industry. However, in most cases temperatures are in degrees celsius, pressures in atmospheres, and energy in kilo joules.

The book chapters have been arranged in a way more or less similar to *From Hydrocarbons to Petrochemicals,* a book I co-authored with the late Professor Hatch and published with Gulf Publishing Company in 1981. Although the book was more addressed to technical personnel and to researchers in the petroleum field, it has been used by many colleges and universities as a reference or as a text for senior and special topics courses. This book is also meant to serve the dual purpose of being a reference as well as a text for chemistry and chemical engineering majors.

In recent years, many learning institutions felt the benefits of one or more technically-related courses such as petrochemicals in their chemistry and chemical engineering curricula. More than forty years ago, Lewis Hatch pioneered such an effort by offering a course in "Chemicals from Petroleum" at the University of Texas. Shortly thereafter, the ter "petrochemicals" was coined to describe chemicals obtained from crude oil or natural gas.

I hope that publishing this book will partially fulfill the objective of continuing the effort of the late Professor Hatch in presenting the state of the art in a simple scientific approach.

At this point, I wish to express my appreciation to the staff of Gulf Publishing Co. for their useful comments.

I wish also to acknowledge the cooperation and assistance I received from my colleagues, the administration of KFUPM, with special mention of Dr. A. Al-Arfaj, chairman of the chemistry department; Dr. M. Z. El-Faer, dean of sciences; and Dr. A. Al-Zakary, vice-rector for graduate studies and research, for their encouragement in completing this work.

Sami Matar, Ph.D.

CHAPTER ONE

Primary Raw Materials
for Petrochemicals

INTRODUCTION

In general, primary raw materials are naturally occurring substances that have not been subjected to chemical changes after being recovered. Natural gas and crude oils are the basic raw materials for the manufacture of petrochemicals. The first part of this chapter deals with natural gas. The second part discusses crude oils and their properties.

Secondary raw materials, or intermediates, are obtained from natural gas and crude oils through different processing schemes. The intermediates may be light hydrocarbon compounds such as methane and ethane, or heavier hydrocarbon mixtures such as naphtha or gas oil. Both naphtha and gas oil are crude oil fractions with different boiling ranges. The properties of these intermediates are discussed in Chapter 2.

Coal, oil shale, and tar sand are complex carbonaceous raw materials and possible future energy and chemical sources. However, they must undergo lengthy and extensive processing before they yield fuels and chemicals similar to those produced from crude oils (substitute natural gas (SNG) and synthetic crudes from coal, tar sand and oil shale). These materials are discussed briefly at the end of this chapter.

NATURAL GAS
(Non-associated and Associated Natural Gases)

Natural gas is a naturally occurring mixture of light hydrocarbons accompanied by some non-hydrocarbon compounds. Non-associated natural gas is found in reservoirs containing no oil (dry wells). Associated gas, on the other hand, is present in contact with and/or dissolved in crude oil and is coproduced with it. The principal component of most

1

Table 1-1
Composition of non-associated and associated natural gases[1]

Component	Non-associated gas		Associated gas	
	Salt Lake US	Kliffside US	Abqaiq Saudi Arabia	North Sea UK
Methane	95.0	65.8	62.2	85.9
Ethane	0.8	3.8	15.1	8.1
Propane	0.2	1.7	6.6	2.7
Butanes	—	0.8	2.4	0.9
Pentane and Heavier	—	0.5	1.1	0.3
Hydrogen sulfide	—	—	2.8	—
Carbon dioxide	3.6	—	9.2	1.6
Nitrogen	0.4	25.6	—	0.5
Helium	—	1.8	—	—

natural gases is methane. Higher molecular weight paraffinic hydrocarbons (C_2-C_7) are usually present in smaller amounts with the natural gas mixture, and their ratios vary considerably from one gas field to another. Non-associated gas normally contains a higher methane ratio than associated gas, while the latter contains a higher ratio of heavier hydrocarbons. Table 1-1 shows the analyses of some selected non-associated and associated gases.[1] In our discussion, both non-associated and associated gases will be referred to as natural gas. However, important differences will be noted.

The non-hydrocarbon constituents in natural gas vary appreciably from one gas field to another. Some of these compounds are weak acids, such as hydrogen sulfide and carbon dioxide. Others are inert, such as nitrogen, helium and argon. Some natural gas reservoirs contain enough helium for commercial production.

Higher molecular weight hydrocarbons present in natural gases are important fuels as well as chemical feedstocks and are normally recovered as natural gas liquids. For example, ethane may be separated for use as a feedstock for steam cracking for the production of ethylene. Propane and butane are recovered from natural gas and sold as liquefied petroleum gas (LPG). Before natural gas is used it must be processed or treated to remove the impurities and to recover the heavier hydrocarbons (heavier than methane). The 1998 U.S. gas consumption was approximately 22.5 trillion ft^3.

NATURAL GAS TREATMENT PROCESSES

Raw natural gases contain variable amounts of carbon dioxide, hydrogen sulfide, and water vapor. The presence of hydrogen sulfide in natural gas for domestic consumption cannot be tolerated because it is poisonous. It also corrodes metallic equipment. Carbon dioxide is undesirable, because it reduces the heating value of the gas and solidifies under the high pressure and low temperatures used for transporting natural gas. For obtaining a sweet, dry natural gas, acid gases must be removed and water vapor reduced. In addition, natural gas with appreciable amounts of heavy hydrocarbons should be treated for their recovery as natural gas liquids.

Acid Gas Treatment

Acid gases can be reduced or removed by one or more of the following methods:

1. Physical absorption using a selective absorption solvent.
2. Physical adsorption using a solid adsorbent.
3. Chemical absorption where a solvent (a chemical) capable of reacting reversibly with the acid gases is used.

Physical Absorption

Important processes commercially used are the Selexol, the Sulfinol, and the Rectisol processes. In these processes, no chemical reaction occurs between the acid gas and the solvent. The solvent, or absorbent, is a liquid that selectively absorbs the acid gases and leaves out the hydrocarbons. In the Selexol process for example, the solvent is dimethyl ether of polyethylene glycol. Raw natural gas passes countercurrently to the descending solvent. When the solvent becomes saturated with the acid gases, the pressure is reduced, and hydrogen sulfide and carbon dioxide are desorbed. The solvent is then recycled to the absorption tower. Figure 1-1 shows the Selexol process.[2]

Physical Adsorption

In these processes, a solid with a high surface area is used. Molecular sieves (zeolites) are widely used and are capable of adsorbing large amounts of gases. In practice, more than one adsorption bed is used for continuous operation. One bed is in use while the other is being regenerated.

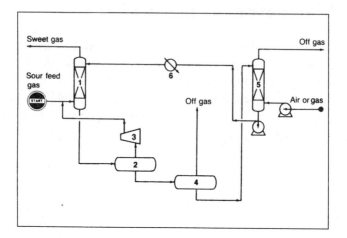

Figure 1-1. The Selexol process for acid gas removal:[2] (1) absorber, (2) flash drum, (3) compressor, (4) low-pressure drum, (5) stripper, (6) cooler.

Regeneration is accomplished by passing hot dry fuel gas through the bed. Molecular sieves are competitive only when the quantities of hydrogen sulfide and carbon disulfide are low.

Molecular sieves are also capable of adsorbing water in addition to the acid gases.

Chemical Absorption (Chemisorption)

These processes are characterized by a high capability of absorbing large amounts of acid gases. They use a solution of a relatively weak base, such as monoethanolamine. The acid gas forms a weak bond with the base which can be regenerated easily. Mono- and diethanolamines are frequently used for this purpose. The amine concentration normally ranges between 15 and 30%. Natural gas is passed through the amine solution where sulfides, carbonates, and bicarbonates are formed.

Diethanolamine is a favored absorbent due to its lower corrosion rate, smaller amine loss potential, fewer utility requirements, and minimal reclaiming needs.[3] Diethanolamine also reacts reversibly with 75% of carbonyl sulfides (COS), while the mono- reacts irreversibly with 95% of the COS and forms a degradation product that must be disposed of.

Diglycolamine (DGA), is another amine solvent used in the Econamine process (Fig 1-2).[4] Absorption of acid gases occurs in an absorber containing an aqueous solution of DGA, and the heated rich

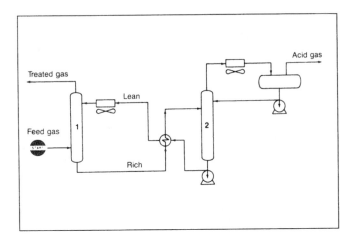

Figure 1-2. The Econamine process:[4] (1) absorption tower, (2) regeneration tower.

solution (saturated with acid gases) is pumped to the regenerator. Diglycolamine solutions are characterized by low freezing points, which make them suitable for use in cold climates.

Strong basic solutions are effective solvents for acid gases. However, these solutions are not normally used for treating large volumes of natural gas because the acid gases form stable salts, which are not easily regenerated. For example, carbon dioxide and hydrogen sulfide react with aqueous sodium hydroxide to yield sodium carbonate and sodium sulfide, respectively.

$$CO_2 + 2NaOH_{(aq)} \rightarrow Na_2 CO_3 + H_2O$$

$$H_2S + 2\,NaOH_{(aq)} \rightarrow Na_2S + 2\,H_2O$$

However, a strong caustic solution is used to remove mercaptans from gas and liquid streams. In the Merox Process, for example, a caustic solvent containing a catalyst such as cobalt, which is capable of converting mercaptans (RSH) to caustic insoluble disulfides (RSSR), is used for streams rich in mercaptans after removal of H_2S. Air is used to oxidize the mercaptans to disulfides. The caustic solution is then recycled for regeneration. The Merox process (Fig. 1-3) is mainly used for treatment of refinery gas streams.[5]

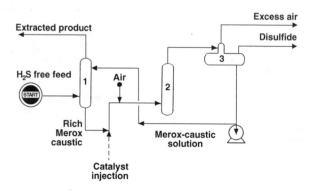

Figure 1-3. The Merox process:[5] (1) extractor, (2) oxidation reactor.

Water Removal

Moisture must be removed from natural gas to reduce corrosion problems and to prevent hydrate formation. Hydrates are solid white compounds formed from a physical-chemical reaction between hydrocarbons and water under the high pressures and low temperatures used to transport natural gas via pipeline. Hydrates reduce pipeline efficiency.

To prevent hydrate formation, natural gas may be treated with glycols, which dissolve water efficiently. Ethylene glycol (EG), diethylene glycol (DEG), and triethylene glycol (TEG) are typical solvents for water removal. Triethylene glycol is preferable in vapor phase processes because of its low vapor pressure, which results in less glycol loss. The TEG absorber normally contains 6 to 12 bubble-cap trays to accomplish the water absorption. However, more contact stages may be required to reach dew points below –40°F. Calculations to determine the number of trays or feet of packing, the required glycol concentration, or the glycol circulation rate require vapor-liquid equilibrium data. Predicting the interaction between TEG and water vapor in natural gas over a broad range allows the designs for ultra-low dew point applications to be made.[6]

A computer program was developed by Grandhidsan et al., to estimate the number of trays and the circulation rate of lean TEG needed to dry natual gas. It was found that more accurate predictions of the rate could be achieved using this program than using hand calculation.[7]

Figure 1-4 shows the Dehydrate process where EG, DEG, or TEG could be used as an absorbent.[8] One alternative to using bubble-cap trays

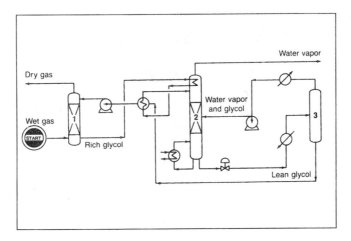

Figure 1-4. Flow diagram of the Dehydrate process[8]: (1) absorption column, (2) glycol sill, (3) vacuum drum.

is structural packing, which improves control of mass transfer. Flow passages direct the gas and liquid flows countercurrent to each other. The use of structural packing in TEG operations has been reviewed by Kean et al.[9]

Another way to dehydrate natural gas is by injecting methanol into gas lines to lower the hydrate-formation temperature below ambient.[10] Water can also be reduced or removed from natural gas by using solid adsorbents such as molecular sieves or silica gel.

Condensable Hydrocarbon Recovery

Hydrocarbons heavier than methane that are present in natural gases are valuable raw materials and important fuels. They can be recovered by lean oil extraction. The first step in this scheme is to cool the treated gas by exchange with liquid propane. The cooled gas is then washed with a cold hydrocarbon liquid, which dissolves most of the condensable hydrocarbons. The uncondensed gas is dry natural gas and is composed mainly of methane with small amounts of ethane and heavier hydrocarbons. The condensed hydrocarbons or natural gas liquids (NGL) are stripped from the rich solvent, which is recycled. Table 1-2 compares the analysis of natural gas before and after treatment.[11] Dry natural gas may then be used either as a fuel or as a chemical feedstock.

Another way to recover NGL is through cryogenic cooling to very low temperatures (−150 to −180°F), which are achieved primarily through

Table 1-2
Typical analysis of natural gas before and after treatment[11]

Component mole %	Feed	Pipeline gas
N_2	0.45	0.62
CO_2	27.85	3.50
H_2S	0.0013	—
C_1	70.35	94.85
C_2	0.83	0.99
C_3	0.22	0.003
C_4	0. 13	0.004
C_5	0.06	0.004
C_{6+}	0.11	0.014

adiabatic expansion of the inlet gas. The inlet gas is first treated to remove water and acid gases, then cooled via heat exchange and refrigeration. Further cooling of the gas is accomplished through turbo expanders, and the gas is sent to a demethanizer to separate methane from NGL. Improved NGL recovery could be achieved through better control strategies and use of on-line gas chromatographic analysis.[12]

NATURAL GAS LIQUIDS (NGL)

Natural gas liquids (condensable hydrocarbons) are those hydrocarbons heavier than methane that are recovered from natural gas. The amount of NGL depends mainly on the percentage of the heavier hydrocarbons present in the gas and on the efficiency of the process used to recover them. (A high percentage is normally expected from associated gas.)

Natural gas liquids are normally fractionated to separate them into three streams:

1. An ethane-rich stream, which is used for producing ethylene.
2. Liquefied petroleum gas (LPG), which is a propane-butane mixture. It is mainly used as a fuel or a chemical feedstock. Liquefied petroleum gas is evolving into an important feedstock for olefin production. It has been predicted that the world (LPG) market for chemicals will grow from 23.1 million tons consumed in 1988 to 36.0 million tons by the year 2000.[13]
3. Natural gasoline (NG) is mainly constituted of $C5^+$ hydrocarbons and is added to gasoline to raise its vapor pressure. Natural gasoline is usually sold according to its vapor pressure.

Natural gas liquids may contain significant amounts of cyclohexane, a precursor for nylon 6 (Chapter 10). Recovery of cyclohexane from NGL by conventional distillation is difficult and not economical because heptane isomers are also present which boil at temperatures nearly identical to that of cyclohexane. An extractive distillation process has been recently developed by Phillips Petroleum Co. to separate cyclohexane.[14]

Liquefied Natural Gas (LNG)

After the recovery of natural gas liquids, sweet dry natural gas may be liquefied for transportation through cryogenic tankers. Further treatment may be required to reduce the water vapor below 10 ppm and carbon dioxide and hydrogen sulfide to less than 100 and 50 ppm, respectively.

Two methods are generally used to liquefy natural gas: the expander cycle and mechanical refrigeration. In the expander cycle, part of the gas is expanded from a high transmission pressure to a lower pressure. This lowers the temperature of the gas. Through heat exchange, the cold gas cools the incoming gas, which in a similar way cools more incoming gas until the liquefaction temperature of methane is reached. Figure 1-5 is a flow diagram for the expander cycle for liquefying natural gas.[15]

In mechanical refrigeration, a multicomponent refrigerant consisting of nitrogen, methane, ethane, and propane is used through a cascade cycle. When these liquids evaporate, the heat required is obtained from

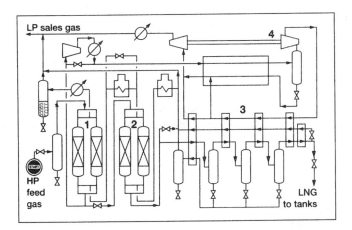

Figure 1-5. Flow diagram of the expander cycle for liquefying natural gas:[15] (1) pretreatment (mol.sieve), (2) heat exchanger, (3) turboexpander.

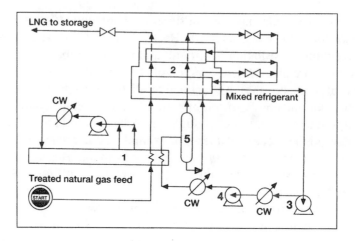

Figure 1-6. The MCR process for liquefying natural gas:[15] (1) coolers, (2) heat exchangers, (3,4) two stage compressors, (5) liquid-vapor phase separator.

natural gas, which loses energy/temperature till it is liquefied. The refrigerant gases are recompressed and recycled. Figure 1-6 shows the MCR natural gas liquefaction process.[15] Table 1-3 lists important properties of a representative liquefied natural gas mixture.

PROPERTIES OF NATURAL GAS

Treated natural gas consists mainly of methane; the properties of both gases (natural gas and methane) are nearly similar. However, natural gas is not pure methane, and its properties are modified by the presence of impurities, such as N_2 and CO_2 and small amounts of unrecovered heavier hydrocarbons.

Table 1-3
Important properties of a representative liquefied natural gas mixture

Density, lb/cf	27.00
Boiling point, °C	−158
Calorific value, Btu/lb	21200
Specific volume, cf/lb	0.037
Critical temperature, °C*	−82.3
Critical pressure, psi*	−673

* Critical temperature and pressure for pure liquid methane.

An important property of natural gas is its heating value. Relatively high amounts of nitrogen and/or carbon dioxide reduce the heating value of the gas. Pure methane has a heating value of 1,009 Btu/ft^3. This value is reduced to approximately 900 Btu/ft3 if the gas contains about 10% N_2 and CO_2. (The heating value of either nitrogen or carbon dioxide is zero.) On the other hand, the heating value of natural gas could exceed methane's due to the presence of higher-molecular weight hydrocarbons, which have higher heating values. For example, ethane's heating value is 1,800 Btu/ft^3, compared to 1,009 Btu/ft^3 for methane. Heating values of hydrocarbons normally present in natural gas are shown in Table 1-4.

Natural gas is usually sold according to its heating values. The heating value of a product gas is a function of the constituents present in the mixture. In the natural gas trade, a heating value of one million Btu is approximately equivalent to 1,000 ft^3 of natural gas.

CRUDE OILS

Crude oil (petroleum) is a naturally occurring brown to black flammable liquid. Crude oils are principally found in oil reservoirs associated with sedimentary rocks beneath the earth's surface. Although exactly how crude oils originated is not established, it is generally agreed that crude oils derived from marine animal and plant debris subjected to high temperatures and pressures. It is also suspected that the transformation may have been catalyzed by rock constituents. Regardless of their origins,

Table 1-4
Heating values of methane and heavier hydrocarbons present in natural gas

Hydrocarbon	Formula	Heating value Btu/ft^3
Methane	CH_4	1,009
Ethane	C_2H_6	1,800
Propane	C_3H_8	2,300
Isobutane	C_4H_{10}	3,253
n-Butane	C_4H_{10}	3,262
Isopentane	C_5H_{12}	4,000
n-Pentane	C_5H_{12}	4,010
n-Hexane	C_6H_{14}	4,750
n-Heptane	C_7H_{16}	5,502

all crude oils are mainly constituted of hydrocarbons mixed with variable amounts of sulfur, nitrogen, and oxygen compounds.

Metals in the forms of inorganic salts or organometallic compounds are present in the crude mixture in trace amounts. The ratio of the different constituents in crude oils, however, vary appreciably from one reservoir to another.

Normally, crude oils are not used directly as fuels or as feedstocks for the production of chemicals. This is due to the complex nature of the crude oil mixture and the presence of some impurities that are corrosive or poisonous to processing catalysts.

Crude oils are refined to separate the mixture into simpler fractions that can be used as fuels, lubricants, or as intermediate feedstock to the petrochemical industries. A general knowledge of this composite mixture is essential for establishing a processing strategy.

COMPOSITION OF CRUDE OILS

The crude oil mixture is composed of the following groups:

1. Hydrocarbon compounds (compounds made of carbon and hydrogen).
2. Non-hydrocarbon compounds.
3. Organometallic compounds and inorganic salts (metallic compounds).

Hydrocarbon Compounds

The principal constituents of most crude oils are hydrocarbon compounds. All hydrocarbon classes are present in the crude mixture, except alkenes and alkynes. This may indicate that crude oils originated under a reducing atmosphere. The following is a brief description of the different hydrocarbon classes found in all crude oils.

Alkanes (Paraffins)

Alkanes are saturated hydrocarbons having the general formula C_nH_{2n+2}. The simplest alkane, methane (CH_4), is the principal constituent of natural gas. Methane, ethane, propane, and butane are gaseous hydrocarbons at ambient temperatures and atmospheric pressure. They are usually found associated with crude oils in a dissolved state.

Normal alkanes (n-alkanes, n-paraffins) are straight-chain hydrocarbons having no branches. Branched alkanes are saturated hydrocarbons with an alkyl substituent or a side branch from the main chain. A branched alkane with the same number of carbons and hydrogens as an n-alkane is called an isomer. For example, butane (C_4H_{10}) has two isomers, n-butane and 2-methyl propane (isobutane). As the molecular weight of the hydrocarbon increases, the number of isomers also increases. Pentane (C_5C_{12}) has three isomers; hexane (C_6H_{14}) has five. The following shows the isomers of hexane:

$$CH_3(CH_2)_4CH_3$$
n-Hexane

$$\overset{\overset{\displaystyle CH_3}{|}}{CH_3CHCH_2CH_2CH_3}$$
2-Methylpentane
(Isohexane)

$$\overset{\overset{\displaystyle CH_3}{|}}{CH_3CH_2CHCH_2CH_3}$$
3-Methylpentane

$$\overset{\overset{\displaystyle CH_3}{|}}{\underset{\underset{\displaystyle CH_3}{|}}{CH_3CCH_2CH_3}}$$
2,2-Dimethylbutane

$$\overset{\overset{\displaystyle CH_3 \quad CH_3}{| \quad |}}{CH_3CH-CHCH_3}$$
2,3-Dimethylbutane

An isoparaffin is an isomer having a methyl group branching from carbon number 2 of the main chain. Crude oils contain many short, medium, and long-chain normal and branched paraffins. A naphtha fraction (obtained as a light liquid stream from crude fractionation) with a narrow boiling range may contain a limited but still large number of isomers.

Cycloparaffins (Naphthenes)

Saturated cyclic hydrocarbons, normally known as naphthenes, are also part of the hydrocarbon constituents of crude oils. Their ratio, however, depends on the crude type. The lower members of naphthenes are cyclopentane, cyclohexane, and their mono-substituted compounds. They are normally present in the light and the heavy naphtha fractions. Cyclohexanes, substituted cyclopentanes, and substituted cyclohexanes are important precursors for aromatic hydrocarbons.

Methylcyclopentane   Cyclohexane   Methylcyclohexane

The examples shown here are for three naphthenes of special importance. If a naphtha fraction contains these compounds, the first two can be converted to benzene, and the last compound can dehydrogenate to toluene during processing. Dimethylcyclohexanes are also important precursors for xylenes (see "Xylenes" later in this section).

Heavier petroleum fractions such as kerosine and gas oil may contain two or more cyclohexane rings fused through two vicinal carbons.

Aromatic Compounds

Lower members of aromatic compounds are present in small amounts in crude oils and light petroleum fractions. The simplest mononuclear aromatic compound is benzene (C_6H_6). Toluene (C_7H_8) and xylene (C_8H_{10}) are also mononuclear aromatic compounds found in variable amounts in crude oils. Benzene, toluene, and xylenes (BTX) are important petrochemical intermediates as well as valuable gasoline components. Separating BTX aromatics from crude oil distillates is not feasible because they are present in low concentrations. Enriching a naphtha fraction with these aromatics is possible through a catalytic reforming process. Chapter 3 discusses catalytic reforming.

Binuclear aromatic hydrocarbons are found in heavier fractions than naphtha. Trinuclear and polynuclear aromatic hydrocarbons, in combination with heterocyclic compounds, are major constituents of heavy crudes and crude residues. Asphaltenes are a complex mixture of aromatic and heterocyclic compounds. The nature and structure of some of these compounds have been investigated.[16] The following are representative examples of some aromatic compounds found in crude oils:

Benzene 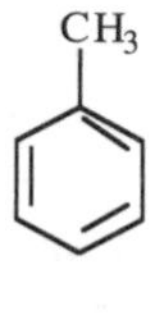   Toluene    p-Xylene

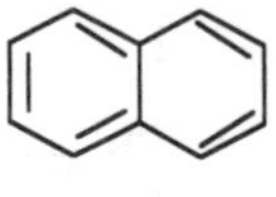

Naphthalene 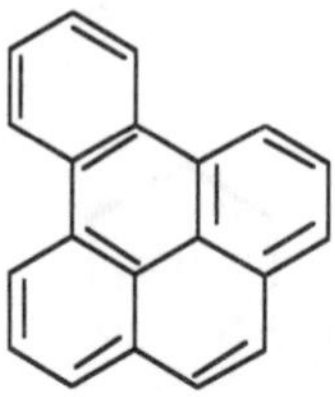   1,2-Benzopyrene 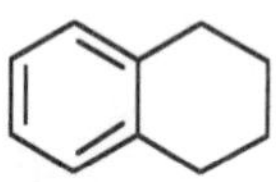   Tetralin

Only a few aromatic-cycloparaffin compounds have been isolated and identified. Tetralin is an example of this class.

Non-hydrocarbon Compounds

Various types of non-hydrocarbon compounds occur in crude oils and refinery streams. The most important are the organic sulfur, nitrogen, and oxygen compounds. Traces of metallic compounds are also found in all crudes. The presence of these impurities is harmful and may cause problems to certain catalytic processes. Fuels having high sulfur and nitrogen levels cause pollution problems in addition to the corrosive nature of their oxidization products.

Sulfur Compounds

Sulfur in crude oils is mainly present in the form of organosulfur compounds. Hydrogen sulfide is the only important inorganic sulfur compound found in crude oil. Its presence, however, is harmful because of its corrosive nature. Organosulfur compounds may generally be classified as acidic and non-acidic. Acidic sulfur compounds are the thiols (mercaptans). Thiophene, sulfides, and disulfides are examples of non-acidic sulfur compounds found in crude fractions. Extensive research has been carried out to identify some sulfur compounds in a narrow light petroleum fraction.[17] Examples of some sulfur compounds from the two types are:

Acidic Sulfur Compounds

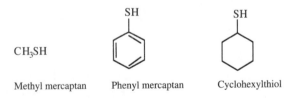

CH₃SH

Methyl mercaptan   Phenyl mercaptan   Cyclohexylthiol

Non-acidic Sulfur Compounds

CH_3SCH_3 $CH_3S\text{-}SCH_3$

Dimethyl sulfide   Dimethyldisulfide

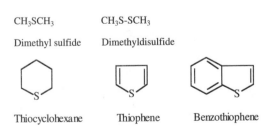

Thiocyclohexane   Thiophene   Benzothiophene

Sour crudes contain a high percentage of hydrogen sulfide. Because many organic sulfur compounds are not thermally stable, hydrogen sulfide is often produced during crude processing. High-sulfur crudes are less desirable because treating the different refinery streams for acidic hydrogen sulfide increases production costs.

Most sulfur compounds can be removed from petroleum streams through hydrotreatment processes, where hydrogen sulfide is produced and the corresponding hydrocarbon released. Hydrogen sulfide is then absorbed in a suitable absorbent and recovered as sulfur (Chapter 4).

Nitrogen Compounds

Organic nitrogen compounds occur in crude oils either in a simple heterocyclic form as in pyridine (C_5H_5N) and pyrrole (C_4H_5N), or in a complex structure as in porphyrin. The nitrogen content in most crudes is very low and does not exceed 0.1 wt%. In some heavy crudes, however, the nitrogen content may reach up to 0.9 wt %.[18] Nitrogen compounds are more thermally stable than sulfur compounds and accordingly are concentrated in heavier petroleum fractions and residues. Light petroleum streams may contain trace amounts of nitrogen compounds, which should be removed because they poison many processing catalysts. During hydrotreatment of petroleum fractions, nitrogen compounds are hydrodenitrogenated to ammonia and the corresponding hydrocarbon. For example, pyridine is denitrogenated to ammonia and pentane:

$+ 5\,H_2 \longrightarrow NH_3 + CH_3CH_2CH_2CH_2CH_3$

Nitrogen compounds in crudes may generally be classified into basic and non-basic categories. Basic nitrogen compounds are mainly those having a pyridine ring, and the non-basic compounds have a pyrrole structure. Both pyridine and pyrrole are stable compounds due to their aromatic nature.

The following are examples of organic nitrogen compounds.

Basic Nitrogen Compounds

Pyridine   Quinoline   Isoquinoline   Acridine

Non-Basic Nitrogen Compounds

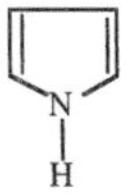

Pyrrole 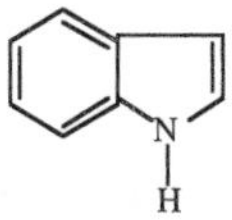   Indole 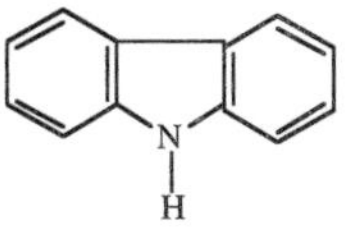   Carbazole 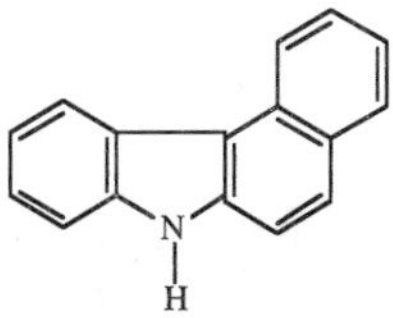   Benzocarbazole

Porphyrins are non-basic nitrogen compounds. The porphyrin ring system is composed of four pyrrole rings joined by =CH-groups. The entire ring system is aromatic. Many metal ions can replace the pyrrole hydrogens and form chelates. The chelate is planar around the metal ion and resonance results in four equivalent bonds from the nitrogen atoms to the metal.[19] Almost all crude oils and bitumens contain detectable amounts of vanadyl and nickel porphyrins. The following shows a porphyrin structure:

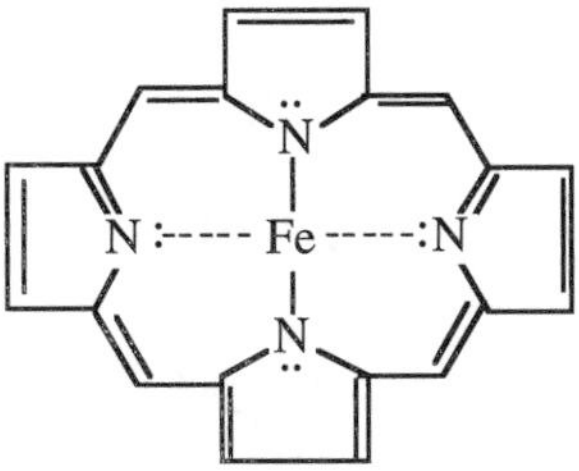

Separation of nitrogen compounds is difficult, and the compounds are susceptible to alteration and loss during handling. However, the basic low-molecular weight compounds may be extracted with dilute mineral acids.

Oxygen Compounds

Oxygen compounds in crude oils are more complex than the sulfur types. However, their presence in petroleum streams is not poisonous to processing catalysts. Many of the oxygen compounds found in crude oils are weakly acidic. They are carboxylic acids, cresylic acid, phenol, and naphthenic acid. Naphthenic acids are mainly cyclopentane and cyclohexane derivatives having a carboxyalkyl side chain.

Naphthenic acids in the naphtha fraction have a special commercial importance and can be extracted by using dilute caustic solutions. The total acid content of most crudes is generally low, but may reach as much as 3%, as in some California crudes.

Non-acidic oxygen compounds such as esters, ketones, and amides are less abundant than acidic compounds. They are of no commercial value. The following shows some of the oxygen compounds commonly found in crude oils:

Acidic Oxygen Compounds

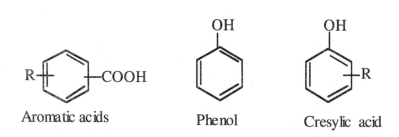

$CH_3(CH_2)_nCOOH$

An aliphatic carboxylic acid

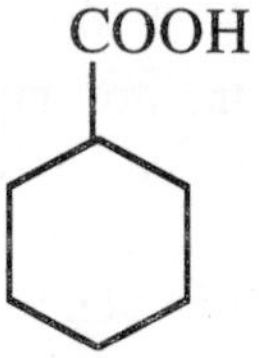

COOH

Cyclohexane
carboxylic acid

R─COOH

Aromatic acids

OH

Phenol

OH
─R

Cresylic acid

Non-Acidic Oxygen Compounds

R-COOR'

Esters

R-CONHR'

Amides

$$R-\overset{\overset{O}{\|}}{C}-R$$

Ketone

Furan

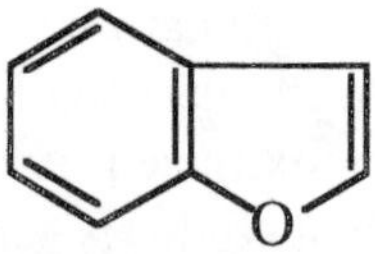

Benzofuran

Metallic Compounds

Many metals occur in crude oils. Some of the more abundant are sodium, calcium, magnesium, aluminium, iron, vanadium, and nickel. They are present either as inorganic salts, such as sodium and magnesium chlorides, or in the form of organometallic compounds, such as those of nickel and vanadium (as in porphyrins). Calcium and magnesium can form salts or soaps with carboxylic acids. These compounds act as emulsifiers, and their presence is undesirable.

Although metals in crudes are found in trace amounts, their presence is harmful and should be removed. When crude oil is processed, sodium and magnesium chlorides produce hydrochloric acid, which is very corrosive. Desalting crude oils is a necessary step to reduce these salts.

Vanadium and nickel are poisons to many catalysts and should be reduced to very low levels. Most of the vanadium and nickel compounds are concentrated in the heavy residues. Solvent extraction processes are used to reduce the concentration of heavy metals in petroleum residues.

PROPERTIES OF CRUDE OILS

Crude oils differ appreciably in their properties according to origin and the ratio of the different components in the mixture. Lighter crudes generally yield more valuable light and middle distillates and are sold at higher prices. Crudes containing a high percent of impurities, such as sulfur compounds, are less desirable than low-sulfur crudes because of their corrosivity and the extra treating cost. Corrosivity of crude oils is a function of many parameters among which are the type of sulfur compounds and their decomposition temperatures, the total acid number, the type of carboxylic and naphthenic acids in the crude and their decomposition temperatures. It was found that naphthenic acids begin to decompose at 600°F. Refinery experience has shown that above 750°F there is no naphthenic acid corrosion. The subject has been reviewed by Kane and Cayard.[20] For a refiner, it is necessary to establish certain criteria to relate one crude to another to be able to assess crude quality and choose the best processing scheme. The following are some of the important tests used to determine the properties of crude oils.

Density, Specific Gravity and API Gravity

Density is defined as the mass of unit volume of a material at a specific temperature. A more useful unit used by the petroleum industry is specific gravity, which is the ratio of the weight of a given volume of a material to the weight of the same volume of water measured at the same temperature.

Specific gravity is used to calculate the mass of crude oils and its products. Usually, crude oils and their liquid products are first measured on a volume basis, then changed to the corresponding masses using the specific gravity.

The API (American Petroleum Institute) gravity is another way to express the relative masses of crude oils. The API gravity could be calculated mathematically using the following equation:

$$°API = \frac{141.5}{Sp.gr. \; 60/60°} - 131.5$$

A low API gravity indicates a heavier crude oil or a petroleum product, while a higher API gravity means a lighter crude or product. Specific gravities of crude oils roughly range from 0.82 for lighter crudes to over 1.0 for heavier crudes (41 - 10 °API scale).

Salt Content

The salt content expressed in milligrams of sodium chloride per liter oil (or in pounds/barrel) indicates the amount of salt dissolved in water. Water in crudes is mainly present in an emulsified form. A high salt content in a crude oil presents serious corrosion problems during the refining process. In addition, high salt content is a major cause of plugging heat exchangers and heater pipes. A salt content higher than 10 lb/1,000 barrels (expressed as NaCl) requires desalting.

Sulfur Content

Determining the sulfur content in crudes is important because the amount of sulfur indicates the type of treatment required for the distillates. To determine sulfur content, a weighed crude sample (or fraction) is burned in an air stream. All sulfur compounds are oxidized to sulfur dioxide, which is further oxidized to sulfur trioxide and finally titrated with a standard alkali.

Identifying sulfur compounds in crude oils and their products is of little use to a refiner because all sulfur compounds can easily be hydrodesulfurized to hydrogen sulfide and the corresponding hydrocarbon.

The sulfur content of crudes, however, is important and is usually considered when determining commercial values.

Pour Point

The pour point of a crude oil or product is the lowest temperature at which an oil is observed to flow under the conditions of the test. Pour point data indicates the amount of long-chain paraffins (petroleum wax) found in a crude oil. Paraffinic crudes usually have higher wax content than other crude types. Handling and transporting crude oils and heavy fuels is difficult at temperatures below their pour points Often, chemical additives known as pour point depressants are used to improve the flow properties of the fuel. Long-chain n-paraffins ranging from 16–60 carbon atoms in particular, are responsible for near-ambient temperature precipitation. In middle distillates, less than 1% wax can be sufficient to cause solidification of the fuel.[21]

Ash Content

This test indicates the amount of metallic constituents in a crude oil. The ash left after completely burning an oil sample usually consists of stable metallic salts, metal oxides, and silicon oxide. The ash could be further analyzed for individual elements using spectroscopic techniques.

CRUDE OIL CLASSIFICATION

Appreciable property differences appear between crude oils as a result of the variable ratios of the crude oil components. For a refiner dealing with crudes of different origins, a simple criterion may be established to group crudes with similar characteristics. Crude oils can be arbitrarily classified into three or four groups depending on the relative ratio of the hydrocarbon classes that predominates in the mixture. The following describes three types of crudes:

1. Paraffinic—the ratio of paraffinic hydrocarbons is high compared to aromatics and naphthenes.
2. Naphthenic—the ratios of naphthenic and aromatic hydrocarbons are relatively higher than in paraffinic crudes.
3. Asphaltic—contain relatively a large amount of polynuclear aromatics, a high asphaltene content, and relatively less paraffins than paraffinic crudes.

A correlation index is a useful criterion for indicating the crude class or type. The following relationship between the mid-boiling point in Kelvin degrees (°K) and the specific gravity of a crude oil or a fraction yields the correlation index (Bureau of Mines Correlation index).[22]

BMCI = 48,640 / K + (473.6d – 456.8)
  K = mid-boiling point in Kelvin degrees (Mid-boiling point is the temperature at which 50 vol % of the crude is distilled.)
  d = specific gravity at 60/60°F

A zero value has been assumed for n-paraffins, 100 for aromatics. A low BMCI value indicates a higher paraffin concentration in a petroleum fraction.

Another relationship used to indicate the crude type is the Watson characterization factor. The factor also relates the mid-boiling point of the crude or a fraction to the specific gravity.

$$\text{Watson characterization factor} = \frac{T^{1/3}}{d}$$

where T = mid-boiling point in °R (°R is the absolute °F, and equals °F + 460)

A value higher than 10 indicates a predominance of paraffins while a value around 10 means a predominance of aromatics.

Table 1-5
Typical analysis of some crude oils

	Arab Extra Light*	Alameen Egypt	Arab Heavy	Bakr-9 Egypt
Gravity, °API	38.5	33.4	28.0	20.9
Carbon residue (wt %)	2.0	5.1	6.8	11.7
Sulfur content (wt %)	1.1	0.86	2.8	3.8
Nitrogen content (wt %)	0.04	0.12	0.15	—
Ash content (wt %)	0.002	0.004	0.012	0.04
Iron (ppm)	0.4	0.0	1.0	—
Nickel (ppm)	0.6	0.0	9.0	108
Vanadium (ppm)	2.2	15	40.0	150
Pour point (°F)	≈Zero	35	–11.0	55
Paraffin wax content (wt %)	—	3.3	—	—

* Ali, M. F et al., Hydrocarbon Processing, Vol. 64, No. 2, 1985 p. 83.

Properties of crude oils vary considerably according to their types. Table 1-5 lists the analyses of some crudes from different origins.

COAL, OIL SHALE, TAR SAND, AND GAS HYDRATES

Coal, oil shale, and tar sand are carbonaceous materials that can serve as future energy and chemical sources when oil and gas are consumed. The H/C ratio of these materials is lower than in most crude oils. As solids or semi-solids, they are not easy to handle or to use as fuels, compared to crude oils. In addition, most of these materials have high sulfur and/or nitrogen contents, which require extensive processing. Changing these materials into hydrocarbon liquids or gaseous fuels is possible but expensive. The following briefly discusses these alternative energy and chemical sources.

COAL

Coal is a natural combustible rock composed of an organic heterogeneous substance contaminated with variable amounts of inorganic compounds. Most coal reserves are concentrated in North America, Europe, and China.

Coal is classified into different ranks according to the degree of chemical change that occurred during the decomposition of plant remains in the prehistoric period. In general, coals with a high heating value and a high fixed carbon content are considered to have been subjected to more severe changes than those with lower heating values and fixed carbon contents. For example, peat, which is considered a young coal, has a low fixed carbon content and a low heating value. Important coal ranks are anthracite (which has been subjected to the most chemical change and is mostly carbon), bituminous coal, sub-bituminous coal, and lignite. Table 1-6 compares the analysis of some coals with crude oil.[23]

During the late seventies and early eighties, when oil prices rose after the 1973 war, extensive research was done to change coal to liquid hydrocarbons. However, coal-derived hydrocarbons were more expensive than crude oils. Another way to use coal is through gasification to a fuel gas mixture of CO and H_2 (medium Btu gas). This gas mixture could be used as a fuel or as a synthesis gas mixture for the production of fuels and chemicals via a Fischer Tropsch synthesis route. This process is

Table 1-6
Typical element analysis of some coals compared with a crude oil[23]

	Weight %					
	C	H	S	N	O	H/C mol ratio
Crude oil	84.6	12.8	1.5	0.4	0.5	1.82
Peat	56.8	5.6	0.3	2.7	34.6	1.18
Lignite	68.8	4.9	0.7	1.1	24.5	0.86
Bitumenous Coal	81.8	5.6	1.5	1.4	9.7	0.82
Anthracite	91.7	3.5	—	—	2.7	0.46

operative in South Africa for the production of hydrocarbon fuels. Fischer Tropsch synthesis is discussed in Chapter 4.

OIL SHALE

Oil shale is a low-permeable rock made of inorganic material interspersed with a high-molecular weight organic substance called "Kerogen." Heating the shale rock produces an oily substance with a complex structure.

The composition of oil shales differs greatly from one shale to another. For example, the amount of oil obtained from one ton of eastern U.S. shale deposit is only 10 gallons, compared to 30 gallons from western U.S. shale deposits.

Retorting is a process used to convert the shale to a high molecular-weight oily material. In this process, crushed shale is heated to high temperatures to pyrolyze Kerogen. The product oil is a viscous, high-molecular weight material. Further processing is required to change the oil into a liquid fuel.

Major obstacles to large-scale production are the disposal of the spent shale and the vast earth-moving operations. Table 1-7 is a typical analysis of a raw shale oil produced from retorting oil shale.

TAR SAND

Tar sands (oil sands) are large deposits of sand saturated with bitumen and water. Tar sand deposits are commonly found at or near the earth's surface entrapped in large sedimentary basins. Large accumulations of tar sand deposits are few. About 98% of all world tar sand is found in

Table 1-7
Typical analysis of shale oil

Test	Result
Gravity	19.7
Nitrogen, wt %	2.18
Conradson Carbon, wt %	4.5
Sulfur, wt %	0.74
Ash, wt %	0.06

seven large tar deposits. The oil sands resources in Western Canada sedimentary basin is the largest in the world. In 1997, it produced 99% of Canada's crude oil. It is estimated to hold 1.7–2.5 trillon barrels of bitumen in place. This makes it one of the largest hydrocarbon deposits in the world.[24] Tar sand deposits are covered by a semifloating mass of partially decayed vegetation approximately 6 meters thick.

Tar sand is difficult to handle. During summer, it is soft and sticky, and during the winter it changes to a hard, solid material.

Recovering the bitumen is not easy, and the deposits are either stripmined if they are near the surface, or recovered *in situ* if they are in deeper beds. The bitumen could be extracted by using hot water and steam and adding some alkali to disperse it. The produced bitumen is a very thick material having a density of approximately 1.05 g/cm^3. It is then subjected to a cracking process to produce distillate fuels and coke. The distillates are hydrotreated to saturate olefinic components. Table 1-8 is a typical analysis of Athabasca bitumen.[25]

GAS HYDRATES

Gas hydrates are an ice-like material which is constituted of methane molecules encaged in a cluster of water molecules and held together by hydrogen bonds. This material occurs in large underground deposits found beneath the ocean floor on continental margins and in places north of the arctic circle such as Siberia. It is estimated that gas hydrate deposits contain twice as much carbon as all other fossil fuels on earth. This source, if proven feasible for recovery, could be a future energy as well as chemical source for petrochemicals.

Due to its physical nature (a solid material only under high pressure and low temperature), it cannot be processed by conventional methods used for natural gas and crude oils. One approach is by dissociating this

Table 1-8
Properties of Athabasca bitumen[25]

Gravity at 60°F (15.6°C)	6.0°API
UOP characterization factor	11.18
Pour point	+50°F (10°C)
Specific heat	0.35 cal/(g)(°C)
Calorific value	17,900 Btu/lb
Viscosity at 60°F (15.6°C)	3,000–300,000 poise
Carbon/hydrogen ratio	8.1
Components, %:	
asphaltenes	20.0
resins	25.0
oils	55.0
Ultimate analysis, %:	
carbon	83.6
hydrogen	10.3
sulfur	5.5
nitrogen	0.4
oxygen	0.2
Heavy metals. ppm:	
nickel	100
vanadium	250
copper	5

cluster into methane and water by injecting a warmer fluid such as sea water. Another approach is by drilling into the deposit. This reduces the pressure and frees methane from water. However, the environmental effects of such drilling must still be evaluated.[26]

REFERENCES

1. Hatch, L. F. and Matar, S., *From Hydrocarbons to Petrochemicals,* Gulf Publishing Company, 1981, p. 5.
2. "Gas Processing Handbook," *Hydrocarbon Processing,* Vol. 69, No.4, 1990, p. 91.
3. Tuttle, R. and Allen, K., *Oil and Gas Journal,* Aug. 9, 1976, pp. 78–82.
4. "Gas Processing Handbook," *Hydrocarbon Processing,* Vol. 69, No. 4, 1990, p. 77.

5. "Gas Processing Handbook" *Hydrocarbon Processing,* Vol. 77, No. 4, 1998, p. 113.

6. Hicks, R. L. and Senules, E. A., "New Gas Water-TEG Equilibria," *Hydrocarbon Processing,* Vol. 70, No. 4, 1991, pp. 55–58.

7. Gandhidasan, P., Al-Farayedhi, A., and Al-Mubarak, A. "A review of types of dessicant dehydrates, solid and liquid," *Oil and Gas Journal,* June 21, 1999, pp. 36–40.

8. "Gas Processing Handbook," *Hydrocarbon Processing,* Vol. 69, No. 4, 1990, p. 76.

9. Kean, J. A., Turner, H. M., and Price, B. C., "How Packing Works in Dehydrators," *Hydrocarbon Processing,* Vol. 70, No. 4, 1991, pp. 47–52.

10. Aggour, M., *Petroleum Economics and Engineering,* edited by Abdel-Aal, H. K., Bakr, B. A., and Al-Sahlawi, M., Marcel Dekker, Inc., 1992, p. 309.

11. *Hydrocarbon Processing,* Vol. 57, No. 4, 1978, p. 122.

12. Jesnen, B. A., "Improve Control of Cryogenic Gas Plants," *Hydrocarbon Processing,* Vol. 70, No. 5, 1991, pp. 109–111.

13. Watters, P. R., "New Partnerships Emerge in LPG and Petrochemicals Trade," *Hydrocarbon Processing,* Vol. 69, No. 6, 1990, pp. 100B–100N.

14. Brown, R. E. and Lee, F. M., "Way to Purify Cyclohexane," *Hydrocarbon Processing,* Vol. 70, No. 5, 1991, pp. 83–84.

15. "Gas Processing Handbook," *Hydrocarbon Processing,* Vol. 71, No. 4, 1992, p. 115.

16. Speight, J. G., *Applied Spectroscopy Reviews,* 5, 1972.

17. Rall, H. C. et al., *Proc. Am. Petrol. Inst.,* Vol. 42, Sec. VIII, 1962, p. 19.

18. Speight, J. G., *The Chemistry and Technology of Petroleum,* Marcel Dekker, Inc. 2nd Ed., 1991, pp. 242–243.

19. Fessenden, R. and Fessenden, J., *Organic Chemistry,* 4th Ed., Brooks/Cole Publishing Company, 1991, p. 793.

20. Kane, R. D., and Cayard, M. S. "Assess crude oil corrosivity," *Hydrocarbon Processing,* Vol. 77, No. 10, 1998, pp. 97–103.

21. Wang, S. L., Flamberg, A., and Kikabhai, T., "Select the optimum pour point depressant," *Hydrocarbon Processing,* Vol. 78, No. 2, 1999, pp. 59–62.

22. Smith, H. M., Bureau of Mines, Technical Paper, 610, 1940.

23. Matar, S., *Synfuels, Hydrocarbons of the Future,* PennWell Publishing Company, 1982, p. 38.

24. Newell, E. P., *Oil and Gas Journal,* June 28, 1999, pp. 44–46.
25. Considine, D. M., *Energy Technology Handbook,* McGraw Hill Book Co., New York, 1977, pp. 3–163.
26. Dagani, R. "Gas hydrates eyed as future energy source," *Chemical and Engineering News,* March 6, 1995, p. 39.

CHAPTER TWO

Hydrocarbon Intermediates

INTRODUCTION

Natural gas and crude oils are the main sources for hydrocarbon intermediates or secondary raw materials for the production of petrochemicals. From natural gas, ethane and LPG are recovered for use as intermediates in the production of olefins and diolefins. Important chemicals such as methanol and ammonia are also based on methane via synthesis gas. On the other hand, refinery gases from different crude oil processing schemes are important sources for olefins and LPG. Crude oil distillates and residues are precursors for olefins and aromatics via cracking and reforming processes. This chapter reviews the properties of the different hydrocarbon intermediates—paraffins, olefins, diolefins, and aromatics. Petroleum fractions and residues as mixtures of different hydrocarbon classes and hydrocarbon derivatives are discussed separately at the end of the chapter.

PARAFFINIC HYDROCARBONS

Paraffinic hydrocarbons used for producing petrochemicals range from the simplest hydrocarbon, methane, to heavier hydrocarbon gases and liquid mixtures present in crude oil fractions and residues.

Paraffins are relatively inactive compared to olefins, diolefins, and aromatics. Few chemicals could be obtained from the direct reaction of paraffins with other reagents. However, these compounds are the precursors for olefins through cracking processes. The C_6–C_9 paraffins and cycloparaffins are especially important for the production of aromatics through reforming. This section reviews some of the physical and chemical properties of C_1–C_4 paraffins. Long-chain paraffins normally present as mixtures with other hydrocarbon types in different petroleum fractions are discussed later in this chapter.

Table 2-1
Selected physical properties of C_1–C_4 paraffins

Name	Formula	Specific gravity	Boiling point °C	Calorific value Btu/ft^3
Methane	CH_4	0.554*	−161.5	1,009
Ethane	CH_3CH_3	1.049*	−88.6	1,800
Propane	$CH_3CH_2CH_3$	1.562*	−42.1	2,300
n-Butane	$CH_3(CH_2)_2CH_3$	0.579	−0.5	3,262
Isobutane	$(CH_3)_2CHCH_3$	0.557	−11.1	3,253

Air= 1.000

METHANE (CH_4)

Methane is the first member of the alkane series and is the main component of natural gas. It is also a by-product in all gas streams from processing crude oils. It is a colorless, odorless gas that is lighter than air. Table 2-1 shows selected physical properties of C_1–C_4 paraffinic hydrocarbon gases.

As a chemical compound, methane is not very reactive. It does not react with acids or bases under normal conditions. It reacts, however, with a limited number of reagents such as oxygen and chlorine under specific conditions. For example, it is partially oxidized with a limited amount of oxygen to a carbon monoxide-hydrogen mixture at high temperatures in presence of a catalyst. The mixture (synthesis gas) is an important building block for many chemicals. (Chapter 5).

Methane is mainly used as a clean fuel gas. Approximately one million BTU are obtained by burning 1,000 ft^3 of dry natural gas (methane). It is also an important source for carbon black.

Methane may be liquefied under very high pressures and low temperatures. Liquefaction of natural gas (methane), allows its transportation to long distances through cryogenic tankers.

ETHANE (CH_3-CH_3)

Ethane is an important paraffinic hydrocarbon intermediate for the production of olefins, especially ethylene. It is the second member of the alkanes and is mainly recovered from natural gas liquids.

Ethane, like methane, is a colorless gas that is insoluble in water. It does not react with acids and bases, and is not very reactive toward many reagents. It can also be partially oxidized to a carbon monoxide and hydrogen mixture or chlorinated under conditions similar to those used for methane. When ethane is combusted in excess air, it produces carbon dioxide and water with a heating value of 1,800 Btu/ft^3 (approximately double that produced from methane).

As a constituent of natural gas, ethane is normally burned with methane as a fuel gas. Ethane's relation with petrochemicals is mainly through its cracking to ethylene. Ethylene is the largest end use of ethane in the U.S. while it is only 5% in Western Europe.[1] Chapter 3 discusses steam cracking of ethane.

PROPANE (CH$_3$CH$_2$CH$_3$)

Propane is a more reactive paraffin than ethane and methane. This is due to the presence of two secondary hydrogens that could be easily substituted (Chapter 6). Propane is obtained from natural gas liquids or from refinery gas streams. Liquefied petroleum gas (LPG) is a mixture of propane and butane and is mainly used as a fuel. The heating value of propane is 2,300 Btu/ft^3. LPG is currently an important feedstock for the production of olefins for petrochemical use.

Liquid propane is a selective hydrocarbon solvent used to separate paraffinic constituents in lube oil base stocks from harmful asphaltic materials. It is also a refrigerant for liquefying natural gas and used for the recovery of condensable hydrocarbons from natural gas.

Chemicals directly based on propane are few, although as mentioned, propane and LPG are important feedstocks for the production of olefins. Chapter 6 discusses a new process recently developed for the dehydrogenation of propane to propylene for petrochemical use. Propylene has always been obtained as a coproduct with ethylene from steam cracking processes. Chapter 6 also discusses the production of aromatics from LPG through the Cyclar process.[2]

BUTANES (C$_4$H$_{10}$)

Like propane, butanes are obtained from natural gas liquids and from refinery gas streams. The C$_4$ acyclic paraffin consists of two isomers: n-butane and isobutane (2-methylpropane). The physical as well as the chemical properties of the two isomers are quite different due to structural differences. For example, the vapor pressure (Reid method) for n-butane is 52 lb/in.^{2}, while it is 71 lb/in.^{2} for isobutane. This makes the former a more favorable gasoline additive to adjust its vapor pressure. However, this use is declining in the United States due to new regulations that reduce the volatility of gasolines to 9 psi, primarily by removing butane.[3]

Isobutane, on the other hand, is a much more reactive compound due to the presence of a tertiary hydrogen.

$CH_3CH_2CH_2CH_3$ $(CH_3)_2CHCH_3$
n-Butane   Isobutane

Butane is primarily used as a fuel gas within the LPG mixture. Like ethane and propane, the main chemical use of butane is as feedstock for steam cracking units for olefin production. Dehydrogenation of n-butane to butenes and to butadiene is an important route for the production of synthetic rubber. n-Butane is also a starting material for acetic acid and maleic anhydride production (Chapter 6).

Due to its higher reactivity, isobutane is an alkylating agent of light olefins for the production of alkylates. Alkylates are a mixture of branched hydrocarbons in the gasoline range having high octane ratings (Chapter 3).

Dehydrogenation of isobutane produces isobutene, which is a reactant for the synthesis of methyl tertiary butyl ether (MTBE). This compound is currently in high demand for preparing unleaded gasoline due to its high octane rating and clean burning properties. (Octane ratings of hydrocarbons are noted later in this chapter.)

OLEFINIC HYDROCARBONS

The most important olefins used for the production of petrochemicals are ethylene, propylene, the butylenes, and isoprene. These olefins are usually coproduced with ethylene by steam cracking ethane, LPG, liquid petroleum fractions, and residues. Olefins are characterized by their higher reactivities compared to paraffinic hydrocarbons. They can easily react with inexpensive reagents such as water, oxygen, hydrochloric acid, and chlorine to form valuable chemicals. Olefins can even add to themselves to produce important polymers such as polyethylene and polypropylene. Ethylene is the most important olefin for producing petrochemicals, and therefore, many sources have been sought for its production. The following discusses briefly, the properties of these olefinic intermediates.

ETHYLENE ($CH_2=CH_2$)

Ethylene (ethene), the first member of the alkenes, is a colorless gas with a sweet odor. It is slightly soluble in water and alcohol. It is a highly active compound that reacts easily by addition to many chemical reagents. For example, ethylene with water forms ethyl alcohol. Addition of chlorine to ethylene produces ethylene dichloride (1,2-dichloroethane), which is cracked to vinyl chloride. Vinyl chloride is an important plastic precursor. Ethylene is also an active alkylating agent. Alkylation of benzene with ethylene produces ethyl benzene, which is dehydrogenated to styrene. Styrene is a monomer used in the manufacture of many commercial polymers and copolymers. Ethylene can be polymerized to different grades of polyethylenes or copolymerized with other olefins.

Catalytic oxidation of ethylene produces ethylene oxide, which is hydrolyzed to ethylene glycol. Ethylene glycol is a monomer for the production of synthetic fibers. Chapter 7 discusses chemicals based on ethylene, and Chapter 12 covers polymers and copolymers of ethylene.

Ethylene is a constituent of refinery gases, especially those produced from catalytic cracking units. The main source for ethylene is the steam cracking of hydrocarbons (Chapter 3). Table 2-2 shows the world ethylene production by source until the year 2000.[4] U.S. production of ethylene was approximately 51 billion lbs in 1997.[5]

PROPYLENE ($CH_3CH=CH_2$)

Like ethylene, propylene (propene) is a reactive alkene that can be obtained from refinery gas streams, especially those from cracking processes. The main source of propylene, however, is steam cracking of hydrocarbons, where it is coproduced with ethylene. There is no special process for propylene production except the dehydrogenation of propane.

$$\text{Catalyst}$$
$$CH_3CH_2\text{–}CH_3 \quad \rightarrow \quad CH_3CH=CH_2+H_2$$

Table 2-2
World ethylene production by feedstock[4] (MMtpd)

Feedstock	1990	1995	2000
Ethane/refinery gas	16	18	20
LPG	6	9	12
Naphtha/condensates	30	36	40
Gasoil/others	4	5	6
Total	56	68	78

Propylene can be polymerized alone or copolymerized with other monomers such as ethylene. Many important chemicals are based on propylene such as isopropanol, allyl alcohol, glycerol, and acrylonitrile. Chapter 8 discusses the production of these chemicals. U.S. production of proplylene was approximately 27.5 billion lbs in 1997.[5]

BUTYLENES (C_4H_8)

Butylenes (butenes) are by-products of refinery cracking processes and steam cracking units for ethylene production.

Dehydrogenation of butanes is a second source of butenes. However, this source is becoming more important because isobutylene (a butene isomer) is currently highly demanded for the production of oxygenates as gasoline additives.

There are four butene isomers: three unbranched, "normal" butenes (n-butenes) and a branched isobutene (2-methylpropene). The three n-butenes are 1-butene and cis- and trans- 2-butene. The following shows the four butylene isomers:

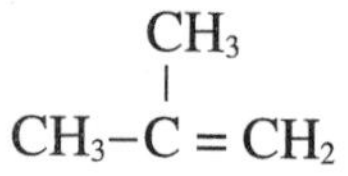

$$CH_3$$
$$|$$
$$CH_3-C = CH_2 \qquad\qquad CH_2 = CHCH_2CH_3$$

Isobutene   1- Butene

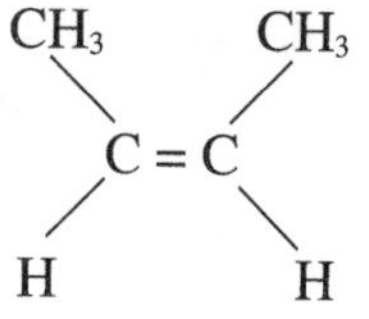

cis-2- Butene 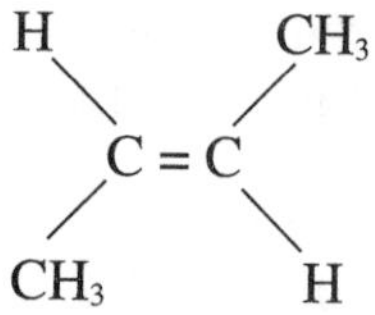   trans-2-Butene

The industrial reactions involving cis- and trans-2-butene are the same and produce the same products. There are also addition reactions where both 1-butene and 2-butene give the same product. For this reason, it is economically feasible to isomerize 1-butene to 2-butene (cis and trans) and then separate the mixture. The isomerization reaction yields two streams, one of 2-butene and the other of isobutene, which are separated by fractional distillation, each with a purity of 80–90%. Table 2-3[6] shows the boiling points of the different butene isomers.

Table 2-3
Structure and boiling points of C_4 olefins[6]

Name	Structure	Boiling Point°C
1-Butene	$CH_2=CHCH_2CH_3$	−6.3
cis-2-Butene	$\begin{array}{c} CH_3 \quad\quad CH_3 \\ \diagdown \quad\quad \diagup \\ C=C \\ \diagup \quad\quad \diagdown \\ H \quad\quad H \end{array}$	+3.7
trans-2-Butene	$\begin{array}{c} H \quad\quad CH_3 \\ \diagdown \quad\quad \diagup \\ C=C \\ \diagup \quad\quad \diagdown \\ CH_3 \quad\quad H \end{array}$	+0.9
Isobutene	$\begin{array}{c} CH_3 \\ | \\ CH_2=C-CH_3 \end{array}$	−6.6

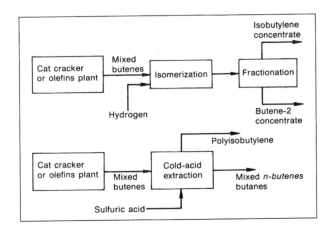

Figure 2-1. The two processes for separating n-butenes and isobutylene.[7]

An alternative method for separating the butenes is by extracting isobutene (due to its higher reactivity) in cold sulfuric acid, which polymerizes it to di- and triisobutylene. The dimer and trimer of isobutene have high octane ratings and are added to the gasoline pool.

Figure 2-1 shows the two processes for the separation of n-butenes from isobutene.[7]

Chemicals based on butenes are discussed in Chapter 9.

THE DIENES

Dienes are aliphatic compounds having two double bonds. When the double bonds are separated by only one single bond, the compound is a conjugated diene (conjugated diolefin). Nonconjugated diolefins have the double bonds separated (isolated) by more than one single bond. This latter class is of little industrial importance. Each double bond in the compound behaves independently and reacts as if the other is not present.[8] Examples of nonconjugated dienes are 1,4-pentadiene and 1,4-cyclo-hexadiene. Examples of conjugated dienes are 1,3-butadiene and 1,3-cyclohexadiene.

$$CH_2=CH-CH_2-CH=CH_2$$

1,4-Pentadiene

1,4-Cyclohexadiene

$$CH_2=CH-CH=CH_2$$

1,3-Butadiene

1,3-Cyclohexadiene

An important difference between conjugated and nonconjugated dienes is that the former compounds can react with reagents such as chlorine, yielding 1,2- and 1,4-addition products. For example, the reaction between chlorine and 1,3-butadiene produces a mixture of 1,4-dichloro-2-butene and 3,4-dichloro-1-butene:

$$CH_2 = CH-CH=CH_2 + Cl_2$$

$ClCH_2CHClCH=CH_2$
3,4-Dichloro-1-butene

$ClCH_2CH=CHCH_2Cl$
1,4-Dichloro-2-butene

When polymerizing dienes for synthetic rubber production, coordination catalysts are used to direct the reaction to yield predominantly 1,4-addition polymers. Chapter 11 discusses addition polymerization. The following reviews some of the physical and chemical properties of butadiene and isoprene.

BUTADIENE (CH$_2$=CH-CH=CH$_2$)

Butadiene is by far the most important monomer for synthetic rubber production. It can be polymerized to polybutadiene or copolymerized with styrene to styrene-butadiene rubber (SBR). Butadiene is an important intermediate for the synthesis of many chemicals such as hexamethylenediamine and adipic acid. Both are monomers for producing nylon. Chloroprene is another butadiene derivative for the synthesis of neoprene rubber.

The unique role of butadiene among other conjugated diolefins lies in its high reactivity as well as its low cost.

Butadiene is obtained mainly as a coproduct with other light olefins from steam cracking units for ethylene production. Other sources of butadiene are the catalytic dehydrogenation of butanes and butenes, and dehydration of 1,4-butanediol. Butadiene is a colorless gas with a mild aromatic odor. Its specific gravity is 0.6211 at 20°C and its boiling temperature is –4.4°C. The U.S. production of butadiene reached 4.1 billion pounds in 1997 and it was the 36th highest-volume chemical.[5]

$$\text{Isoprene } (CH_2{=}\overset{\overset{\displaystyle CH_3}{\displaystyle |}}{C}-CH{=}CH_2)$$

Isoprene (2-methyl-1,3-butadiene) is a colorless liquid, soluble in alcohol but not in water. Its boiling temperature is 34.1°C.

Isoprene is the second important conjugated diene for synthetic rubber production. The main source for isoprene is the dehydrogenation of C$_5$ olefins (tertiary amylenes) obtained by the extraction of a C$_5$ fraction from catalytic cracking units. It can also be produced through several synthetic routes using reactive chemicals such as isobutene, formaldehyde, and propene (Chapter 3).

The main use of isoprene is the production of polyisoprene. It is also a comonomer with isobutene for butyl rubber production.

AROMATIC HYDROCARBONS

Benzene, toluene, xylenes (BTX), and ethylbenzene are the aromatic hydrocarbons with a widespread use as petrochemicals. They are important precursors for many commercial chemicals and polymers such as phenol, trinitrotoluene (TNT), nylons, and plastics. Aromatic compounds are characterized by having a stable ring structure due to the overlap of the π-orbitals (resonance).

Accordingly, they do not easily add to reagents such as halogens and acids as do alkenes. Aromatic hydrocarbons are susceptible, however, to electrophilic substitution reactions in presence of a catalyst.

Aromatic hydrocarbons are generally nonpolar. They are not soluble in water, but they dissolve in organic solvents such as hexane, diethyl ether, and carbon tetrachloride.

EXTRACTION OF AROMATICS

Benzene, toluene, xylenes (BTX), and ethylbenzene are obtained mainly from the catalytic reforming of heavy naphtha. The product reformate is rich in C_6, C_7, and C_8 aromatics, which could be extracted by a suitable solvent such as sulfolane or ethylene glycol.

These solvents are characterized by a high affinity for aromatics, good thermal stability, and rapid phase separation. The Tetra extraction process by Union Carbide (Figure 2-2) uses tetraethylene glycol as a solvent.[9] The feed (reformate), which contains a mixture of aromatics, paraffins,

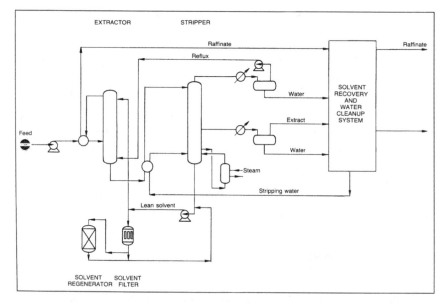

Figure 2-2. The Union Carbide aromatics extraction process using tetraethylene glycol.[9]

and naphthenes, after heat exchange with hot raffinate, is countercurrentIy contacted with an aqueous tetraethylene lycol solution in the extraction column. The hot, rich solvent containing BTX aromatics is cooled and introduced into the top of a stripper column. The aromatics extract is then purified by extractive distillation and recovered from the solvent by steam stripping. Extractive distillation has been reviewed by Gentry and Kumar.[10] The raffinate (constituted mainly of paraffins, isoparaffins and cycloparaffins) is washed with water to recover traces of solvent and then sent to storage. The solvent is recycled to the extraction tower.

The extract, which is composed of BTX and ethylbenzene, is then fractionated. Benzene and toluene are recovered separately, and ethylbenzene and xylenes are obtained as a mixture (C_8 aromatics).

Due to the narrow range of the boiling points of C_8 aromatics (Table 2-4), separation by fractional distillation is difficult. A superfractionation technique is used to segregate ethylbenzene from the xylene mixture.

Because p-xylene is the most valuable isomer for producing synthetic fibers, it is usually recovered from the xylene mixture. Fractional crystallization used to be the method for separating the isomers, but the yield was only 60%. Currently, industry uses continuous liquid-phase adsorption separation processes.[11] The overall yield of p-xylene is increased

Table 2-4
Boiling and freezing points of C_8 aromatics

Name	Structure	Boiling point °C	Freezing point °C
o-Xylene	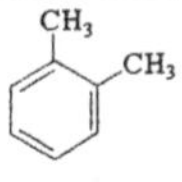	144.4	−25.2
p-Xylene	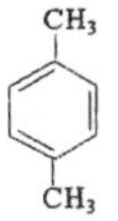	138.4	+13.3
m-Xylene	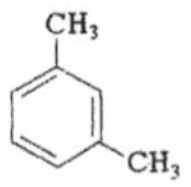	139.1	−46.8
Ethylbenzene	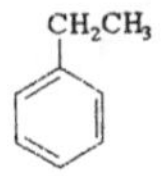	136.2	−94.9

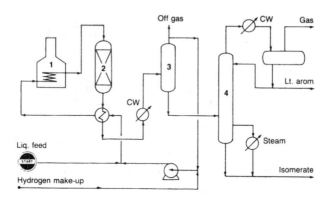

Figure 2-3. Flow diagram of the Mobil xylene isomerization process.[12]

by incorporating an isomerization unit to isomerize o- and m-xylenes to p-xylene. An overall yield of 90% p-xylene could be achieved. Figure 2-3 is a flow diagram of the Mobil isomerization process. In this process, partial conversion of ethylbenzene to benzene also occurs. The catalyst used is shape selective and contains ZSM-5 zeolite.[12]

Benzene

Benzene (C_6H_6) is the simplest aromatic hydrocarbon and by far the most widely used one. Before 1940, the main source of benzene and substituted benzene was coal tar. Currently, it is mainly obtained from catalytic reforming. Other sources are pyrolysis gasolines and coal liquids.

Benzene has a unique structure due to the presence of six delocalized π electrons that encompass the six carbon atoms of the hexagonal ring.

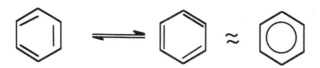

Benzene could be represented by two resonating Kekule structures.

It may also be represented as a hexagon with a circle in the middle. The circle is a symbol of the π cloud encircling the benzene ring. The delocalized electrons associated with the benzene ring impart very special properties to aromatic hydrocarbons. They have chemical properties of single-bond compounds such as paraffin hydrocarbons and double-bond compounds such as olefins, as well as many properties of their own.

Aromatic hydrocarbons, like paraffin hydrocarbons, react by substitution, but by a different reaction mechanism and under milder conditions. Aromatic compounds react by addition only under severe conditions. For example, electrophilic substitution of benzene using nitric acid produces nitrobenzene under normal conditions, while the addition of hydrogen to benzene occurs in presence of catalyst only under high pressure to

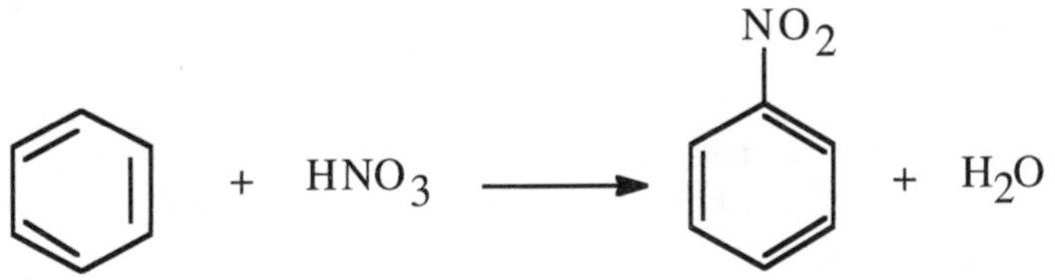

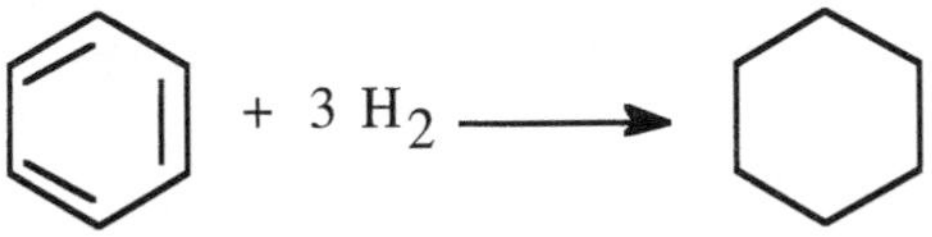

give cyclohexane:

Monosubstitution can occur at any one of the six equivalent carbons of the ring. Most of the monosubstituted benzenes have common names such as toluene (methylbenzene), phenol (hydroxybenzene), and aniline (aminobenzene).

When two hydrogens in the ring are substituted by the same reagent, three isomers are possible. The prefixes ortho, meta, and para are used to indicate the location of the substituents in 1,2-; 1,3-; or 1,4-positions. For

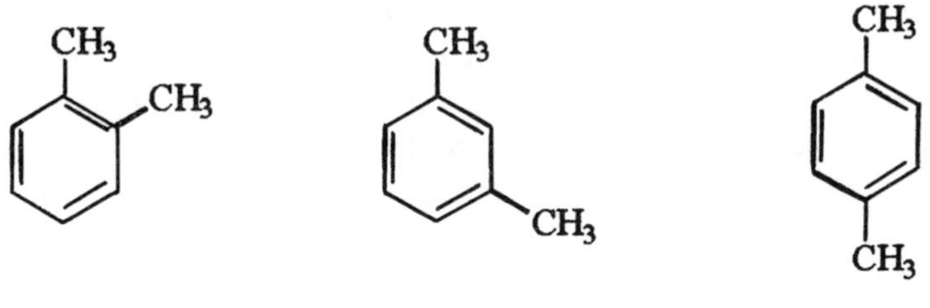

o-Xylene	m-Xylene	p-Xylene
(1,2-Dimethyl-	(1,3-Dimethyl-	(1,4-Dimethyl-
benzene)	benzene)	benzene)

example, there are three xylene isomers:

Benzene is an important chemical intermediate and is the precursor for many commercial chemicals and polymers such as phenol, styrene for polystyrenics, and caprolactom for nylon 6. Chapter 10 discusses chemicals based on benzene. The U.S. production of benzene was approximately 15 billion pounds in 1994.

Ethylbenzene

Ethylbenzene ($C_6H_5CH_2CH_3$) is one of the C_8 aromatic constituents in reformates and pyrolysis gasolines. It can be obtained by intensive fractionation of the aromatic extract, but only a small quantity of the demanded ethylbenzene is produced by this route. Most ethylbenzene is obtained by the alkylation of benzene with ethylene. Chapter 10 discusses conditions for producing ethylbenzene with benzene chemicals. The U.S. production of ethylbenzene was approximately 12.7 billion pounds in 1997. Essentially, all of it was directed for the production of styrene.

Methylbenzenes (Toluene and Xylenes)

Methylbenzenes occur in small quantities in naphtha and higher boiling fractions of petroleum. Those presently of commercial importance are toluene, o-xylene, p-xylene, and to a much lesser extent m-xylene.

The primary sources of toluene and xylenes are reformates from catalytic reforming units, gasoline from catcracking, and pyrolysis gasoline from steam reforming of naphtha and gas oils. As mentioned earlier, solvent extraction is used to separate these aromatics from the reformate mixture.

Only a small amount of the total toluene and xylenes available from these sources is separated and used to produce petrochemicals.

Toluene and xylenes have chemical characteristics similar to benzene, but these characteristics are modified by the presence of the methyl substituents. Although such modification activates the ring, toluene and xylenes have less chemicals produced from them than from benzene. Currently, the largest single use of toluene is to convert it to benzene.

para-Xylene is mainly used to produce terephthalic acid for polyesters. o-Xylene is mainly used to produce phthalic anhydride for plasticizers.

In 1997, the U.S. produced approximately 7.8 billion pounds of p-xylene and only one billion pounds of o-xylene.[5]

LIQUID PETROLEUM FRACTIONS AND RESIDUES

Liquid Petroleum fractions are light naphtha, heavy naphtha, kerosine and gas oil. The bottom product from distillation units is the residue. These mixtures are intermediates through which other reactive intermediates are obtained. Heavy naphtha is a source of aromatics via catalytic reforming and of olefins from steam cracking units. Gas oils and residues are sources of olefins through cracking and pyrolysis processes. The composition and the properties of these mixtures are reviewed in the following sections.

Naphtha

Naphtha is a generic term normally used in the petroleum refining industry for the overhead liquid fraction obtained from atmospheric distillation units. The approximate boiling range of light straight-run naphtha (LSR) is 35–90°C, while it is about 80–200°C for heavy straight-run naphtha (HSR) .

Naphtha is also obtained from other refinery processing units such as catalytic cracking, hydrocracking, and coking units. The composition of naphtha, which varies appreciably, depends mainly on the crude type and whether it is obtained from atmospheric distillation or other processing units.

Naphtha from atmospheric distillation is characterized by an absence of olefinic compounds. Its main constituents are straight and branched-chain paraffins, cycloparaffins (naphthenes), and aromatics, and the ratios of these components are mainly a function of the crude origin.

Naphthas obtained from cracking units generally contain variable amounts of olefins, higher ratios of aromatics, and branched paraffins. Due to presence of unsaturated compounds, they are less stable than straight-run naphthas. On the other hand, the absence of olefins increases the stability of naphthas produced by hydrocracking units. In refining operations, however, it is customary to blend one type of naphtha with another to obtain a required product or feedstock.

Selecting the naphtha type can be an important processing procedure. For example, a paraffinic-base naphtha is a better feedstock for steam cracking units because paraffins are cracked at relatively lower temperatures than cycloparaffins. Alternately, a naphtha rich in cycloparaffins would be a better feedstock to catalytic reforming units because cycloparaffins are easily dehydrogenated to aromatic compounds. Table 2-5 is a typical analysis of naphtha from two crude oil types.

The main use of naphtha in the petroleum industry is in gasoline production. Light naphtha is normally blended with reformed gasoline (from catalytic reforming units) to increase its volatility and to reduce the aromatic content of the product gasoline.

Heavy naphtha from atmospheric distillation units or hydrocracking

Table 2-5
**Typical analyses of two straight-run naphtha fractions from
two crude types**

Test	Marine Balayem Egypt	Bakr-9 Egypt
Boiling range °C	58–170	71–182
Specific gravity 60/60°F	0.7485	0.7350
°API	57.55	
Sulfur content wt %	0.055	0.26
Hydrocarbon types vol %		
Paraffins	62.7	80.2
Naphthenes	29.1	11.0
Aromatics	8.2	8.8

units has a low octane rating, and it is used as a feedstock to catalytic reforming units. Catalytic reforming is a process of upgrading low-octane naphtha to a high-octane reformate by enriching it with aromatics and branched paraffins. The octane rating of gasoline fuels is a property related to the spontaneous ignition of unburned gases before the flame front and causes a high pressure. A fuel with a low octane rating produces a strong knock, while a fuel with a high octane rating burns smoothly without detonation. Octane rating is measured by an arbitrary scale in which isooctane (2,2,4-trimethylpentane) is given a value of 100 and n-heptane a value of zero. A fuel's octane number equals the percentage of isooctane in a blend with n-heptane.[13]

The octane number is measured using a single-cylinder engine (CFR engine) with a variable compression ratio. The octane number of a fuel is a function of the different hydrocarbon constituents present. In general, aromatics and branched paraffins have higher octane ratings than straight-chain paraffins and cycloparaffins. Table 2-6 shows the octane rating of different hydrocarbons in the gasoline range. Chapter 3 discusses the reforming process.

Reformates are the main source for extracting C_6-C_8 aromatics used for petrochemicals. Chapter 10 discusses aromatics-based chemicals.

Naphtha is also a major feedstock to steam cracking units for the production of olefins. This route to olefins is especially important in places such as Europe, where ethane is not readily available as a feedstock because most gas reservoirs produce non-associated gas with a low ethane content.

Naphtha could also serve as a feedstock for steam reforming units for

Table 2-6
Boiling points and octane ratings of different hydrocarbons in the gasoline range

| | | Octane number clear | |
Hydrocarbon	Boiling point, °F	Research method F-1	Motor method F-2
n-Butane	0.5	...	...
n-Pentane	97	61.7	61.9
2-Methylbutane	82	92.3	90.3
2,2-Dimethylbutane	122	91.8	93.4
2,3 Dimethylbutane	137	103.5	94.3
n-Hexane	156	24.8	26.0
2-Methylpentane	146	73.4	73.5
3-Methylpentane	140	74.5	74.3
n-Heptane	208	0.0	0.0
2-Methylhexane	194	42.4	46.4
n-Octane	258	−19.0*	−15.0*
2,2,4-Trimethyl pentane (isooctane)	211	100.0	100.0
Benzene	176	...	114.8
Toluene	231	120.1	103.5
Ethylbenzene	278	107.4	97.9
Isopropylbenzene	306	...	...
o-Xylene	292	120.0*	103.0*
m-Xylene	283	145.0	124.0*
p-Xylene	281	146.0*	127.0*

Blending value of 20% in 60 octane number reference fuel.

the production of synthesis gas for methanol (Chapter 4).

KEROSINE

Kerosine, a distillate fraction heavier than naphtha, is normally a product from distilling crude oils under atmospheric pressures. It may also be obtained as a product from thermal and catalytic cracking or hydrocracking units. Kerosines from cracking units are usually less stable than those produced from atmospheric distillation and hydrocracking units due to presence of variable amounts of olefinic constituents.

Kerosine is usually a clear colorless liquid which does not stop flowing except at very low temperature (normally below −30°C). However, kerosine containing high olefin and nitrogen contents may develop some color (pale yellow) after being produced.

The main constituents of kerosines obtained from atmospheric and hydrocracking units are paraffins, cycloparaffins, and aromatics. Kerosines with a high normal-paraffin content are suitable feedstocks for extracting C_{12}-C_{14} n-paraffins, which are used for producing biodegradable detergents (Chapter 6). Currently, kerosine is mainly used to produce jet fuels, after it is treated to adjust its burning quality and freezing point. Before the widespread use of electricity, kerosine was extensively used to fuel lamps, and is still used for this purpose in remote areas. It is also used as a fuel for heating purposes.

Gas Oil

Gas oil is a heavier petroleum fraction than kerosine. It can be obtained from the atmospheric distillation of crude oils (atmospheric gas oil, AGO), from vacuum distillation of topped crudes (vacuum gas oil, VGO), or from cracking and hydrocracking units.

Atmospheric gas oil has a relatively lower density and sulfur content than vacuum gas oil produced from the same crude. The aromatic content of gas oils varies appreciably, depending mainly on the crude type and the process to which it has been subjected. For example, the aromatic content is approximately 10% for light gas oil and may reach up to 50% for vacuum and cracked gas oil. Table 2-7 is a typical analysis of atmospheric and vacuum gas oils.[14]

A major use of gas oil is as a fuel for diesel engines. Another important use is as a feedstock to cracking and hydrocracking units. Gases produced from these units are suitable sources for light olefins and LPG. Liquefied petroleum gas LPG may be used as a fuel, as a feedstock to

Table 2-7
Characteristics of typical atmospheric gas oil (AGO) and vacuum gas oil (VGO)[14]

| | Gas oil | |
Properties	Atmospheric AGO	Vacuum VGO
Specific gravity, °API	38.6	30.0
Specific gravity, 15/15°C	0.832	0.876
Boiling range, °C	232–327	299–538
Hydrogen, wt %	13.7	13.0
Aromatics, wt %	24.0	28.0

steam cracking units for olefin production, or as a feedstock for a Cyclar unit for the production of aromatics.

Residual Fuel Oil

Residual fuel oil is generally known as the bottom product from atmospheric distillation units. Fuel oils produced from cracking units are unstable. When used as fuels, they produce smoke and deposits that may block the burner orifices.

The constituents of residual fuels are more complex than those of gas oils. A major part of the polynuclear aromatic compounds, asphaltenes, and heavy metals found in crude oils is concentrated in the residue.

The main use of residual fuel oil is for power generation. It is burned in direct-fired furnaces and as a process fuel in many petroleum and chemical companies. Due to the low market value of fuel oil, it is used as a feedstock to catalytic and thermal cracking units.

Residues containing high levels of heavy metals are not suitable for catalytic cracking units. These feedstocks may be subjected to a demetallization process to reduce their metal contents. For example, the metal content of vacuum residues could be substantially reduced by using a selective organic solvent such as pentane or hexane, which separates the residue into an oil (with a low metal and asphaltene content) and asphalt (with high metal content). Demetallized oils could be processed by direct hydrocatalysis.[15]

Another approach used to reduce the harmful effects of heavy metals in petroleum residues is metal passivation. In this process an oil-soluble treating agent containing antimony is used that deposits on the catalyst surface in competition with contaminant metals, thus reducing the catalytic activity of these metals in promoting coke and gas formation. Metal passivation is especially important in fluid catalytic cracking (FCC) processes. Additives that improve FCC processes were found to increase catalyst life and improve the yield and quality of products.[16]

Residual fuels with high heavy metal content can serve as feedstocks for thermal cracking units such as delayed coking. Low-metal fuel oils are suitable feedstocks to catalytic cracking units. Product gases from cracking units may be used as a source for light olefins and LPG for petrochemical production. Residual fuel oils are also feedstocks for steam cracking units for the production of olefins.

REFERENCES

1. *Chemical Industries Newsletter,* October–December 1998, pp. 9–10.
2. "Petrochemical Handbook," *Hydrocarbon Processing,* Vol. 70, No. 3, 1991, p. 142.
3. Yepsen, G. and Witoshkin, T., "Refiners Have Options to Deal with Reformulated Gasoline," *Oil and Gas Journal,* April 8, 1991, pp. 68–71.
4. DiCintio, R. et al., "Separate Ethylene Efficiently," *Hydrocarbon Processing,* Vol. 70, No. 7, 1991, pp. 83–86.
5. *Chemical and Engineering News,* June 29, 1998, pp. 43–47.
6. Hatch, L. F. and Matar, S. "Chemicals from C_4," *Hydrocarbon Processing,* Vol. 57, No. 8, 1978, pp. 153–165.
7. *Chemical Week,* Nov. 16, 1977, p. 49.
8. Fessenden, R. J. and Fessenden, J. S., *Organic Chemistry,* 4th Ed., Brooks/Cole Publishing Co., Pacific Grove, California, 1991, p. 70.
9. "Petrochemical Handbook," *Hydrocarbon Processing,* Vol. 61, No. 11, 1982, p. 195.
10. Gentry, J. C. and Kumar, C. S. "Improve BTX Processing Economics" *Hydrocarbon Processing,* Vol. 77, No. 3, 1998, pp. 69–82.
11. Biesser, H. J. and Winter, G. R., *Oil and Gas Journal,* Aug. 11, 1975, pp. 74–75.
12. "Petrochemical Handbook," *Hydrocarbon Processing,* Vol. 70, No. 3, 1991, pp. 166.
13. Matar, S., *Synfuels; Hydrocarbons of the Future,* PennWell Publishing Co., Tulsa, Okla, 1982, p. 10.
14. Barwell, J. and Martin, S. R., International Seminar on Petrochemical Industries, No. 9 (P-2) Iraq, Oct. 25–30, 1975.
15. *Oil and Gas Journal,* March 20, 1978, p. 94.
16. Krishna, A. S. et al., "Additives Improve FCC Process," *Hydrocarbon Processing,* Vol. 70, No. 11, 1991, pp. 59–66.

CHAPTER THREE

Crude Oil Processing and Production of Hydrocarbon Intermediates

INTRODUCTION

The hydrocarbon intermediates referred to in the previous chapter are produced by subjecting crude oils to various processing schemes. These include a primary distillation step to separate the crude oil complex mixture into simpler fractions. These fractions are primarily used as fuels. However, a small percentage of these streams are used as secondary raw materials or intermediates for obtaining olefins, diolefins, and aromatics for petrochemicals production. Further processing of these fractions may be required to change their chemical composition to the required products. These new products may also be used as fuels of improved qualities or as chemical feedstocks. For example, reforming a naphtha fraction catalytically produces a reformate rich in aromatics. The major use of the reformate is to supplement the gasoline pool due to its high octane rating. However, the reformate is also used to extract the aromatics for petrochemicals use. At this point, the production of intermediates for petrochemicals is not separable from the production of fuels. In this chapter, the production of hydrocarbon intermediates is discussed in conjunction with different crude oil processing schemes. These include physical separation techniques and chemical conversion processes. The production of olefins is also discussed in the last section.

PHYSICAL SEPARATION PROCESSES

Physical separation techniques separate a mixture such as a crude oil without changing the chemical characteristics of the components. The separation is based on differences of certain physical properties of the constituents such as the boiling and melting points, adsorption affinities on a certain solid, and diffusion through certain membranes.

The important physical separation processes, discussed here, are distillation, absorption, adsorption, and solvent extraction.

ATMOSPHERIC DISTILLATION

Atmospheric distillation separates the crude oil complex mixture into different fractions with relatively narrow boiling ranges. In general, separation of a mixture into fractions is based primarily on the difference in the boiling points of the components. In atmospheric distillation units, one or more fractionating columns are used.

Distilling a crude oil starts by preheating the feed by exchange with the hot product streams. The feed is further heated to about 320°C as it passes through the heater pipe (pipe still heater).

The hot feed enters the fractionator, which normally contains 30–50 fractionation trays. Steam is introduced at the bottom of the fractionator to strip off light components. The efficiency of separation is a function of the number of theoretical plates of the fractionating tower and the reflux ratio. Reflux is provided by condensing part of the tower overhead vapors. Reflux ratio is the ratio of vapors condensing back to the still to vapors condensing out of the still (distillate). The higher the reflux ratio, the better the separation of the mixture.

Products are withdrawn from the distillation tower as side streams, while the reflux is provided by returning a portion of the cooled vapors from the tower overhead condenser. Additional reflux could be obtained by returning part of the cold side stream products to the tower. In practice, the reflux ratio varies over a wide range according to the specific separations desired. From the overhead condenser, the uncondensed gases are separated, and the condensed light naphtha liquid is withdrawn to storage. Heavy naphtha, kerosine, and gas oil are withdrawn as side stream products. Table 3-1 shows the approximate boiling ranges for crude oil fractions. The residue (topped crude) is removed from the bottom of the distillation tower and may be used as a fuel oil. It may also be charged to a vacuum distillation unit, a catalytic cracking or steam cracking process. Figure 3-1 is a flow diagram for atmospheric and vacuum distillation units.[1]

Table 3-1
Approximate ASTM boiling point ranges for crude oil fractions

| | Boiling range | |
Fractions	°F	°C
Light naphtha	85–210	30–99
Heavy naphtha	190–400	88–204
Kerosine	340–520	171–271
Atmospheric gas oil	540–820	288–438
Vacuum gas oil	750–1,050	399–566
Vacuum residue	1,000+	538+

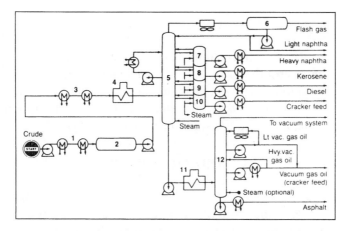

Figure 3-1. Flow diagram of atmospheric and vacuum distillation units:[1] (1,3) heat exchangers; (2) desalter, (3,4) heater; (5) distillation column, (6) overhead condenser, (7–10) pump around streams, (11) vacuum distillation heater; (12) vacuum tower.

VACUUM DISTILLATION

Vacuum distillation increases the amount of the middle distillates and produces lubricating oil base stocks and asphalt. The feed to the unit is the residue from atmospheric distillation. In vacuum distillation, reduced pressures are applied to avoid cracking long-chain hydrocarbons present in the feed.

The feed is first preheated by exchange with the products, charged to the vacuum unit heater, and then passed to the vacuum tower in an atmosphere of superheated steam. Using superheated steam is important: it decreases the partial pressure of the hydrocarbons and reduces coke formation in the furnace tubes. Distillation normally occurs at a temperature range of 400–440°C and an absolute pressure of 25–40 mmHg. The top tower temperature is adjusted by refluxing part of the gas oil product (top product). The size (diameter) of the vacuum distillation tower is much larger than atmospheric towers because the volume of the vapor/unit-volume of the feed is much larger than in atmospheric distillation.[2]

Products obtained as side streams are vacuum gas oil (VGO), lube oil base stocks, and asphalt. Asphalt may be used for paving roads or may be charged to a delayed coking unit.

ABSORPTION PROCESS

This process selectively removes a certain gas from a gas mixture using a liquid absorbent. In the refining industry, this process is used extensively to free the product gas streams from acid gases (mainly H_2S) either by using a physical or a chemical absorbent. Absorption of acid gases from natural gas are discussed in Chapter 1.

ADSORPTION PROCESS

Adsorption processes use a solid material (adsorbent) possessing a large surface area and the ability to selectively adsorb a gas or a liquid on its surface. Examples of adsorbents are silica (SiO_2), anhydrous alumina (Al_2O_3), and molecular sieves (crystalline silica/alumina). Adsorption processes may be used to remove acid gases from natural gas and gas streams. For example, molecular sieves are used to dehydrate natural gas and to reduce its acid gases.

Adsorption processes are also used to separate liquid mixtures. For example, molecular sieve 5A selectively adsorbs n-paraffins from a low-octane naphtha fraction. Branched paraffins and aromatics in the mixture are not adsorbed on the solid surface. The collected fraction containing mainly aromatics and branched paraffins have a higher octane number than the feed. Desorbing n-paraffins is effected by displacement with another solvent or by using heat. The recovered n-paraffins in this range are good steam cracking feedstocks for olefin production.

Adsorption of n-paraffins (C_{10}-C_{14}) from a kerosine or a gas oil fraction can be achieved in a liquid or a vapor phase adsorption process.

Normal paraffins in this range are important intermediates for alkylating benzene for synthetic detergents production (Chapter 10). They are also good feedstocks for single-cell protein (SCP).

The IsoSiv process is an isobaric, isothermal adsorption technique used to separate n-paraffins from gas oils. The operation conditions are approximately 370°C and 100 psi.[3] Desorption is achieved using n-pentane or n-hexane. The solvent is easily distilled from the heavier n-paraffins and then recycled.

SOLVENT EXTRACTION

Liquid solvents are used to extract either desirable or undesirable compounds from a liquid mixture. Solvent extraction processes use a liquid solvent that has a high solvolytic power for certain compounds in the feed mixture. For example, ethylene glycol has a greater affinity for aromatic hydrocarbons and extracts them preferentially from a reformate mixture (a liquid paraffinic and aromatic product from catalytic reforming). The raffinate, which is mainly paraffins, is freed from traces of ethylene glycol by distillation. Other solvents that could be used for this purpose are liquid sulfur dioxide and sulfolane (tetramethylene sulfone).

The sulfolane process is a versatile extractant for producing high purity BTX aromatics (benzene, toluene, and xylenes). It also extracts aromatics from kerosines to produce low-aromatic jet fuels.

On the other hand, liquid propane also has a high affinity for paraffinic hydrocarbons. Propane deasphalting removes asphaltic materials from heavy lube oil base stocks. These materials reduce the viscosity index of lube oils. In this process, liquid propane dissolves mainly paraffinic hydrocarbons and leaves out asphaltic materials. Higher extraction temperatures favor better separation of the asphaltic components. Deasphalted oil is stripped to recover propane, which is recycled.

Solvent extraction may also be used to reduce asphaltenes and metals from heavy fractions and residues before using them in catalytic cracking. The organic solvent separates the resids into demetallized oil with lower metal and asphaltene content than the feed, and asphalt with high metal content. Figure 3-2 shows the IFP deasphalting process and Table 3-2 shows the analysis of feed before and after solvent treatment.[4]

Solvent extraction is used extensively in the petroleum refining industry. Each process uses its selective solvent, but, the basic principle is the same as above.

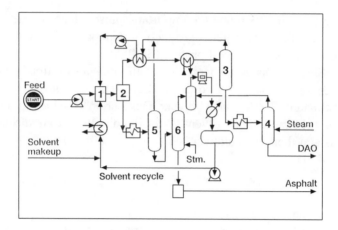

Figure 3-2. The IFP deasphalting process:[4] (1,2) extractor, (3-6) solvent recovery towers.

Table 3-2
Typical analysis of light Arabian vacuum resid before and after solvent treatment using once C_4 and another C_5 hydrocarbon solvent[4]

	Feed	DAO	
Solvent	—	C_4	C_5
Yield, wt %	—	70.1	85.5
Sp. gr.	1.003	0.959	0.974
Visc., cSt @ 210°F	345	63	105
Conradson carbon, wt %	16.4	5.3	7.9
Asphaltenes (C_7 insol.), wt %	4.20	<0.05	<0.05
Ni, ppm	19	2.0	7.0
V, ppm	61	2.6	15.5
S, wt%	4.05	3.3	3.65
N_2, ppm	2,875	1,950	2,170

CONVERSION PROCESSES

Conversion processes in the petroleum industry are generally used to:

1. Upgrade lower-value materials such as heavy residues to more valuable products such as naphtha and LPG. Naphtha is mainly used to supplement the gasoline pool, while LPG is used as a fuel or as a petrochemical feedstock.

2. Improve the characteristics of a fuel. For example, a lower octane naphtha fraction is reformed to a higher octane reformate product. The reformate is mainly blended with naphtha for gasoline formulation or extracted for obtaining aromatics needed for petrochemicals production.
3. Reduce harmful impurities in petroleum fractions and residues to control pollution and to avoid poisoning certain processing catalysts. For example, hydrotreatment of naphtha feeds to catalytic reformers is essential because sulfur and nitrogen impurities poison the catalyst.

Conversion processes are either thermal, where only heat is used to effect the required change, or catalytic, where a catalyst lowers the reaction activation energy. The catalyst also directs the reaction toward a desired product or products (selective catalyst).

THERMAL CONVERSION PROCESSES

Thermal cracking was the first process used to increase gasoline production. After the development of catalytic cracking, which improved yields and product quality, thermal cracking was given other roles in refinery operations. The three important thermal cracking techniques are coking, viscosity breaking, and steam cracking.

Steam cracking is of special importance as a major process designed specifically for producing light olefins. It is discussed separately later in this chapter.

Coking Processes

Coking is a severe thermal cracking process designed to handle heavy residues with high asphaltene and metal contents. These residues cannot be fed to catalytic cracking units because their impurities deactivate and poison the catalysts.

Products from coking processes vary considerably with feed type and process conditions. These products are hydrocarbon gases, cracked naphtha, middle distillates, and coke. The gas and liquid products are characterized by a high percentage of unsaturation. Hydrotreatment is usually required to saturate olefinic compounds and to desulfurize products from coking units.

Thermal Cracking Reactions

The first step in cracking is the thermal decomposition of hydrocarbon molecules to two free radical fragments. This initiation step can occur by a homolytic carbon-carbon bond scission at any position along the hydrocarbon chain. The following represents the initiation reaction:

$$RCH_2CH_2CH_2R' \rightarrow RCH_2\dot{C}H_2 + R'\dot{C}H_2$$

The radicals may further crack, yielding an olefin and a new free radical. Cracking usually occurs at a bond beta to the carbon carrying the unpaired electron.

$$RCH_2\dot{C}H_2 \rightarrow \dot{R} + CH_2=CH_2$$

Further β bond scission of the new free radical $\dot{R}$ can continue to produce ethylene until the radical is terminated.

Free radicals may also react with a hydrocarbon molecule from the feed by abstracting a hydrogen atom. In this case the attacking radical is terminated, and a new free radical is formed. Abstraction of a hydrogen atom can occur at any position along the chain. However, the rate of hydrogen abstraction is faster from a tertiary position than from a secondary, which is faster than from a primary position.

$$\dot{R} + RCH_2CH_2CH_2R' \rightarrow RCH_2\dot{C}HCH_2R' + RH$$

The secondary free radical can crack on either side of the carbon carrying the unpaired electron according to the beta scission rule, and a terminal olefin is produced.

$$RCH_2\dot{C}HCH_2R' \underbrace{\begin{array}{l} \longrightarrow \dot{R} + \overset{\cdot}{R'}CH_2CH=CH_2 \\ \longrightarrow \dot{R'} + RCH_2CH=CH_2 \end{array}}$$

Free radicals, unlike carbocations, do not normally undergo isomerization by methyl or hydrogen migration. However, hydrogen transfer (chain transfer) occurs when a free radical reacts with other hydrocarbons.

There are two major commercial thermal cracking processes, delayed coking and fluid coking. Flexicoking is a fluid coking process in which the coke is gasified with air and steam. The resulting gas mixture partially provides process heat.

Delayed Coking

In delayed coking, the reactor system consists of a short contact-time heater coupled to a large drum in which the preheated feed "soaks" on a batch basis. Coke gradually forms in the drum. A delayed coking unit has at least a pair of drums. When the coke reaches a predetermined level in one drum, flow is diverted to the other so that the process is continuous.

Vapors from the top of the drum are directed to the fractionator where they are separated into gases, naphtha, kerosine, and gas oil. Table 3-3 shows products from a delayed coker using different feeds.[5]

Decoking the filled drum can be accomplished by a hydraulic system using several water jets under at least 3,000 pounds per square inch gauge.

Operating conditions for delayed coking are 25–30 psi at 480–500°C, with a recycle ratio of about 0.25 based on equivalent feed. Improved liquid yields could be obtained by operating at lower pressures. Coking at approximately 15 psi with ultra low recycle produced about 10% more gas oil.[6] Operating at too-low temperature produces soft spongy coke. On the other hand, operating at a higher temperature produces more coke and gas but less liquid products. Mochida et al. reviewed the chemistry and different options for the production of delayed coke.[7] It is the chemistry of the pyrolysis system which controls the properties of the semi

Table 3-3
Feeds and products from a delayed coker unit
(using different feeds)[5]

Operating conditions:

Heater outlet temperature, °F	900–950
Coke drum pressure, psig	15–90
Recycle ratio, vol/vol feed, %	10–100

Yields:

Feedstock	Middle East vac. residue	Vacuum residue of hydrotreated bottoms	Coal tar pitch
Gravity, °API	7.4	1.3	–11.0
Sulfur, wt %	4.2	2.3	0.5
Conradson carbon, wt %	20.0	27.6	—
Products, wt %			
Gas + LPG	7.9	9.0	3.9
Naphtha	12.6	11.1	—
Gas oil	50.8	44.0	31.0
Coke	28.7	35.9	65.1

and final coke structure. Factors that govern the reactions are the coke drum size, the heating rate, the soak time, the pressure, and the final reaction temperature.[8] However, if everything is equal (temperature, pressure, soak time, etc.), the quality of coke produced by delayed coking is primarily a function of the feed quality. Figure 3-3 shows a delayed coking unit.[5]

Coke produced from delayed coking is described as delayed sponge, shot, or needle coke depending on its physical structure. Shot coke is the most common when running the unit under severe conditions with sour crude residues. Needle coke is produced from selected aromatic feedstocks. Sponge coke is more porous and has a high surface area. The properties and markets for petroleum cokes have been reviewed by Dymond.[9] Table 3-4 shows the types of petroleum cokes and their uses.[9]

Fluid Coking

In the fluid coking process, part of the coke produced is used to provide the process heat. Cracking reactions occur inside the heater and the fluidized-bed reactor. The fluid coke is partially formed in the heater. Hot coke slurry from the heater is recycled to the fluid reactor to provide the heat required for the cracking reactions. Fluid coke is formed by spraying the hot feed on the already-formed coke particles. Reactor temperature is about 520°C, and the conversion into coke is immediate, with

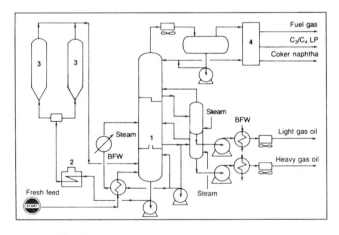

Figure 3-3. Flow diagram of a delayed coking unit:[5] (1) coker fractionator, (2) coker heater, (3) coke drum, (4) vapor recovery column.

Table 3-4
Types of petroleum cokes and their end uses[9]

Application	Type coke	State	End use
Carbon source	Needle	Calcined	Electrodes
			Synthetic graphite
	Sponge	Calcined	Aluminum anodes
			TiO_2 pigments
			Carbon raiser
	Sponge	Green	Silicon carbide
			Foundries
			Coke ovens
Fuel use	Sponge	Green lump	Europe/Japan space heating
	Sponge	Green	Industrial boilers
	Shot	Green	Utilities
	Fluid	Green	Cogeneration
	Flexicoke	Green	Lime
			Cement

complete disorientation of the crystallites of product coke. The burning process in fluid coking tends to concentrate the metals, but it does not reduce the sulfur content of the coke.

Fluid coking has several characteristics that make it undesirable for most petroleum coke markets. These characteristics are high sulfur content, low volatility, poor crystalline structure, and low grindability index.[10]

Flexicoking, on the other hand, integrates fluid coking with coke gasification. Most of the coke is gasified. Flexicoking gasification produces a substantial concentration of the metals in the coke product. Figure 3-4 shows an Exxon flexicoking process.[5]

Viscosity Breaking (Vis-breaking)

Viscosity breaking aims to thermally crack long-chain feed molecules to shorter ones, thus reducing the viscosity and the pour point of the product.

In this process, the feed is usually a high viscosity, high pour point fuel oil that cannot be used or transported, especially in cold climates, due to the presence of waxy materials. Wax is a complex mixture of long-chain paraffins mixed with aromatic compounds having long paraffinic side chains. Vis-breaking is a mild cracking process that operates at approximately 450°C using short residence times. Long paraffinic chains break to

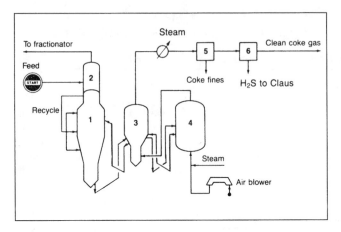

Figure 3-4. Flow diagram of an Exxon flexicoking unit:[5] (1) reactor, (2) scrubber, (3) heater, (4) gasifier, (5) coke fines removal, (6) H_2S removal.

shorter ones, and dealkylation of the aromatic side chains occurs. Table 3-5 shows the analysis of feed and products from a vis-breaking unit.[11]

CATALYTIC CONVERSION PROCESSES

Catalytic conversion processes include naphtha catalytic reforming, catalytic cracking, hydrocracking, hydrodealkylation, isomerization, alkylation, and polymerization. In these processes, one or more catalyst is used. A common factor among these processes is that most of the reactions are initiated by an acid-type catalyst that promotes carbonium ion formation.

Other important catalytic processes are those directed toward improving the product quality through hydrotreatment. These processes use heterogeneous hydrogenation catalysts.

Catalytic Reforming

The aim of this process is to improve the octane number of a naphtha feedstock by changing its chemical composition. Hydrocarbon compounds differ greatly in their octane ratings due to differences in structure. In general, aromatics have higher octane ratings than paraffins and cycloparaffins. Similar to aromatics, branched paraffins have high octane ratings. The octane number of a hydrocarbon mixture is a function of the octane numbers of the different components and their ratio in the mixture. (See octane ratings of different hydrocarbons in Chapter 2.)

Table 3-5
Analysis of feed and products from viscosity breaking[11]

Charge inspections	Libyan residue
Gravity, °API	24.4
Vacuum Engler, corrected °F	
IBP	510
5%	583
10%	608
20%	650
Pour point (max.), °F	75
Visc. SUS @ 122°F	175.8
Product yield, vol %	
Gasoline, 100% C_4, 330 EP	10.8
Furnace oil, 805°F EP	42.7
Fuel oil	46.3
Gas, C_3 & Lighter (wt %)	2.1
Properties of products	
Furnace oil	
Pour point (max.), °F	+5
Flash (PMCO), °F	150
Fuel oil	
Pour point (max.), °F	+40
Flash (PMCC), °F	150
Visc., SFS @ 122°F	67.5
Stability (ASTM D-1661)	No. 1

Increasing the octane number of a low-octane naphtha fraction is achieved by changing the molecular structure of the low octane number components. Many reactions are responsible for this change, such as the dehydrogenation of naphthenes and the dehydrocyclization of paraffins to aromatics. Catalytic reforming is considered the key process for obtaining benzene, toluene, and xylenes (BTX). These aromatics are important intermediates for the production of many chemicals.[12]

Reformer Feeds

The feed to a catalytic reformer is normally a heavy naphtha fraction produced from atmospheric distillation units. Naphtha from other sources such as those produced from cracking and delayed coking may also be used. Before using naphtha as feed for a catalytic reforming unit, it must be hydrotreated to saturate the olefins and to hydrodesulfurize and hydrodenitrogenate sulfur and nitrogen compounds. Olefinic compounds are undesirable because they are precursors for coke, which deactivates the catalyst. Sulfur and nitrogen compounds poison the reforming catalyst. The reducing atmosphere in catalytic reforming promotes forming of hydrogen sulfide and ammonia. Ammonia reduces the acid sites of the catalyst, while platinum becomes sulfided with H_2S.

Types of hydrocarbons in the feed have significant effects on the operation severity. Feeds with a high naphthene content are easier to aromatize than feeds with a high ratio of paraffins (see "Reforming reactions"). The boiling range of the feeds is also an effective parameter. Feeds with higher end points ($\approx200°C$) are favorable because some of the long-chain molecules are hydrocracked to molecules in the gasoline range. These molecules can isomerize and dehydrocyclize to branched paraffins and to aromatics, respectively.

Reforming Catalysts

The catalysts generally used in catalytic reforming are dual functional to provide two types of catalytic sites, hydrogenation-dehydrogenation sites and acid sites. The former sites are provided by platinum, which is the best known hydrogenation-dehydrogenation catalyst and the latter (acid sites) promote carbonium ion formation and are provided by an alumina carrier. The two types of sites are necessary for aromatization and isomerization reactions.

Bimetallic catalysts such as Pt/Re were found to have better stability, increased catalyst activity, and selectivity. Trimetallic catalysts of noble metal alloys are also used for the same purpose. The increased stability of these catalysts allowed operation at lower pressures. A review of reforming catalysts by Al-Kabbani manifests the effect of the ratio of the metallic components of the catalyst. A ratio of 0.5 or less for Pt/Re in the new generation catalysts versus 1.0 for the older ones can tolerate much higher coke levels. Reforming units can perform similarly with higher coke levels (20–25% versus 15–20%). These catalysts can tolerate higher sulfer naphtha feeds (>1 ppm). Higher profitability may be realized by increasing the cycle length.[13]

Reforming Reactions

Many reactions occur in the reactor under reforming conditions. These are aromatization reactions, which produce aromatics; isomerization reactions, which produce branched paraffins; and other reactions, which are not directly involved in aromatics formation (hydrocracking and hydrodealkylation).

Aromatization. The two reactions directly responsible for enriching naphtha with aromatics are the dehydrogenation of naphthenes and the dehydrocyclization of paraffins. The first reaction can be represented by the dehydrogenation of cyclohexane to benzene.

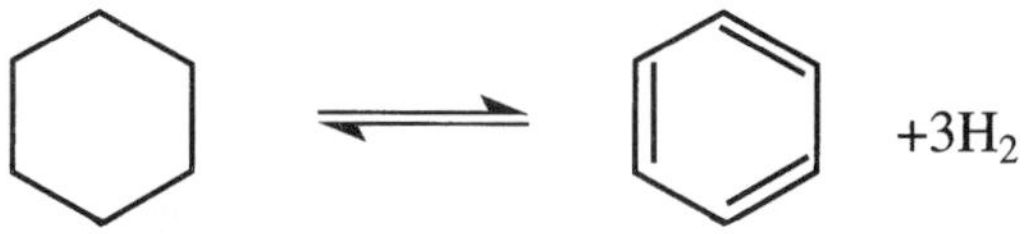

$$+3H_2 \qquad \Delta H = +221 \text{ KJ/mol}$$
$$K_p = 6 \times 10^5 @ 500°C$$

This reaction is fast; it reaches equilibrium quickly. The reaction is also reversible, highly endothermic, and the equilibrium constant is quite large (6×10^5 @ 500°C).

It is evident that the yield of aromatics (benzene) is favored at higher temperatures and lower pressures. The effect of decreasing H_2 partial pressure is even more pronounced in shifting the equilibrium to the right.

The second aromatization reaction is the dehydrocyclization of paraffins to aromatics. For example, if n-hexane represents this reaction, the first step would be to dehydrogenate the hexane molecule over the platinum surface, giving 1-hexene (2- or 3-hexenes are also possible isomers, but cyclization to a cyclohexane ring may occur through a different mechanism). Cyclohexane then dehydrogenates to benzene.

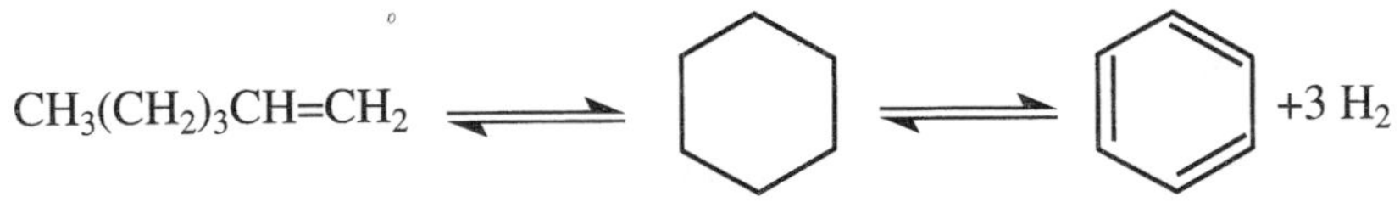

$$CH_3(CH_2)_3CH=CH_2$$

$$\Delta H = +266 \text{ KJ/mol}$$
$$K_p = 7.8 \times 104 @ 500°C$$

This is also an endothermic reaction, and the equilibrium production of aromatics is favored at higher temperatures and lower pressures. However, the relative rate of this reaction is much lower than the dehydrogenation of cyclohexanes. Table 3-6 shows the effect of temperature on the selectivity to benzene when reforming n-hexane using a platinum catalyst.[14]

Table 3-6
Selectivity to benzene from reforming n-hexane over a
platinum catalyst[14]

LHSV	Temp.,°F	% Conversion	Selectivity to Benzene	Selectivity to Isohexane
2	885	80.2	16.6	58
2	932	86.8	24.1	36.9
2	977	90.4	27.4	23.4

More often than what has been mentioned above regarding the cyclization of paraffins over the platinum catalyst, the formed olefin species reacts with the acid catalyst forming a carbocation. Carbocation formation may occur by abstraction of a hydride ion from any position along the hydrocarbon chain. However, if the carbocation intermediate has the right configuration, cyclization occurs. For example, cyclization of 1-heptene over the alumina catalyst can occur by the following successive steps:

$$CH_3CH_2(CH_2)_3CH{=}CH_2 \longrightarrow CH_3\overset{+}{C}H(CH_2)_3CH{=}CH_2$$

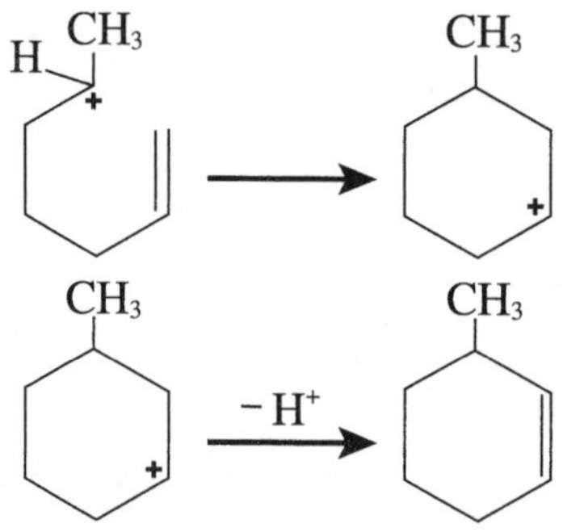

The formed methylcyclohexane carbocation eliminates a proton, yielding 3-methylcyclohexene. 3-Methylcyclohexene can either dehydrogenate over the platinum surface or form a new carbocation by losing H⁻ over the acid catalyst surface. This step is fast, because an allylic carbonium ion is formed. Losing a proton on a Lewis base site produces methyl cyclohexadiene. This sequence of carbocation formation, followed by loss of a proton, continues till the final formation of toluene.

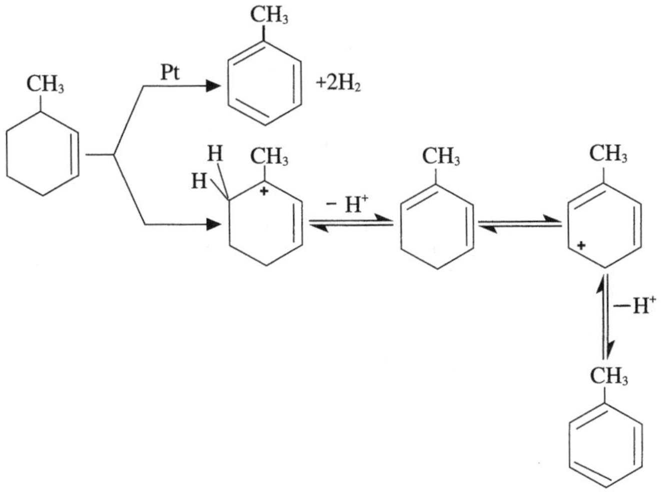

It should be noted that both reactions leading to aromatics (dehydrogenation of naphthenes and dehydrocyclization of paraffins) produce hydrogen and are favored at lower hydrogen partial pressure.

Isomerization. Reactions leading to skeletal rearrangement of paraffins and cycloparaffins in a catalytic reactor are also important in raising the octane number of the reformate product. Isomerization reactions may occur on the platinum catalyst surface or on the acid catalyst sites. In the former case, the reaction is slow. Most isomerization reactions, however, occur through formation of a carbocation. The formed carbocation could rearrange through a hydride-methide shift that would lead to branched isomers. The following example illustrates the steps for the isomerization of n-heptane to 2-methylhexane through 1,2-methide-hydride shifts:

Carbocation Formation:

$$CH_3CH_2CH_2(CH_2)_3CH_3 \longrightarrow CH_3CH_2\overset{+}{C}H(CH_2)_3CH_3$$

1,2-Methide-Hydride Shift:

$$CH_3CH_2\overset{+}{C}H(CH_2)_3CH_3 \longrightarrow CH_3-\underset{\underset{+}{|}}{\overset{\overset{CH_3}{|}}{C}}(CH_2)_3CH_3$$

Hydride Abstraction:

$$\underset{\overset{\displaystyle |}{+}}{\overset{\overset{\displaystyle CH_3}{\displaystyle |}}{CH_3-C(CH_2)_3CH_3}} + RH \longrightarrow R^+ + \underset{}{\overset{\overset{\displaystyle CH_3}{\displaystyle |}}{CH_3-CH(CH_2)_3CH_3}}$$

RH = Acyclic or cyclic hydrocarbon molecule

Isomerization of alkylcyclopentanes may also occur on the platinum catalyst surface or on the silica/alumina. For example, methylcyclopentane isomerizes to cyclohexane:

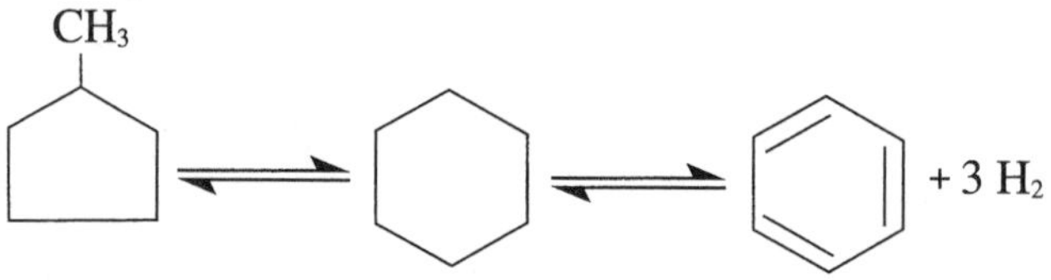

The formed cyclohexane can dehydrogenate to benzene.

Hydrocracking. Hydrocracking is a hydrogen-consuming reaction that leads to higher gas production and lower liquid yield. This reaction is favored at high temperatures and high hydrogen partial pressure. The following represents a hydrocracking reaction:

$$RCH_2CH_2CH_2R' + H_2 \rightarrow RCH_2CH_3 + R'CH_3$$

Bond breaking can occur at any position along the hydrocarbon chain. Because the aromatization reactions mentioned earlier produce hydrogen and are favored at high temperatures, some hydrocracking occurs also under these conditions. However, hydrocracking long-chain molecules can produce C_6, C_7, and C_8 hydrocarbons that are suitable for hydrodecyclization to aromatics.

For more aromatics yield, the end point of the feed may be raised to include higher molecular weight hydrocarbons in favor of hydrocracking and dehydrocyclization. However, excessive hydrocracking is not desirable because it lowers liquid yields.

Hydrodealkylation. Hydrodealkylation is a cracking reaction of an aromatic side chain in presence of hydrogen. Like hydrocracking, the reaction consumes hydrogen and is favored at a higher hydrogen partial pressure. This reaction is particularly important for increasing benzene yield when methylbenzenes and ethylbenzene are dealkylated. Although the overall reaction is slightly exothermic, the cracking step is favored at higher temperatures. Hydrodealkylation may be represented by the reaction of toluene and hydrogen.

As in hydrocracking, this reaction increases the gas yield and changes the relative equilibrium distribution of the aromatics in favor of benzene. Table 3-7 shows the properties of feed and products from Chevron Rheiniforming process.[15]

Table 3-7
Properties of feed and products from Chevron Rheiniforming process[15]

Yields: Typical yields for severe reforming:

Naphtha Feed	Hydrotreated		Hydrocracked
Feed type	Paraffinic		Naphthenic
Boiling range, °F	200–330		200–390
Paraffins, LV%	68.6		32.6
Naphthenes, LV%	23.4		55.5
Aromatics, LV%	8.0		11.9
Sulfur, ppm	<0.2		<0.2
Nitrogen, ppm	<0.5		<0.5
Reactor outlet press., psig	90	200	200
Products			
Hydrogen, scf/bbl feed	1,510	1,205	1,400
C_1-C_3, scf/bbl feed	160	355	160
C_5^+ reformate			
Yield, LV%	80.1	73.5	84.7
Research octane clear	98	99	100
Paraffins, LV%	32.4	31.2	27.5
Naphthenes, LV%	1.1	0.9	2.6
Aromatics, LV%	66.5	67.9	69.9

Reforming Process

Catalytic reformers are normally designed to have a series of catalyst beds (typically three beds). The first bed usually contains less catalyst than the other beds. This arrangement is important because the dehydrogenation of naphthenes to aromatics can reach equilibrium faster than the other reforming reactions. Dehydrocyclization is a slower reaction and may only reach equilibrium at the exit of the third reactor. Isomerization and hydrocracking reactions are slow. They have low equilibrium constants and may not reach equilibrium before exiting the reactor.

The second and third reactors contain more catalyst than the first one to enhance the slow reactions and allow more time in favor of a higher yield of aromatics and branched paraffins. Because the dehydrogenation of naphthenes and the dehydrocyclization of paraffins are highly endothermic, the reactor outlet temperature is lower than the inlet temperature. The effluent from the first and second reactors are reheated to compensate for the heat loss.

Normally, catalytic reformers operate at approximately 500–525°C and 100–300 psig, and a liquid hourly space velocity range of 2–4 hr^{-1}. Liquid hourly space velocity (LHSV) is an important operation parameter expressed as the volume of hydrocarbon feed per hour per unit volume of the catalyst. Operating at lower LHSV gives the feed more contact with the catalyst.

Regeneration of the catalyst may be continuous for certain processes that are designed to permit the removal and replacement of the catalyst during operation. In certain other processes, an additional reactor is used (Swing reactor). When the activity of the catalyst is decreased in one of the reactors on stream, it is replaced with the stand-by (Swing) reactor.

In many processes, regeneration occurs by shutting down the unit and regenerating the catalyst (Semi-regenerative). Figure 3-5 shows a Chevron Rheiniforming semiregenerative fixed three-bed process.[15]

Products from catalytic reformers (the reformate) is a mixture of aromatics, paraffins and cycloparaffins ranging from C_6-C_8. The mixture has a high octane rating due to presence of a high percentage of aromatics and branched paraffins. Extraction of the mixture with a suitable solvent produces an aromatic-rich extract, which is further fractionated to separate the BTX components. Extraction and extractive distillation of reformate have been reviewed by Gentray and Kumar.[16]

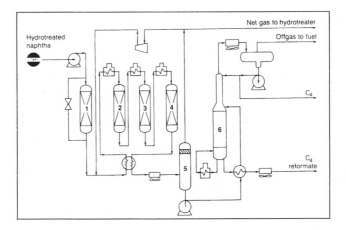

Figure 3-5. Flow diagram of a Chevron Rheiniforming unit:[15] (1) sulfur sorber, (2–4) reactors, (5) separator, (6) stabilizer.

Catalytic Cracking

Catalytic cracking (Cat-cracking) is a remarkably versatile and flexible process. Its principal aim is to crack lower-value stocks and produce higher-value light and middle distillates. The process also produces light hydrocarbon gases, which are important feedstocks for petrochemicals. Catalytic cracking produces more gasoline of higher octane than thermal cracking. This is due to the effect of the catalyst, which promotes isomerization and dehydrocyclization reactions.

Products from catalytic cracking units are also more stable due to a lower olefin content in the liquid products. This reflects a higher hydrogen transfer activity, which leads to more saturated hydrocarbons than in thermally cracked products from delayed coking units, for example.

The feeds to catalytic cracking units vary from gas oils to crude residues. Heavier feeds contain higher concentrations of basic and polar molecules as well as asphaltenes. Examples are basic nitrogen compounds, which are readily adsorbed on the catalyst acid sites and lead to instantaneous albeit temporary deactivation. Polycyclic aromatics and asphaltenes contribute strongly to coke formation. FCC (fluid catalytic cracking) catalyst deactivation in resid processing have been reviewed by O'Connor et al.[17] and Occelli.[18] These feedstocks are often pretreated to decrease the metallic and asphaltene contents. Hydrotreatment, solvent extraction, and propane deasphalting are important treatment processes.

Excessive asphaltene and aromatics in the feed are precursors to carbon formation on the catalyst surface, which substantially reduces its activity and produces gasolines of lower quality.

Residium fluid catalytic cracking (RFCC) has gained wide acceptance due to a larger production of gasoline with only small amounts of low-value products. Pretreating the feed in a low-severity residue desulfurization (RDS) increased the gasoline yield by 7.4%.[19] Table 3-8 compares the effect of RDS pretreatment on product yields from RFCC (with and without RDS).[19] Other resid treatment approaches to passivate heavy metals in catalytic cracking feeds are noted in the following section "Cracking Catalysts."

Cracking Catalysts

Acid-treated clays were the first catalysts used in catalytic cracking processes, but have been replaced by synthetic amorphous silica-alumina, which is more active and stable. Incorporating zeolites (crystalline alumina-silica) with the silica/alumina catalyst improves selectivity towards aromatics. These catalysts have both Lewis and Bronsted acid sites that promote carbonium ion formation. An important structural feature of zeolites is the presence of holes in the crystal lattice, which are formed by the silica-alumina tetrahedra. Each tetrahedron is made of four oxygen anions with either an aluminum or a silicon cation in the center. Each oxygen anion with a -2 oxidation state is shared between either two silicon, two aluminum, or an aluminum and a silicon cation.

The four oxygen anions in the tetrahedron are balanced by the $+4$ oxidation state of the silicon cation, while the four oxygen anions connecting the aluminum cation are not balanced. This results in -1 net charge, which should be balanced. Metal cations such as Na^+, Mg^{2+}, or protons (H^+) balance the charge of the alumina tetrahedra. A two-dimensional representation of an H-zeolite tetrahedra is shown:

Bronsted acid sites in HY-zeolites mainly originate from protons that neutralize the alumina tetrahedra. When HY-zeolite (X- and Y-zeolites

Table 3-8
Effect of RDS pretreatment on product yields from RFCC
(with and without RDS)[19]

	Arabian light RDS feed	Arabian light RDS product
RFCC feed properties		
Boiling range, °C	370+	370+
API	15.1	20.1
CCR, wt %	8.9	4.9
Sulfur, wt %	3.30	0.48
Nitrogen, wt %	0.17	0.13
Nickel + vanadium, ppm	51	7
RFCC yields, %		
H_2S, wt	1.7	0.2
C_2, wt	4.0	4.0
C_3, LV	8.4	10.1
C_4, LV	12.4	15.2
Gasoline (C_5–221°C), LV	50.6	58.0
LCO (221°C to 360°C), LV	21.4	18.2
Bottoms (360°C$^+$), LV	9.7	7.2
Coke, wt	10.3	7.0
Catalyst makeup, lb/bbl	1.72	0.23
Catalyst cooler required	Yes	No

are cracking catalysts) is heated to temperatures in the range of 400–500°C, Lewis acid sites are formed.

A Lewis acid site

Zeolites as cracking catalysts are characterized by higher activity and better selectivity toward middle distillates than amorphous silica-alumina catalysts. This is attributed to a greater acid sites density and a higher adsorption power for the reactants on the catalyst surface.

The higher selectivity of zeolites is attributed to its smaller pores, which allow diffusion of only smaller molecules through their pores, and to the higher rate of hydrogen transfer reactions. However, the silica-alumina matrix has the ability to crack larger molecules. Hayward and Winkler have recently demonstrated the importance of the interaction of the zeolite with the silica-alumina matrix. In a set of experiments using gas oil and rare earth zeolite/silica-alumina, the yield of gasoline increased when the matrix was used before the zeolite. This was explained by the mechanism of initial matrix cracking of large feedstock molecules to smaller ones and subsequent zeolite cracking of the smaller molecules to products.[20]

Aluminum distribution in zeolites is also important to the catalytic activity. An inbalance in charge between the silicon atoms in the zeolite framework creates active sites, which determine the predominant reactivity and selectivity of FCC catalyst. Selectivity and octane performance are correlated with unit cell size, which in turn can be correlated with the number of aluminum atoms in the zeolite framework.[21]

Deactivation of zeolite catalysts occurs due to coke formation and to poisoning by heavy metals. In general, there are two types of catalyst deactivation that occur in a FCC system, reversible and irreversible. Reversible deactivation occurs due to coke deposition. This is reversed by burning coke in the regenerator. Irreversible deactivation results as a combination of four separate but interrelated mechanisms: zeolite dealumination, zeolite decomposition, matrix surface collapse, and contamination by metals such as vanadium and sodium.[22]

Pretreating the feedstocks with hydrogen is not always effective in reducing heavy metals, and it is expensive. Other means that proved successful are modifying the composition and the microporous structure of the catalyst or adding metals like Sb, Bi or Sn, or Sb-Sn combination.[23] Antimony organics have been shown to reduce by 50% gas formation due to metal contaminants, especially nickel.[24]

Cracking Reactions

A major difference between thermal and catalytic cracking is that reactions through catalytic cracking occur via carbocation intermediate, compared to the free radical intermediate in thermal cracking. Carbocations are longer lived and accordingly more selective than free radicals. Acid catalysts such as amorphous silica-alumina and crystalline zeolites promote the formation of carbocations. The following illustrates the different ways by which carbocations may be generated in the reactor:

1. Abstraction of a hydride ion by a Lewis acid site from a hydrocarbon

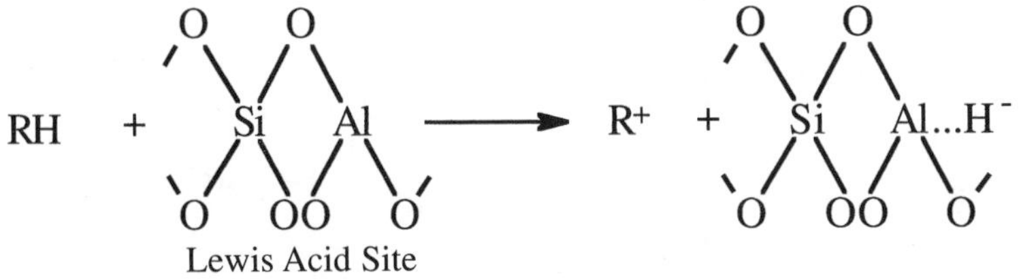

$$RH \ + \ \text{(Lewis Acid Site)} \longrightarrow R^+ \ + \ \text{(Si Al...H}^-)$$

Lewis Acid Site

2. Reaction between a Bronsted acid site (H+) and an olefin

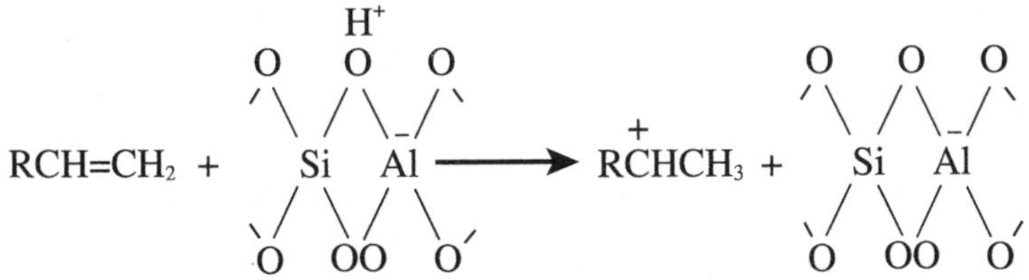

$$RCH{=}CH_2 \ + \ \text{(Si Al with H}^+) \longrightarrow R\overset{+}{C}HCH_3 \ + \ \text{(Si Al)}$$

3. Reaction of a carbonium ion formed from step 1 or 2 with another hydrocarbon by abstraction of a hydride ion

$$R^+ + RCH_2CH_3 \rightarrow RH + R\overset{+}{C}HCH_3$$

Abstraction of a hydride ion from a tertiary carbon is easier than from a secondary, which is easier than from a primary position. The formed carbocation can rearrange through a methide-hydride shift similar to what has been explained in catalytic reforming. This isomerization reaction is responsible for a high ratio of branched isomers in the products.

The most important cracking reaction, however, is the carbon-carbon beta bond scission. A bond at a position beta to the positively-charged carbon breaks heterolytically, yielding an olefin and another carbocation. This can be represented by the following example:

$$RCH_2\overset{+}{C}HCH_3 \rightarrow R^+ + CH_2{=}CHCH_3$$

The new carbocation may experience another beta scission, rearrange to a more stable carbonium ion, or react with a hydrocarbon molecule in the mixture and produce a paraffin.

The carbon-carbon beta scission may occur on either side of the carbocation, with the smallest fragment usually containing at least three carbon atoms. For example, cracking a secondary carbocation formed from a long chain paraffin could be represented as follows:

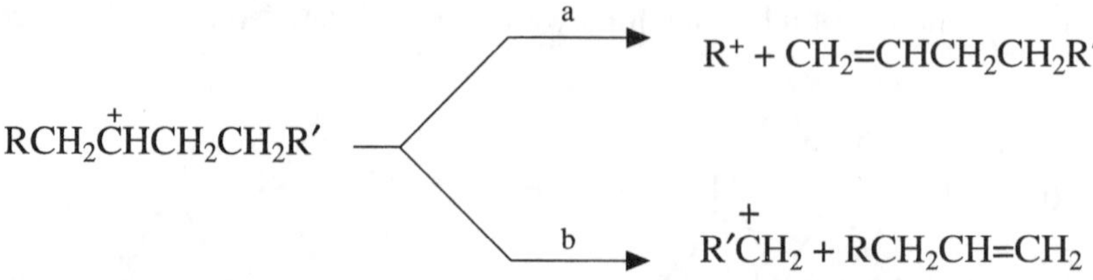

$$RCH_2\overset{+}{C}HCH_2CH_2R' \underset{b}{\overset{a}{\longleftarrow}} \begin{cases} R^+ + CH_2=CHCH_2CH_2R' \\ \\ \overset{+}{R'CH_2} + RCH_2CH=CH_2 \end{cases}$$

If R = H in the above example, then according to the beta scission rule (an empirical rule) only route b becomes possible, and propylene would be a product:

$$CH_3\overset{+}{C}HCH_2CH_2R' \rightarrow R'\overset{+}{C}H_2 + CH_3CH=CH_2$$

The propene may be protonated to an isopropyl carbocation:

$$CH_2=CHCH_3 + H^+ \rightarrow CH_3\overset{+}{C}HCH_3$$

An isopropyl carbocation cannot experience a beta fission (no C-C bond beta to the carbon with the positive charge).[25] It may either abstract a hydride ion from another hydrocarbon, yielding propane, or revert back to propene by eliminating a proton. This could explain the relatively higher yield of propene from catalytic cracking units than from thermal cracking units.

Aromatization of paraffins can occur through a dehydrocyclization reaction. Olefinic compounds formed by the beta scission can form a carbocation intermediate with the configuration conducive to cyclization. For example, if a carbocation such as that shown below is formed (by any of the methods mentioned earlier), cyclization is likely to occur.

$$RCH_2CH_2CH_2CH_2CH=CH_2 \overset{zeolite}{\longrightarrow} R\overset{+}{C}HCH_2CH_2CH_2CH=CH_2$$

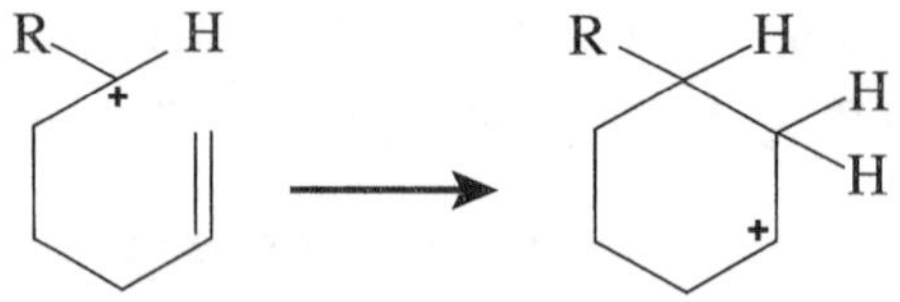

Once cyclization has occurred, the formed carbocation can lose a proton, and a cyclohexene derivative is obtained. This reaction is aided by the presence of an olefin in the vicinity ($R-CH=CH_2$).

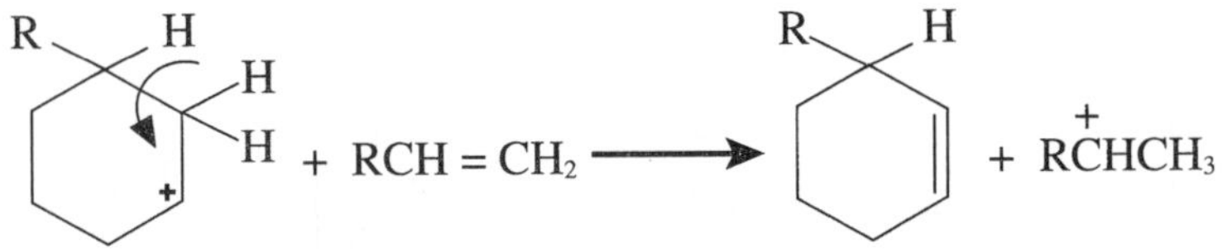

The next step is the abstraction of a hydride ion by a Lewis acid site from the zeolite surface to form the more stable allylic carbocation. This is again followed by a proton elimination to form a cyclohexadiene intermediate. The same sequence is followed until the ring is completely aromatized.

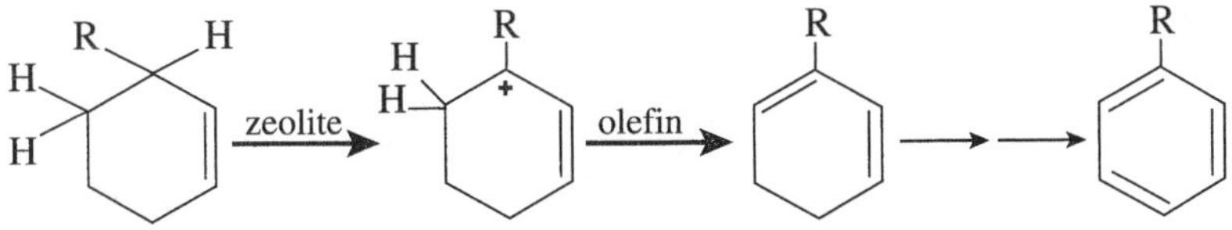

During the cracking process, fragmentation of complex polynuclear cyclic compounds may occur, leading to formation of simple cycloparaffins. These compounds can be a source of C_6, C_7, and C_8 aromatics through isomerization and hydrogen transfer reactions.

Coke formed on the catalyst surface is thought to be due to polycondensation of aromatic nuclei. The reaction can also occur through a carbonium ion intermediate of the benzene ring. The polynuclear aromatic structure has a high C/H ratio.

Cracking Process

Most catalytic cracking reactors are either fluid bed or moving bed. In the more common fluidized bed process (FCC), the catalyst is an extremely porous powder with an average particle size of 60 microns. Catalyst size is important, because it acts as a liquid with the reacting hydrocarbon mixture. In the process, the preheated feed enters the reactor section with hot regenerated catalyst through one or more risers where cracking occurs. A riser is a fluidized bed where a concurrent upward flow of the reactant gases and the catalyst particles occurs. The reactor temperature is usually held at about 450–520°C, and the pressure is approximately 10–20 psig. Gases leave the reactor through cyclones to remove the powdered catalyst, and pass to a fractionator for separation of the product streams. Catalyst regeneration occurs by combusting carbon deposits to carbon dioxide and the regenerated catalyst is then returned

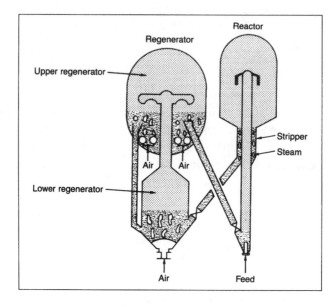

Figure 3-6. Typical FCC reactor/regenerator.[26]

to the bottom of the riser. Figure 3-6 is a typical FCC reactor/regeneration system.[26]

Fluid catalytic cracking produces unsaturates, especially in the light hydrocarbon range C_3–C_5, which are used as petrochemical feedstocks and for alkylate production. In addition to hydrocarbon gases, FCC units produce gasolines with high octane numbers (due to the high aromatic content, branched paraffins and olefins), gas oils, and tar. The ratio of these products depends greatly on the different process variables. In general, higher conversions increase gas and gasoline yields. Higher conversion also increases coke formation. Process variables that increase conversion are higher temperatures, longer residence times, and higher catalyst/oil ratio. Table 3-9 shows the analysis of the feed and the products from an FCC unit.[27]

In the moving bed processes, the preheated feed meets the hot catalyst, which is in the form of beads that descend by gravity to the regeneration zone. As in fluidized bed cracking, conversion of aromatics is low, and a mixture of saturated and unsaturated light hydrocarbon gases is produced. The gasoline product is also rich in aromatics and branched paraffins.

Table 3-9
Analysis of feed and products from a fluid catalytic cracking process[27]

Yields: Typical examples

	North slope vac. resid	Maya crude	P.R. Springs bitumen
Feed			
Gravity, °API	10.7	23.5	2.1
Sulfur, wt %	2.0	3.0	1.0
Nitrogen, wt %	0.48	0.3	0.76
Con carb resid, wt %	11.8	11.2	18.0
Ni + V, ppm	73	264	89
Product yields			
H_2S, wt %	0.3	0.3	0.8
Light-C_2, wt %	5.1	2.9	1.6
LPG, vol %	7.8	4.2	3.0
Naphtha, whole, vol %	18.7	26.5	14.0
Light gas oil, vol %	13.7	29.1	17.9
Heavy gas oil, vol %	54.3	334.9	55.4
Coke, burned, wt %	9.5	8.7	17.1
Heavy gas oil cut			
Gravity, °API	11.5	17.0	14.9
Sulfur, wt %	2.2	3.1	0.5
Nitogren, wt %	0.44	0.22	0.48
Ni + V, ppm	3.0	20.7	12.0
Visc, cSt @ 210°F	18	12	—

Deep Catalytic Cracking

Deep catalytic cracking (DCC) is a catalytic cracking process which selectively cracks a wide variety of feedstocks into light olefins. The reactor and the regenerator systems are similar to FCC. However, innovation in the catalyst development, severity, and process variable selection enables DCC to produce more olefins than FCC. In this mode of operation, propylene plus ethylene yields could reach over 25%. In addition, a high yield of amylenes (C_5 olefins) is possible. Figure 3-7 shows the DCC process and Table 3-10 compares olefins produced from DCC and FCC processes.[28]

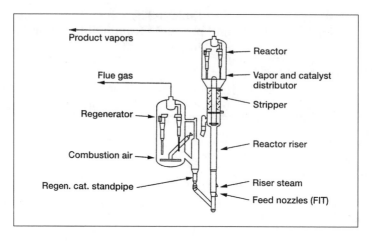

Figure 3-7. Deep catalytic cracking process.[28]

Table 3-10
Comparison of products from DCC with those from FCC[28]

Products: wt % FF	DCC Type I	DCC Type II	FCC
Ethylene	6.1	2.3	0.9
Propylene	20.5	14.3	6.8
Butylene	14.3	14.6	11.0
in which $IC_4^=$	5.4	6.1	3.3
Amylene	—	9.8	8.5
in which $IC_5^=$	—	6.5	4.3

Hydrocracking Process

Hydrocracking is essentially catalytic cracking in the presence of hydrogen. It is one of the most versatile petroleum refining schemes adapted to process low value stocks. Generally, the feedstocks are not suitable for catalytic cracking because of their high metal, sulfur, nitrogen, and asphaltene contents. The process can also use feeds with high aromatic content.

Products from hydrocracking processes lack olefinic hydrocarbons. The product slate ranges from light hydrocarbon gases to gasolines to residues. Depending on the operation variables, the process could

Table 3-11
Analysis of feed and products from hydrocracking process[29]

Yields: Typical from various feeds:

Feed	Naphtha	LCCO	VGO	VGO
Catalyst stages	1	2	2	2
Gravity, °API	72.5	24.6	25.8	21.6
Aniline pt, °F	145	92	180	180
ASTM 10%/EP, °F	154/290	478/632	740/1,050	740/1,100
Sulfur, wt %	0.005	0.6	1.0	2.5
Nitrogen, ppm	0.1	500	1,000	900
Yields, vol %				
Propane	55	3.4	—	—
iso-Butane	29	9.1	3.0	2.5
n-Butane	19	4.5	3.0	2.5
Light naphtha	23	30.0	11.9	7.0
Heavy naphtha	—	78.7	14.2	7.0
Kerosine	—	—	86.8	48.0
Diesel	—	—	—	50.0
Product quality				
Lt naphtha RON cl	85	76	77	76
Hv. naphtha RON cl	—	65	61	61
Kerosine freeze pt, °F	—	—	–65	–75
Diesel pour pt, °F	—	—	—	–10

be adapted for maximizing gasoline, jet fuel, or diesel production. Table 3-11 shows the feed and the products from a hydrocracking unit.[29]

Hydrocracking Catalysts and Reactions

The dual-function catalysts used in hydrocracking provide high surface area cracking sites and hydrogenation-dehydrogenation sites. Amorphous silica-alumina, zeolites, or a mixture of them promote carbonium ion formation. Catalysts with strong acidic activity promote isomerization, leading to a high iso/normal ratios.[30] The hydrogenation-dehydrogenation activity, on the other hand, is provided by catalysts such as cobalt, molybdenum, tungsten, vanadium, palladium, or rare earth elements. As with catalytic cracking, the main reactions occur by carbonium ion and beta scission, yielding two fragments that could be hydrogenated on the catalyst surface. The main hydro-cracking reaction could be illustrated by the first-step formation of a carbocation over the catalyst surface:

$$RCH_2CH_2R' \xrightarrow{\text{zeolite}} RCH_2\overset{+}{C}HR'$$

The carbocation may rearrange, eliminate a proton to produce an olefin, or crack at a beta position to yield an olefin and a new carbocation. Under an atmosphere of hydrogen and in the presence of a catalyst with hydrogenation-dehydrogenation activity, the olefins are hydrogenated to paraffinic compounds. This reaction sequence could be represented as follows:

$$RCH_2\overset{+}{C}HR' \xrightarrow{-H^+} RCH{=}CHR'$$

$$RCH_2\overset{+}{C}HR' \xrightarrow{\beta \text{ scission}} R'CH{=}CH_2 + R^+$$

$$R'CH{=}CH_2 + H_2 \xrightarrow{\text{catalyst}} R'CH_2CH_3$$

As can be anticipated, most products from hydrocracking are saturated. For this reason, gasolines from hydrocracking units have lower octane ratings than those produced by catalytic cracking units; they have a lower aromatic content due to high hydrogenation activity. Products from hydrocracking units are suitable for jet fuel use. Hydrocracking also produces light hydrocarbon gases (LPG) suitable as petrochemical feedstocks.

Other reactions that occur during hydrocracking are the fragmentation followed by hydrogenation (hydrogenolysis) of the complex asphaltenes and heterocyclic compounds normally present in the feeds.

Dealkylation, fragmentation, and hydrogenation of substituted polynuclear aromatics may also occur. The following is a representative example of hydrocracking of a substituted anthracene.

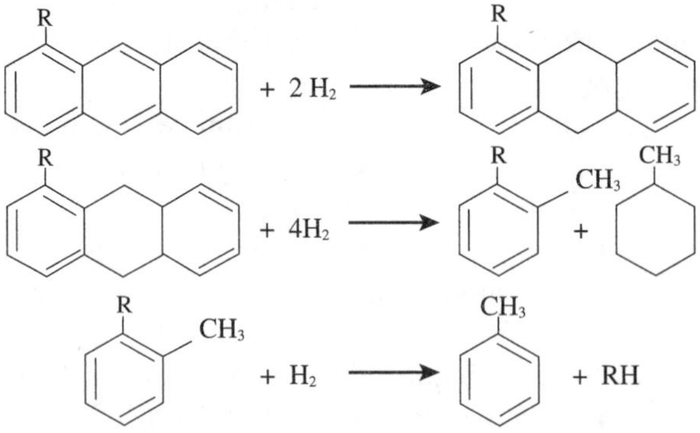

It should be noted, however, that this reaction sequence may be different from what may actually be occurring in the reactor. The reactions proceed at different rates depending on the process variables. Hydrodesulfurization of complex sulfur compounds such as dibenzothiophene also occurs under these conditions. The desulfurized product may crack to give two benzene molecules:

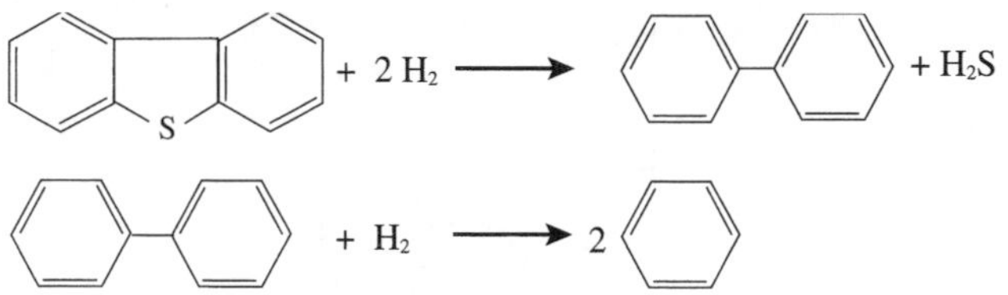

Process

Most commercial hydrocracking operations use a single stage for maximum middle-distillate optimization despite the flexibility gained by having more than one reactor. In the single stage process two operation modes are possible, a once-through mode and a total conversion of the fractionator bottoms through recycling.

In the once-through operation low sulfur fuels are produced and the fractionator bottoms are not recycled. In the total conversion mode the fractionator bottoms are recycled to the inlet of the reactor to obtain more middle distillates.

In the two-stage operation, the feed is hydrodesulfurized in the first reactor with partial hydrocracking. Reactor effluent goes to a high-pressure separator to separate the hydrogen-rich gas, which is recycled and mixed with the fresh feed. The liquid portion from the separator is fractionated, and the bottoms of the fractionator are sent to the second stage reactor.

Hydrocracking reaction conditions vary widely, depending on the feed and the required products. Temperature and pressure range from 400 to 480°C and 35 to 170 atmospheres. Space velocities in the range of 0.5 to 2.0 hr^{-1} are applied. Figure 3-8 shows the Chevron two-stage hydrocracking process.[29]

Hydrodealkylation Process

This process is designed to hydrodealkylate methylbenzenes, ethylbenzene and C$_9^+$ aromatics to benzene. The petrochemical demand for benzene is greater than for toluene and xylenes. After separating benzene

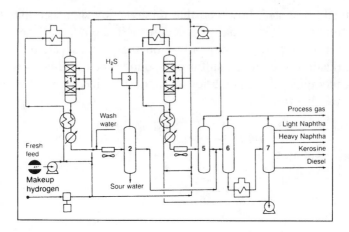

Figure 3-8. Flow diagram of a Cheveron hydocracking unit:[29] (1,4) reactors, (2,5) HP separators, (3) recycle scrubber (optional), (6) LP separator, (7) fractionator.

from the reformate, the higher aromatics are charged to a hydrodealkylation unit. The reaction is a hydrocracking one, where the alkyl side chain breaks and is simultaneously hydrogenated. For example, toluene dealkylates to methane and benzene, while ethylbenzene produces ethane and benzene. In each case one mole of H_2 is consumed:

$$\underset{CH_3}{\bigcirc} + H_2 \longrightarrow \bigcirc + CH_4$$

$$\underset{C_2H_5}{\bigcirc} + H_2 \longrightarrow \bigcirc + CH_3CH_3$$

Consuming hydrogen is mainly a function of the number of benzene substituents. Dealkylation of polysubstituted benzene increases hydrogen consumption and gas production (methane). For example, dealkylating one mole xylene mixture produces two methane moles and one mole of benzene; it consumes two moles of hydrogen.

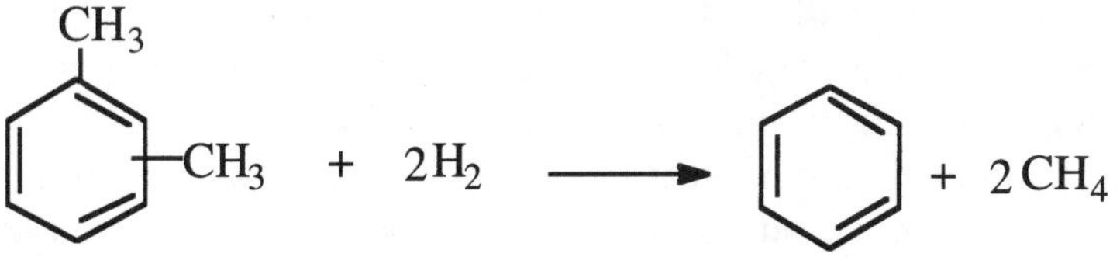

Unconverted toluene and xylenes are recycled.

Hydrotreatment Processes

Hydrotreating is a hydrogen-consuming process primarily used to reduce or remove impurities such as sulfur, nitrogen, and some trace metals from the feeds. It also stabilizes the feed by saturating olefinic compounds.

Feeds to hydrotreatment units vary widely; they could be any petroleum fraction, from naphtha to crude residues. The process is relatively simple: choosing the desulfurization process depends largely on the feed type, the level of impurities present, and the extent of treatment needed to suit the market requirement. Table 3-12 shows the feed and product properties from a hydrotreatment unit.[31]

In this process, the feed is mixed with hydrogen, heated to the proper temperature, and introduced to the reactor containing the catalyst. The

Table 3-12
Products from hydrodesulfurization of feeds with
different sulfur levels[31]

Process	VGO*	VRDS**	VGO+ VRDS	RDS***
Feed sulfur, wt %	2.3	4.1	2.9	2.9
Product sulfur, wt %	0.1	1.28	0.5	0.5
Product yields				
C_1-C_4 wt %	0.59	0.56	0.58	0.58
H_2S, NH_3, wt %	2.44	3.00	2.55	2.55
C_5^+, wt %	97.51	97.34	97.46	97.67
C_5^+, LV %	100.6	102.0	101.0	101.5
Hydrogen consumption				
scf/bbl	330	720	450	550
scf/lb sulfur	47	71	56	69

* *Vacuum gas oil hydrotreater*
** *Vacuum residuum hydrotreater*
*** *Atmospheric residuum desulfurization hydrotreating*

conditions are usually adjusted to minimize hydrocracking. Typical reactor temperatures range from 260 to 425°C. Hydrogen partial pressure and space velocity are important process variables. Increasing the temperature and hydrogen partial pressure increases the hydrogenation and hydrodesulfurization reactions. Lower space velocities are used with feeds rich in polyaromatics. Total pressure varies widely—from 100 to 3,000 psi—depending on the type of feed, level of impurities, and the extent of hydrotreatment required. Figure 3-9 shows an Exxon hydrotreatment unit.[32]

Hydrotreatment Catalysts and Reactions

Catalysts used in hydrotreatment (hydrodesulfurization, HDS) processes are the same as those developed in Germany for coal hydrogenation during World War II. The catalysts should be sulfur-resistant. The cobalt-molybdenum system supported on alumina was found to be an effective catalyst.

The catalyst should be reduced and sulfided during the initial stages of operation before use. Other catalyst systems used in HDS are NiO/MoO_3 and NiO/WO_3. Because mass transfer has a significant influence on the reaction rates, catalyst performance is significantly affected by the particle size and pore diameter.

Reactions occurring in hydrotreatment units are mainly hydrodesulfurization and hydrodenitrogenation of sulfur and nitrogen compounds. In

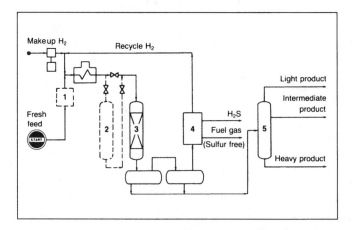

Figure 3-9. Flow diagram of an Exxon hydrotreating unit[32]: (1) filter, (2) guard vessel to protect reactor, (3) main reactor, (4) gas treatment, (5) fractionator.

the first case H_2S is produced along with the hydrocarbon. In the latter case, ammonia is released. The following examples are hydrodesulfurization reactions of some representative sulfur compounds present in petroleum fractions and coal liquids.

$$R\text{-}SH + H_2 \rightarrow RH + H_2S$$

$$R\text{-}S\text{-}R + 2H_2 \rightarrow 2RH + H_2S$$

$$RS\text{-}SR + 3H_2 \rightarrow 2RH + 2H_2S$$

$+ \ 4\,H_2 \longrightarrow CH_3(CH_2)_2CH_3 + H_2S$

Examples of hydrodenitrogenation of two types of nitrogen compounds normally present in some light and middle crude distillates are shown as follows:

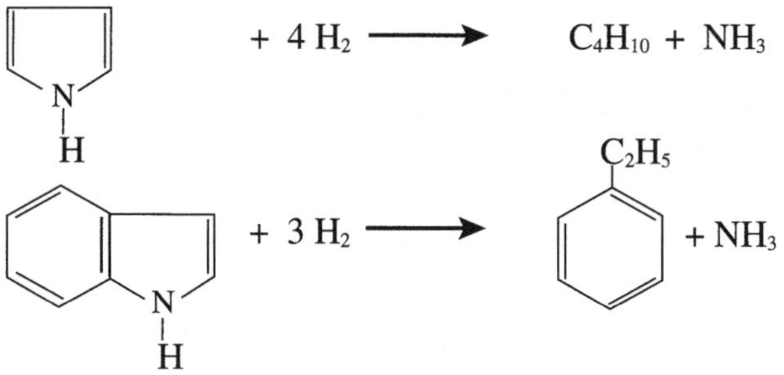

$+ \ 4\,H_2 \longrightarrow C_4H_{10} + NH_3$

$+ \ 3\,H_2 \longrightarrow$ $+ NH_3$

More complex sulfur and nitrogen compounds are present in heavy residues. These are hyrodesulfurized and hydrodenitrogenated, but under more severe conditions than normally used for lighter distillates. For example, for light petroleum distillates the approximate temperature and pressure ranges of 300–400°C and 35–70 atm. are used, versus 340–425°C and 55–170 atm. for heavy petroleum residua. Liquid hourly space velocities (LHSV) in the range of 2–10 hr^{-1} are used for light products, while it is 0.2–10 hr^{-1} for heavy residues.[33]

Alkylation Process

Alkylation in petroleum processing produces larger hydrocarbon molecules in the gasoline range from smaller molecules. The products are branched hydrocarbons having high octane ratings.

The term alkylation generally is applied to the acid catalyzed reaction between isobutane and various light olefins, and the product is known as the alkylate. Alkylates are the best of all possible motor fuels, having both excellent stability and a high octane number.

Either concentrated sulfuric acid or anhydrous hydrofluoric acid is used as a catalyst for the alkylation reaction. These acid catalysts are capable of providing a proton, which reacts with the olefin to form a carbocation. For example, when propene is used with isobutane, a mixture of C_5 isomers is produced. The following represents the reaction steps:

$$CH_2{=}CHCH_3 + H^+ \longrightarrow CH_3{-}\overset{\overset{\displaystyle CH_3}{|}}{\underset{\underset{\displaystyle H}{|}}{C}}{}^+$$

$$CH_3{-}\overset{\overset{\displaystyle CH_3}{|}}{\underset{\underset{\displaystyle H}{|}}{C}}{-}CH_3 + CH_3{-}\overset{\overset{\displaystyle CH_3}{|}}{\underset{\underset{\displaystyle H}{|}}{C}}{}^+ \longrightarrow CH_3CH_2CH_3 + CH_3{-}\overset{\overset{\displaystyle CH_3}{|}}{\underset{\underset{\displaystyle CH_3}{|}}{C}}{}^+$$

$$(CH_3)_3\overset{+}{C} + CH_2{=}CHCH_3 \longrightarrow (CH_3)_3CCH_2\overset{+}{C}HCH_3$$

The formed carbocation from the last step may abstract a hydride ion from an isobutane molecule and produce 2,2-dimethylpentane, or it may rearrange to another carbocation through a hydride shift.

$$CH_3{-}\overset{\overset{\displaystyle CH_3}{|}}{\underset{\underset{\displaystyle CH_3}{|}}{C}}{-}CH_2\overset{+}{C}HCH_3 \longrightarrow$$

$$\xrightarrow{+H^-} CH_3{-}\overset{\overset{\displaystyle CH_3}{|}}{\underset{\underset{\displaystyle CH_3}{|}}{C}}{-}CH_2CH_2CH_3$$

$$\xrightarrow{\boxed{\text{Rearrange}}} CH_3{-}\overset{\overset{\displaystyle CH_3}{|}}{\underset{\underset{\displaystyle CH_3}{|}}{C}}{-}\overset{+}{C}HCH_2CH_3$$

The new carbocation can rearrange again through a methide/hydride shift as shown in the following equation:

$$
\underset{\underset{CH_3}{|}}{\overset{\overset{CH_3}{|}}{CH_3-C-\overset{+}{C}HCH_2CH_3}} \longrightarrow \underset{\underset{CH_3}{|}}{\overset{\overset{CH_3}{|}}{CH_3-C-\overset{+}{C}(CH_3)_2}}
$$

The rearranged carbocation finally reacts with isobutane to form 2,2,3-trimethylbutane.

$$
\underset{\underset{CH_3}{|}}{\overset{\overset{CH_3}{|}}{CH_3C-C(CH_3)_2}} + \underset{\underset{H}{|}}{\overset{\overset{CH_3}{|}}{CH_3-C-CH_3}} \rightarrow \underset{\underset{CH_3}{|}}{\overset{\overset{CH_3}{|}}{CH_3C-CH(CH_3)_2}} + \underset{\underset{CH_3}{|}}{\overset{\overset{CH_3}{|}}{CH_3-C^+}}
$$

The final product contains approximately 60–80% 2,2-dimethylpentane and varying amounts of 2,2,3-trimethylbutane and 2-methylhexane.

The primary process variables affecting the economics of sulfuric acid alkylation are the reaction temperature, isobutane recycle rate, reactor space velocity, and spent acid strength. To control fresh acid makeup, spent acid could be monitored by continuously measuring its density, the flow rate, and its temperature. This can reduce the acid usage in alkylation units.[34]

The presence of impurities such as butadiene affects the product yield and properties. Butadiene tends to polymerize and form acid-soluble oils, which increases acid makeup requirements. For every pound of butadiene in the feed, ten pounds of additional make-up acid will be required.[35]

Other olefins that are commercially alkylated are isobutene and 1- and 2-butenes. Alkylation of isobutene produces mainly 2,2,4-trimethylpentane (isooctane).

Both sulfuric acid and hydrofluoric acid catalyzed alkylations are low temperature processes. Table 3-13 gives the alkylation conditions for HF and H_2SO_4 processes.[36] One drawback of using H_2SO_4 and HF in alkylation is the hazards associated with it. Many attempts have been tried to use solid catalysts such as zeolites, alumina and ion exchange resins. Also strong solid acids such as sulfated zirconia and SbF_5/sulfonic acid resins were tried. Although they were active, nevertheless they lack stability.[37] No process yet proved successful due to the fast deactivation of the catalyst. A new process which may have commercial possibility, uses

Table 3-13
Ranges of operating conditions for H_2SO_4 and HF alkylation[36]

Process catalysts	H_2SO_4	HF
Temperature, °C	2–16	16–52
Isobutane/olefin feed	3–12	3–12
Olefin space velocity, vo/hr./vo	0.1–0.6	—
Olefin contact time. min	20–30	8–20
Catalysts acidity, wt %	88–95	80–95
Acid in emulsion, vol %	40–60	25–80

liquid trifilic acid (CF_3-SO_2OH) on a porous solid bed. Using isobutane and light olefins, the intermediates are: isopropyl, sec-butyl, 2-pentyl, and 3-pentyl esters of trifilic acid.[38]

Isomerization Process

Isomerization is a small-volume but important refinery process. Like alkylation, it is acid catalyzed and intended to produce highly-branched hydrocarbon mixtures. The low octane C_5/C_6 fraction obtained from natural gasoline or from a light naphtha fraction may be isomerized to a high octane product.

Dual-function catalysts activated by either inorganic or organic chlorides are the preferred isomerization catalysts. A typical catalyst is platinum with a zeolite base. These catalysts serve the dual purpose of promoting carbonium ion formation and hydrogenation-dehydrogenation reactions. The reaction may start by forming a carbocation via abstraction of a hydride ion by a catalyst acid site. Alternatively, an olefin formed on the catalyst surface could be protonated to form the carbocation. The carbocation isomerizes by a 1,2-hydride/methide shift as mentioned earlier (see this chapter, "Reforming Reactions"). Figure 3-10 shows the vapor phase equilibrium of hexane isomers.[39]

Oligomerization of Olefins (Dimerization)

This process produces polymer gasoline with a high octane. Dimerization was first used (1935) to dimerize isobutylene to diisobutylene, constituted of 2,4,4-trimethyl-1-pentene (80%) and 2,4,4-trimethyl-2-pentene (20%). Both phosphoric and sulfuric acid were used as catalysts.

At present, the feedstock is either a propylene-propane mixture or propylene-butane mixture where propane and butane are diluents. The

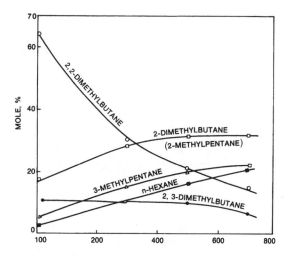

Figure 3-10. Vapor phase equilibrium for hexanes.[39]

product is an olefin having a high octane number. When propylene is used, a trimer or a tetramer is formed. The polymerization reaction is highly exothermic, so the temperature has to be controlled. The presence of propane and butane in the mixture acts as a heat sink to absorb part of the reaction heat. Typical reaction conditions are 170–250°C and 25–100 atm.

The polymerization reaction starts by protonating the olefin and forming a carbocation. For example, protonating propene gives isopropyl carbocation. The proton is provided by the ionization of phosphoric acid:

$$CH_3CH{=}CH_2 + H^+ \longrightarrow CH_3\overset{+}{C}HCH_3$$

The next step is the reaction of the carbocation with the olefin (propene or butene).

$$CH_3\overset{+}{C}HCH_3 + CH_3CH{=}CH_2 \longrightarrow CH_3\underset{\underset{CH_3}{|}}{C}HCH_2\overset{+}{C}HCH_3$$

The newly-formed carbocation either eliminates a proton and forms a dimer or attacks another propene molecule and eliminates a proton, giving the trimer.

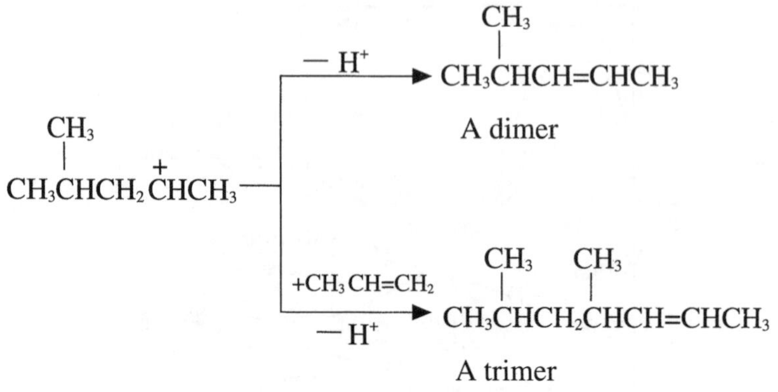

$$CH_3CHCH_2\overset{+}{C}HCH_3 \begin{cases} \xrightarrow[-H^+]{} \quad \underset{\displaystyle \text{A dimer}}{\overset{\displaystyle CH_3}{CH_3CHCH=CHCH_3}} \\ \\ \xrightarrow[-H^+]{+CH_3CH=CH_2} \underset{\displaystyle \text{A trimer}}{\overset{\displaystyle CH_3 \quad CH_3}{CH_3CHCH_2CHCH=CHCH_3}} \end{cases}$$

Further protonation of the trimer produces a C_9 carbocation which may further react with another propene molecule and eventually produce propylene tetramer.

The product is a mixture of dimers, trimers, tetramers, and pentamers having an average RON (Research Octane Number) = 95. Table 3-14 shows the analysis of feed and products from dimerization of propylene.[40]

Table 3-14
Typical feed and products from the dimerization of propylene[40]

			Vol. %	Total	wt %	Total
Feed						
Propylene			71	—	—	—
Propane			29	100	—	—
Products						
LPG						
Propylene			4.2	—	—	—
Propane			34.6	—	—	—
Isohexanes*			61.2	100	—	—
Isohexenes			—	—	92.0	—
Isononenes			—	—	6.5	—
Heavier			—	—	1.5	100
ASTM distillation (°F)	IBP	133				
	10	136				
	50	140				
	90	160				
	95	320				
	EP	370				

* *"Dimersol isohexenes"*

PRODUCTION OF OLEFINS

The most important olefins and diolefins used to manufacture petrochemicals are ethylene, propylene, butylenes, and butadiene. Butadiene, a conjugated diolefin, is normally coproduced with C_2–C_4 olefins from different cracking processes. Separation of these olefins from catalytic and thermal cracking gas streams could be achieved using physical and chemical separation methods. However, the petrochemical demand for olefins is much greater than the amounts these operations produce. Most olefins and butadienes are produced by steam cracking hydrocarbons.

Butadiene can be alternatively produced by other synthetic routes discussed with the synthesis of isoprene, the second major diolefin for rubber production.

STEAM CRACKING OF HYDROCARBONS
(Production of Olefins)

The main route for producing light olefins, especially ethylene, is the steam cracking of hydrocarbons. The feedstocks for steam cracking units range from light paraffinic hydrocarbon gases to various petroleum fractions and residues. The properties of these feedstocks are discussed in Chapter 2.

The cracking reactions are principally bond breaking, and a substantial amount of energy is needed to drive the reaction toward olefin production.

The simplest paraffin (alkane) and the most widely used feedstock for producing ethylene is ethane. As mentioned earlier, ethane is obtained from natural gas liquids. Cracking ethane can be visualized as a free radical dehydrogenation reaction, where hydrogen is a coproduct:

$$CH_3CH_3 \rightarrow CH_2{=}CH_2 + H_2 \qquad \Delta H_{590°C} = +143 \text{ KJ}$$

The reaction is highly endothermic, so it is favored at higher temperatures and lower pressures. Superheated steam is used to reduce the partial pressure of the reacting hydrocarbons' (in this reaction, ethane). Superheated steam also reduces carbon deposits that are formed by the pyrolysis of hydrocarbons at high temperatures. For example, pyrolysis of ethane produces carbon and hydrogen:

$$CH_3CH_3 \rightarrow 2C + 3H_2$$

Ethylene can also pyrolyse in the same way. Additionally, the presence of steam as a diluent reduces the hydrocarbons' chances of being in contact with the reactor tube-wall. Deposits reduce heat transfer through the reactor tubes, but steam reduces this effect by reacting with the carbon deposits (steam reforming reaction).

$$C + H_2O \rightarrow CO + H_2$$

Many side reactions occur when ethane is cracked. A probable sequence of reactions between ethylene and a formed methyl or an ethyl free radical could be represented:

$$CH_2= CH_2 + \dot{C}H_3 \rightarrow CH_3CH_2\dot{C}H_2 \rightarrow CH_3CH= CH_2 + \dot{H}$$

$$CH_2{=}CH_2 + CH_3\dot{C}H_2 \rightarrow CH_3CH_2CH_2\dot{C}H_2$$

$$\rightarrow CH_3CH_2CH{=}CH_2 + \dot{H}$$

Propene and l-butene, respectively, are produced in this free radical reaction. Higher hydrocarbons found in steam cracking products are probably formed through similar reactions.

When liquid hydrocarbons such as a naphtha fraction or a gas oil are used to produce olefins, many other reactions occur. The main reaction, the cracking reaction, occurs by a free radical and beta scission of the C-C bonds. This could be represented as:

$$RCH_2CH_2CH_2R \rightarrow RCH_2CH_2\dot{C}H_2 + \dot{R}$$

$$RCH_2CH_2\dot{C}H_2 \rightarrow R\dot{C}H_2 + CH_2{=}CH_2$$

The newly formed free radical may terminate by abstraction of a hydrogen atom, or it may continue cracking to give ethylene and a free radical. Aromatic compounds with side chains are usually dealkylated. The produced free radicals further crack to yield more olefins.

In the furnace and in the transfer line exchanger, coking is a significant problem. Catalytic coking occurs on clean metal surfaces when nickel and other transition metals used in radiant tube alloys catalyze dehydrogenation and formation of coke. Coke formation reduces product yields, increases energy consumption, and shortens coil service life. Coking is related to feedstock, temperature, and steam dilution. The radiant tubes gradually become coated with an internal layer of coke, thus raizing the tube metal temperature and increasing pressure drop through the radiant coils. When coke reaches an allowable limit as indicated by a high pressure drop, it should be removed.[41] Coke could be reduced by adding antifoulants, which passivate the catalytic coking mechanism.

The subject has been reviewed by Burns et al.[42] Over the past 20 years, significant improvements have been made in the design and operation of high severity pyrolysis furnaces. Using better alloys for tubing has enabled raising the temperature, shortening residence time and lowering pressure drop in the cracking coils. The use of cast alloys with a higher alloy content increases their long-term strength. Figure 3-11 shows the effect of alloy content on the long-term rupture stress for modified Ni-Cr-Fe alloys.[41]

Steam Cracking Process

A typical ethane cracker has several identical pyrolysis furnaces in which fresh ethane feed and recycled ethane are cracked with steam as a diluent. Figure 3-12 is a block diagram for ethylene from ethane. The outlet temperature is usually in the 800°C range. The furnace effluent is quenched in a heat exchanger and further cooled by direct contact in a water quench tower where steam is condensed and recycled to the pyrolysis furnace. After the cracked gas is treated to remove acid gases, hydrogen and methane are separated from the pyrolysis products in the demethanizer. The effluent is then treated to remove acetylene, and ethylene is separated from ethane and heavier in the ethylene fractionator. The bottom fraction is separated in the deethanizer into ethane and C_3^+ fraction. Ethane is then recycled to the pyrolysis furnace.

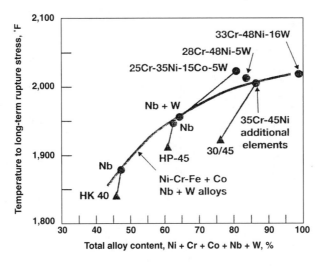

Figure 3-11. Effect of alloy content on long-term rupture stress for cast modified Ni-Cr-Fe alloys.[41]

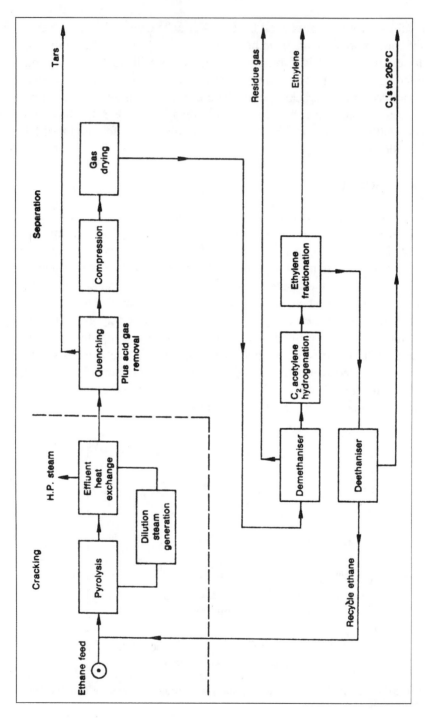

Figure 3-12. Block diagram for producing ethylene from ethane.

An olefin plant that uses liquid feeds requires an additional pyrolysis furnace, an effluent quench exchanger, and a primary fractionator for fuel oil separation.

Process Variables

The important process variables are reactor temperature, residence time, and steam/hydrocarbon ratio. Feed characteristics are also considered, since they influence the process severity.

Temperature

Steam cracking reactions are highly endothermic. Increasing temperature favors the formation of olefins, high molecular weight olefins, and aromatics. Optimum temperatures are usually selected to maximize olefin production and minimize formation of carbon deposits.

Reactor temperature is also a function of the feedstock used. Higher molecular weight hydrocarbons generally crack at lower temperatures than lower molecular weight compounds. For example, a typical furnace outlet temperature for cracking ethane is approximately 800°C, while the temperature for cracking naphtha or gas oil is about 675–700°C.

Residence Time

In steam cracking processes, olefins are formed as primary products. Aromatics and higher hydrocarbon compounds result from secondary reactions of the formed olefins. Accordingly, short residence times are required for high olefin yield. When ethane and light hydrocarbon gases are used as feeds, shorter residence times are used to maximize olefin production and minimize BTX and liquid yields; residence times of 0.5–1.2 sec are typical. Cracking liquid feedstocks for the dual purpose of producing olefins plus BTX aromatics requires relatively longer residence times than for ethane. However, residence time is a compromise between the reaction temperature and other variables.

A fairly new development in cracking liquid feeds that improves ethylene yield is the Millisecond furnace, which operates between 0.03–0.1 sec with an outlet temperature range of 870–925°C. "The Millisecond furnace probably represents the last step that can be taken with respect to this critical variable because contact times below the .01 sec range lead to production of acetylenes in large quantities."[43]

Steam/Hydrocarbon Ratio

A higher steam/hydrocarbon ratio favors olefin formation. Steam reduces the partial pressure of the hydrocarbon mixture and increases the yield of olefins. Heavier hydrocarbon feeds require more steam than gaseous feeds to additionally reduce coke deposition in the furnace tubes. Liquid feeds such as gas oils and petroleum residues have complex polynuclear aromatic compounds, which are coke precursors. Steam to hydrocarbon weight ratios range between 0.2–1 for ethane and approximately 1–1.2 for liquid feeds.

Feedstocks

Feeds to steam cracking units vary appreciably, from light hydrocarbon gases to petroleum residues. Due to the difference in the cracking rates of the various hydrocarbons, the reactor temperature and residence time vary. As mentioned before, long chain hydrocarbons crack more easily than shorter chain compounds and require lower cracking temperatures. For example, it was found that the temperature and residence time that gave 60% conversion for ethane yielded 90% conversion for propane.[44]

Feedstock composition also determines operation parameters. The rates of cracking hydrocarbons differ according to structure. Paraffinic hydrocarbons are easier to crack than cycloparaffins, and aromatics tend to pass through unaffected. Isoparaffins such as isobutane and isopentane give high yields of propylene. This is expected, because cracking at a tertiary carbon is easier:

$$\underset{\displaystyle H_3C\overset{\displaystyle CH_3}{\overset{\displaystyle |}{C}}HCH_3}{} \longrightarrow CH_3CH{=}CH_2 + CH_4$$

As feedstocks progress from ethane to heavier fractions with lower H/C ratios, the yield of ethylene decreases, and the feed per pound ethylene product ratio increases markedly. Table 3-15 shows yields from steam cracking of different feedstocks,[45] and how the liquid by-products and BTX aromatics increase dramatically with heavier feeds.

Cracking Gas Feeds

The main gas feedstock for ethylene production is ethane. Propane and butane or their mixture, LPG, are also used, but to a lesser extent. They

Table 3-15
Ultimate yields from steam cracking various feedstocks[45]

	Feedstock					
Yield, wt %	Ethane	Propane	Butane	Naphtha	Gas oil	Saudi NGL
$H_2 + CH_4$	13	28	24	26	18	23
Ethylene	80	45	37	30	25	50
Propylene	2.4	15	18	13	14	12
Butadiene	1.4	2	2	4.5	5	2.5
Mixed butenes	1.6	1	6.4	8	6	3.5
C_5^+	1.6	9	12.6	18.5	32	9

are specially used when coproduct propylene, butadiene, and the butenes are needed. The advantage of using ethane as a feed to cracking units is a high ethylene yield with minimal coproducts. For example, at 60% per pass conversion level, the ultimate yield of ethylene is 80% based on ethane being recycled to extinction.

The following are typical operating conditions for an ethane cracking unit and the products obtained:

Conditions:

Temperature, °C	750–850
Pressure, Kg/cm^2	1–1.2
Steam/HC	0.5

Yield wt %

Hydrogen + methane	12.9
Ethylene	80.9
Propylene	1.8
Butadiene	1.9
Other*	2.5

* Other: Propane 0.3, butanes 0.4, butenes 0.4, C_5^+ 1.4

Propane cracking is similar to ethane except for the furnace temperature, which is relatively lower (longer chain hydrocarbons crack easier). However, more by-products are formed than with ethane, and the separation section is more complex. Propane gives lower ethylene yield, higher propylene and butadiene yields, and significantly more aromatic pyrolysis gasoline. Residual gas (mainly H_2 and methane) is about two and half times that produced when ethane is used. Increasing the severity of a propane cracking unit increases ethylene and residual gas yields and decreases propylene yield. Figure 3-13 shows the influence of conversion severity on the theoretical product yield for cracking propane.[46]

Cracking n-butane is also similar to ethane and propane, but the yield of ethylene is even lower. It has been noted that cracking either propane or butanes at nearly similar severity produced approximately equal liquid yields. Mixtures of propane and butane LPG are becoming important steam cracker feedstocks for C_2–C_4 olefin production. It has been forecasted that world LPG markets will grow from 114.7 million metric tons/day in 1988 to 136.9 MMtpd in the year 2000, and the largest portion of growth will be in the chemicals field.[47]

Cracking Liquid Feeds

Liquid feedstocks for olefin production are light naphtha, full range naphtha, reformer raffinate, atmospheric gas oil, vacuum gas oil, residues, and crude oils. The ratio of olefins produced from steam cracking of these feeds depends mainly on the feed type and, to a lesser extent, on the operation variables. For example, steam cracking light naphtha produces about twice the amount of ethylene obtained from steam cracking vacuum gas oil under nearly similar conditions. Liquid feeds are usually

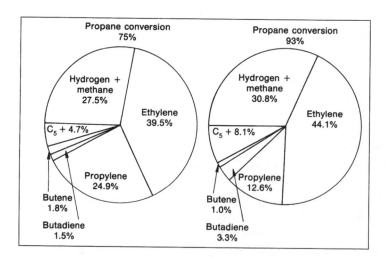

Figure 3-13. The influence of conversion severity on the theoretical product yield for the cracking of propane. Acetylene, methyl acetylene, and propadiene are hydrogenated and both ethane and propane are recycled to extinction (wt%).[46]

cracked with lower residence times and higher steam dilution ratios than those used for gas feedstocks. The reaction section of the plant is essentially the same as with gas feeds, but the design of the convection and the quenching sections are different. This is necessitated by the greater variety and quantity of coproducts. An additional pyrolysis furnace for cracking coproduct ethane and propane and an effluent quench exchanger are required for liquid feeds. Also, a propylene separation tower and a methyl acetylene removal unit are incorporated in the process. Figure 3-14 is a flow diagram for cracking naphtha or gas oil for ethylene production.[42]

As with gas feeds, maximum olefin yields are obtained at lower hydrocarbon partial pressures, pressure drops, and residence times. These variables may be adjusted to obtain higher BTX at the expense of higher olefin yield.

One advantage of using liquid feeds over gas feedstocks for olefin production is the wider spectrum of coproducts. For example, steam cracking naphtha produces, in addition to olefins and diolefins, pyrolysis gasoline rich in BTX. Table 3-16 shows products from steam cracking naphtha at low and at high severities.[44, 48] It should be noted that operation at a higher severity increased ethylene product and by-product methane and decreased propylene and butenes. The following conditions are typical for naphtha cracking:

Temperature °C:	800
Pressure Atm.:	Atmospheric
Steam/HC Kg/Kg:	0.6–0.8
Residence time sec:	0.35

Steam cracking raffinate from aromatic extraction units is similar to naphtha cracking. However, because raffinates have more isoparaffins, relatively less ethylene and more propylene is produced.

Cracking gas oils for olefin production has been practiced since 1930. However, due to the simplicity of cracking gas feeds, the use of gas oil declined. Depending on gas feed availability and its price, which is increasing relative to crude prices, gas oil cracking may return as a potential source for olefins. Gas oils in general are not as desirable feeds for olefin production as naphtha because they have higher sulfur and aromatic contents. The presence of a high aromatic content in the feed affects the running time of the system and the olefin yield; gas oils generally produce less ethylene and more heavy fuel oil. Although high sulfur gas oils could be directly cracked, it is preferable to hydrodesulfurize these feeds before cracking to avoid separate treatment schemes for each product.

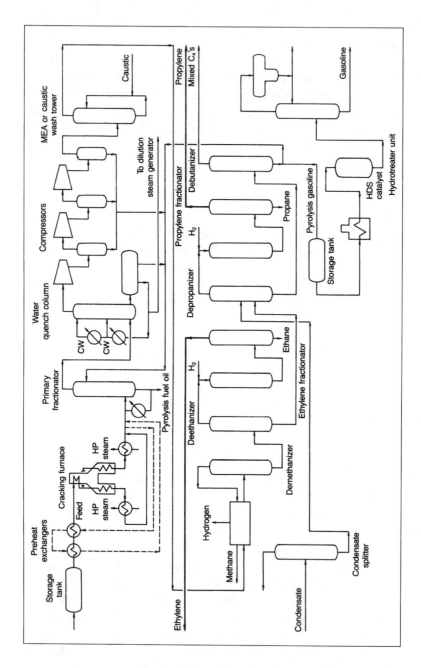

Figure 3-14. Flow diagram of an ethylene plant using liquid feeds.[42]

Table 3-16
Products from steam cracking naphtha at high severities[44,48]

Products**	Cracking severity	
	Low	High
Methane	10.3	15
Ethylene	25.8	31.3
Propylene	16.0	12.1
Butadiene	4.5	4.2
Butenes	7.9	2.8
BTX	10	13
C_5^+	17	9
Fuel oil	3	6
Other***	5.5	6.6

Feed:

Sp. gr 60/60°F	0.713	
Boiling range °C	32–170	
Aromatics	7	

**Weight percent

***Ethane (3.3 and 3.4%), acetylene, methylacetylene, propane, hydrogen.

Processes used to crack gas oils are similar to those for naphtha. However, gas oil throughput is about 20–25% higher than that for naphtha. The ethylene cracking capacity for AGO is about 15% lower than for naphtha. There must be a careful balance between furnace residence time, hydrocarbon partial pressure, and other factors to avoid problems inherent in cracking gas oils.[49] Table 3-17 shows the product composition from cracking AGO and VGO at low and high severities.[44,48,50] Figure 3-15 shows the effect of cracking severity when using gas oil on the product composition.[51]

PRODUCTION OF DIOLEFINS

Diolefins are hydrocarbon compounds that have two double bonds. Conjugated diolefins have two double bonds separated by one single bond. Due to conjugation, these compounds are more stable than monoolefins and diolefins with isolated double bonds. Conjugated diolefins also have different reactivities than monoolefins. The most important industrial diolefinic hydrocarbons are butadiene and isoprene.

Table 3-17
**Product composition from cracking atmospheric gas oil
and vacuum gas oil[44,48,50]**

Products*	AGO		VGO	
	Severity		Severity	
	Low	High	Low	High
Methane	8.0	13.7	6.6	9.4
Ethylene	19.5	26.0	19.4	23.0
Ethane	3.3	3.0	2.8	3.0
Propylene	14.0	9.0	13.9	13.7
Butadiene	4.5	4.2	5.0	6.3
Butenes	6.4	2.0	7.0	4.9
BTX	10.7	12.6		
C_5-205°C**	10.0	8.0	18.9	16.9
Fuel oil	21.8	19.0	25.0	21.0
Other***	1.8	2.5	1.4	1.8

*Weight %.
**Other than BTX.
***Acetylene, methylacetylene, propane, hydrogen.

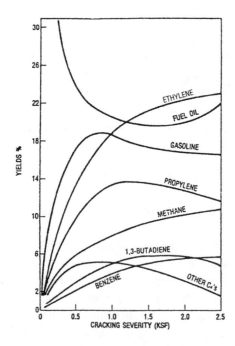

Figure 3-15. Component yields vs cracking severity for a typical gas oil.[51]

Butadiene ($CH_2 = CH\text{-}CH = CH_2$)

Butadiene is the raw material for the most widely used synthetic rubber, a copolymer of butadiene and styrene (SBR). In addition to its utility in the synthetic rubber and plastic industries (over 90% of butadiene produced), many chemicals could also be synthesized from butadiene.

Production

Butadiene is obtained as a by-product from ethylene production. It is then separated from the C_4 fraction by extractive distillation using furfural.

Butadiene could also be produced by the catalytic dehydrogenation of butanes or a butane/butene mixture.

$$CH_3CH_2CH_2CH_3 \rightarrow CH_2\text{=}CH\text{-}CH\text{=}CH_2 + 2H_2$$

The first step involves dehydrogenation of the butanes to a mixture of butenes which are then separated, recycled, and converted to butadiene. Figure 3-16 is the Lummus fixed-bed dehydrogenation of C_4 mixture to butadiene.[52] The process may also be used for the dehydrogenation of mixed amylenes to isoprene. In the process, the hot reactor effluent is quenched, compressed, and cooled. The product mixture is extracted: unreacted butanes are separated and recycled, and butadiene is recovered.

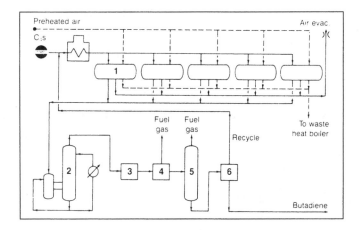

Figure 3-16. Flow diagram of the Lummus process for producing butadiene:[52] (1) reactor, (2) quenching, (3) compressor, (4) cryogenic recovery, (5) stabilizer, (6) extraction.

The Phillips process uses an oxidative-dehydrogenation catalyst in the presence of air and steam. The C_4 mixture is passed over the catalyst bed at 900 to 1100°C. Hydrogen released from dehydrogenation reacts with oxygen, thus removing it from the equilibrium mixture and shifting the reaction toward the formation of more butadiene. An in-depth study of the oxidative dehydrogenation process was made by Welch et al. They concluded that conversion and overall energy costs are favorable for butadiene production via this route.[53]

In some parts of the world, as in Russia, fermented alcohol can serve as a cheap source for butadiene. The reaction occurs in the vapor phase under normal or reduced pressures over a zinc oxide/alumina or magnesia catalyst promoted with chromium or cobalt. Acetaldehyde has been suggested as an intermediate: two moles of acetaldehyde condense and form crotonaldehyde, which reacts with ethyl alcohol to give butadiene and acetaldehyde.

Butadiene could also be obtained by the reaction of acetylene and formaldehyde in the vapor phase over a copper acetylide catalyst. The produced 1,4-butynediol is hydrogenated to 1,4-butanediol. Dehydration of 1,4-butanediol yields butadiene.

$$HC\equiv CH + 2\,H\!-\!\overset{\displaystyle O}{\overset{\|}{C}}\!-\!H \longrightarrow HOCH_2C\equiv CCH_2OH$$

1,4 butynediol

$$HOH_2CC\equiv CCH_2OH + 2\,H_2 \longrightarrow HOCH_2CH_2CH_2CH_2OH$$

$$HOCH_2CH_2CH_2CH_2OH \rightarrow CH_2=CH\!-\!CH=CH_2 + 2H_2O$$

Isoprene $(H_2C{=}\overset{\displaystyle CH_3}{\overset{|}{C}}\!-\!CH{=}CH_2)$

Isoprene (2-methyl 1,3-butadiene) is the second most important conjugated diolefin after butadiene. Most isoprene production is used for the manufacture of cis-polyisoprene, which has a similar structure to natural rubber. It is also used as a copolymer in butyl rubber formulations.

Production

There are several different routes for producing isoprene. The choice of one process over the other depends on the availability of the raw materials and the economics of the selected process.

While most isoprene produced today comes from the dehydrogenation of C_5 olefin fractions from cracking processes, several schemes are used for its manufacture via synthetic routes. The following reviews the important approaches for isoprene production.

Dehydrogenation of Tertiary Amylenes (Shell Process)

t-Amylenes (2-methyl-1-butene and 2-methyl-2-butene) are produced in small amounts with olefins from steam cracking units. The amylenes are extracted from a C_5 fraction with aqueous sulfuric acid.

Dehydrogenation of t-amylenes over a dehydrogenation catalyst produces isoprene. The overall conversion and recovery of t-amylenes is approximately 70%.

The C_5 olefin mixture can also be produced by the reaction of ethylene and propene using an acid catalyst.

$$2CH_2{=}CH_2 + 2CH_3CH{=}CH_2 \longrightarrow \overset{\overset{\textstyle CH_3}{\textstyle |}}{CH_3C}{=}CHCH_3 + \overset{\overset{\textstyle CH_3}{\textstyle |}}{CH_2{=}C}CH_2CH_3$$

The C_5 olefin mixture is then dehydrogenated to isoprene.

From Acetylene and Acetone

A three-step process developed by Snamprogetti is based on the reaction of acetylene and acetone in liquid ammonia in the presence of an alkali metal hydroxide. The product, methylbutynol, is then hydrogenated to methylbutenol followed by dehydration at 250–300°C over an acidic heterogeneous catalyst.

$$HC{\equiv}CH + CH_3\overset{\overset{\textstyle O}{\textstyle ||}}{C}CH_3 \longrightarrow HC{\equiv}C-\overset{\overset{\textstyle OH}{\textstyle |}}{\underset{\underset{\textstyle CH_3}{\textstyle |}}{C}}-CH_3$$

Methylbutynol

$$\xrightarrow{\text{H}_2} \text{CH}_2 = \text{CH} - \overset{\overset{\displaystyle \text{OH}}{|}}{\underset{\underset{\displaystyle \text{CH}_3}{|}}{\text{C}}} - \text{CH}_3 \xrightarrow{-\text{H}_2\text{O}} \text{CH}_2 = \text{CH} - \overset{\overset{\displaystyle \text{CH}_3}{|}}{\text{C}} = \text{CH}_2$$

From Isobutylene and Formaldehyde (IFP Process)

The reaction between isobutylene (separated from C_4 fractions from cracking units or from cracking isobutane to isobutene) and formaldehyde produces a cyclic ether (dimethyl dioxane). Pyrolysis of dioxane gives isoprene and formaldehyde. The formaldehyde is recovered and recycled to the reactor.

$$\underset{\underset{\displaystyle \text{CH}_3}{|}}{\text{CH}_3} - \text{C} = \text{CH}_2 + 2\text{H} - \overset{\overset{\displaystyle \text{O}}{\|}}{\text{C}} - \text{H} \longrightarrow$$

$$\text{CH}_2 = \overset{\overset{\displaystyle \text{CH}_3}{|}}{\text{C}} - \text{CH} = \text{CH}_2 + \text{H} - \overset{\overset{\displaystyle \text{O}}{\|}}{\text{C}} - \text{H} + \text{H}_2\text{O}$$

From Isobutylene and Methylal (Sun Oil Process)

In this process, methylal (dimethoxymethane) is used instead of formaldehyde. The advantage of using methylal over formaldehyde is its lower reactivity toward 1-butene than formaldehyde, thus allowing mixed feedstocks to be used. Also, unlike formaldehyde, methylal does not decompose to CO and H_2.

The first step in this process is to produce methylal by the reaction of methanol and formaldehyde using an acid catalyst.

$$\overset{\overset{\displaystyle \text{O}}{\|}}{\text{HCH}} + 2\text{CH}_3\text{OH} \xrightarrow{\text{H}^+} \text{CH}_3\text{OCH}_2\text{OCH}_3 + \text{H}_2\text{O}$$

$$\underset{\underset{\displaystyle \text{CH}_3}{|}}{\text{CH}_3} - \text{C} = \text{CH}_2 + \text{CH}_3\text{OCH}_2\text{OCH}_3 \longrightarrow \text{CH}_2 = \overset{\overset{\displaystyle \text{CH}_3}{|}}{\text{C}} - \text{CH} = \text{CH}_2 + 2\text{CH}_3\text{OH}$$

The second step is the vapor phase reaction of methylal with isobutene to produce isoprene.

2-Butene in the C_4 mixture also reacts with methylal but at a slower rate to give isoprene. 1-Butene reacts slowly to give 1,3-pentadiene.

From Propylene (Goodyear Process)

Another approach for producing isoprene is the dimerization of propylene to 2-methyl-1-pentene. The reaction occurs at 200°C and about 200 atmospheres in the presence of a tripropyl aluminum catalyst combined with nickel or platinum.

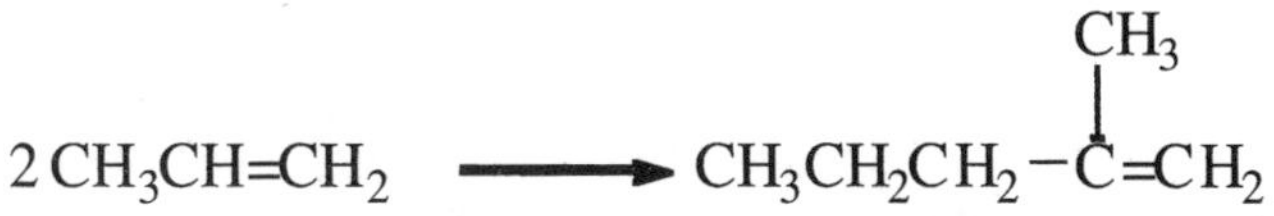

$$2 \, CH_3CH{=}CH_2 \longrightarrow CH_3CH_2CH_2{-}\underset{\underset{CH_3}{|}}{C}{=}CH_2$$

The next step is the isomerization of 2-methyl-1-pentene to 2-methyl-2-pentene using an acid catalyst.

$$CH_3CH_2CH_2\underset{\underset{CH_3}{|}}{C}{=}CH_2 \longrightarrow CH_3CH_2CH{=}\underset{\underset{CH_3}{|}}{C}{-}CH_3$$

2-Methyl-2-pentene is finally pyrolyzed to isoprene.

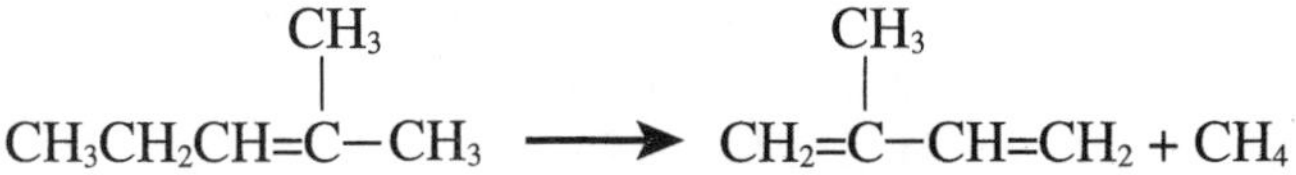

$$CH_3CH_2CH{=}\underset{\underset{CH_3}{|}}{C}{-}CH_3 \longrightarrow CH_2{=}\underset{\underset{CH_3}{|}}{C}{-}CH{=}CH_2 + CH_4$$

REFERENCES

1. "Refining Handbook," *Hydrocarbon Processing,* Vol. 59, No. 11, 1990, p. 86.
2. Gary, J. H. and Handwerk, G. E., *Petroleum Refining, Technology and Economics,* Second Edition, Marcell Dekker, Inc., 1984, p. 45.
3. Reber, R. A. and Symoniak, M. F., Ind. Eng. Chem. Div., 169th ACS National Meeting, paper 75, April 1975.
4. "Refining Handbook" *Hydrocarbon Processing,* Vol. 77, No. 11, 1998, p. 68.

5. "Refining Handbook," *Hydrocarbon Processing,* Vol. 69, No. 11, 1990, p. l06.

6. Elliot, J. D., "Maximize Distillate Liquid Products," *Hydrocarbon Processing,* Vol. 71, No. 1, 1992, pp. 75–82.

7. Mochida, I., Fugimato, K. and Oyama, T. Thrower, P. A. editor, *Chemistry and Physics of Carbon,* Vol. 24, Marcell Dekker, 1994.

8. Martinez-Escandell, M. et al., "Pyrolysis of Petroleum Residues," *Carbon,* Vol. 37, No. 10, 1999, pp. 1567–1582.

9. Dymond, R. E., "World Markets for Petroleum Coke," *Hydrocarbon Processing,* Vol. 70, No. 9, 1991, pp. 162C–162J.

10. Gotshall, W. W., Reprints, Division of Petroleum Chemistry, A.C.S., No. 20, Nov. 3, 1975.

11. "Refining Handbook," *Hydrocarbon Processing,* Vol. 53, No. 11, 1974, p. 123.

12. Matar, S., "Aromatics Production and Chemicals," *The Arabian Journal for Science and Engineering,* Vol. 11, No.1,1986, pp. 23–32.

13. Al-Kabbani, A. S. "Reforming Catalyst Optimization," *Hydrocarbon Processing,* Vol. 78, No. 7, 1999, pp. 61–67.

14. Pollitzer, E. L., Hayes, J. C., and Haensel, V., "The Chemistry of Aromatics Production via Catalytic Reforming," *Refining Petroleum for Chemicals,* Advances in Chemistry Series No. 97, American Chemical Society, 1970, pp. 20–23.

15. "Refining Handbook," *Hydrocarbon Processing,* Vol. 69, No. 11, 1990, p. 118.

16. Gentry, J. C. and Kumar, C. S., "Improve BTX Processing Economics," *Hydrocarbon Processing,* Vol. 77, No. 3, 1998, pp. 69–82.

17. O'Connor, P. et al. "Improve Resid Processing," *Hydrocarbon Processing,* Vol . 70, No. 11, 1991, pp. 76–84.

18. Ocelli, M. L., "Metal-Resistant Fluid Cracking Catalyst: Thirty Years Of Research," ACS Symposium Series, No. 52, Washington, DC., 1990, p. 343.

19. Reynolds, B. E., Brown, E. C., and Silverman, M. A., "Clean Gasoline via VRDS/RFCC," *Hydrocarbon Processing,* Vol. 71, No. 4, 1992, pp. 43–51.

20. Hayward, C. M., and Winkler, W .S. "FCC: Matrix/zeolite Interactions," *Hydrocarbon Processing,* Vol. 69, No. 2, 1990, pp. 55–56.

21. Humphries, A. et al. "Catalyst Helps Reformulation," *Hydrocarbon Processing,* Vol. 70, No. 4, 1991, pp. 69–72.

22. McLean, J. B. and Moorehead, E. L. "Steaming Affects FCC Catalyst," *Hydrocarbon Processing,* Vol. 70, No. 2, 1991, pp. 41–45.
23. Occelli, M. L., (ed.) *Fluid Catalytic Cracking, Role in Modern Refining,* ACS Symposium Series, American Chemical Society, Washington DC, 1988, pp. 1–16.
24. Gall, J. W. et al., NPRA Annual Meeting, AM 82–50, 5, 1982.
25. Hatch, L. F. and Matar, S., "Refining Processes and Petrochemicals" (Part I), *Hydrocarbon Processing,* Vol. 56, No. 7, 1977, pp. 191–201.
26. Jazayeri, B., "Optimize FCC Riser Design," *Hydrocarbon Processing,* Vol. 70, No. 5, 1991, pp. 93–95.
27. "Refining Handbook," *Hydrocarbon Processing,* Vol. 75, No. 11, 1996, p. 121.
28. "Petrochemical Handbook," *Hydrocarbon Processing,* Vol. 78, No. 3, 1999, p. 124.
29. "Refining Handbook," *Hydrocarbon Processing,* Vol. 69, No. 11, 1990, p. 1 00.
30. Scott, J. W. and Bridge, A. G., "Origin and Refining of Petroleum," No. 7, Washington D.C., American Chemical Society, 1971, p. 116.
31. Bridge, A. G., Scott, J. W., and Reed A. M., *Hydrocarbon Processing,* Vol. 54, No. 5, 1975, pp. 74–81.
32. "Refining Handbook," *Hydrocarbon Processing,* Vol. 69, No. 11, 1990, p. 116.
33. Gates, B. C., Katzer, J. R., and Schuit, G. C., Chemistry of Catalytic Processes, McGraw-Hill Book Company, 1979, p. 394.
34. Jensen, B. et al., "Reduce Acid Usage on Alkylation" *Hydrocarbon Processing,* Vol. 77, No. 7, 1998, p. 101.
35. Lerner, H. and Citarella, V. A., "Improve Alkylation Efficiency," *Hydrocarbon Processing,* Vol. 70, No. 11, 1991, pp. 89–92.
36. Lafferty, W. L. and Stokeld, R. W., *Origin and Refining of Petroleum, Advances in Chemistry Series* 103, ACS, Washington D.C., 1971 , p. 134
37. Cheung, T. and Gates, B., "Strong Acid Catalyst for Paraffin Conversion," *CHEMTECH,* Vol. 27, No. 9, 1997, pp. 28–34.
38. Albright, L. F., Improving Alkylate Gasoline Technology," *CHEMTECH,* Vol. 28, No. 7, 1998, pp. 46–53.
39. Lawrance, P. A., and Rawlings A. A., Proceedings 7th World Pet. Congress, 1967, p. 137.
40. Andrews, J. W. et al., *Hydrocarbon Processing,* Vol. 54, No. 5, 1975, pp. 69–73.

41. Wysiekierski, A. G. et al., "Control Coking for Olefin Production *Hydrocarbon Processing,* Vol. 78, No. 1, 1999, pp. 97–100.

42. Burns, K. G., et al., "Chemicals Increase Ethylene Plant Efficiency," *Hydrocarbon Processing,* Vol. 70, No. 1, 1991, pp. 83–87.

43. Belgian Patent 840–343 to Continental Oil (Houston).

44. Barwell, J. and Martin, S. R., International Seminar on Petrochemicals, paper No. 9 (p. 2) Baghdad Oct. 25–30, 1975.

45. Lee, A. K. K and Aitani, A. M., "Saudi Ethylene Plants Move Toward More Feed Flexibility," *Oil and Gas Journal* (Special), Sept. 10, 1990, pp. 60–64.

46. Nahas, R. S. and Nahas, M. R., Second Arab Conference on Petrochemicals paper No. 6 (P-1) Abu Dhabi, March 15–22, 1976.

47. Watters, P. R., "New Partnership Emerge in LPG and Petrochemicals trade," *Hydrocarbon Processing,* Vol. 69, No. 6, 1990, pp. 100B–100N.

48. El-Enany, N. M. and Abdel Rahman O. F., Second Arab Conference on Petrochemicals, paper No. 9 (p.2) Abu Dhabi, March 15–23, 1976.

49. Smith. J, *Chemical Engineering,* Sept. 15, 1975, pp. 131–136.

50. Bassler, E. J., *Oil and Gas Journal,* March 17, 1975, pp. 93–96.

51. Zdonik, S. B., Potter, W. S., and Hayward, G. L., *Hydrocarbon Processing,* Vol. 55, No. 4, 1976, pp. 161–166.

52. Petrochemical Handbook, *Hydrocarbon Processing,* Vol. 70, No. 3, 1991, p. 141.

53. Welch, L. M., Croce, L. J. and Christmann, H. F., *Hydrocarbon Processing,* Vol. 57, No. 11, 1978, pp. 131–136.

CHAPTER FOUR

Nonhydrocarbon Intermediates

INTRODUCTION

From natural gas, crude oils, and other fossil materials such as coal, few intermediates are produced that are not hydrocarbon compounds. The important intermediates discussed here are hydrogen, sulfur, carbon black, and synthesis gas.

Synthesis gas consists of a nonhydrocarbon mixture (H_2,CO) obtainable from more than one source. It is included in this chapter and is further noted in Chapter 5 in relation to methane as a major feedstock for this mixture. This chapter discusses the use of synthesis gas obtained from coal gasification and from different petroleum sources for producing gaseous as well as liquid hydrocarbons (Fischer Tropsch synthesis).

Naphthenic acids and cresylic acid, which are extracted from certain crude oil fractions, are briefly reviewed at the end of the chapter.

HYDROGEN

Hydrogen is the lightest known element. Although only found in the free state in trace amounts, it is the most abundant element in the universe and is present in a combined form with other elements. Water, natural gas, crude oils, hydrocarbons, and other organic fossil materials are major sources of hydrogen.

Hydrogen has been of great use to theoretical investigation. The structure of the atom developed by Bohr (Nobel Prize Winner 1922) was based on a model of the hydrogen atom. Chemically, hydrogen is a very reactive element. Obtaining hydrogen from its compounds is an energy-extensive process. To decompose water into hydrogen and oxygen, an energy input equal to an enthalpy change of +286 KJ/mol is required[1]:

$$H_2O \rightarrow H_2 + \frac{1}{2}O_2 \qquad \Delta H = +286 \text{ KJ/mol}$$

Electrolysis, and thermochemical and photochemical decomposition of water followed by purification through diffusion methods are expensive processes to produce hydrogen.

The most economical way to produce hydrogen is by steam reforming petroleum fractions and natural gas (Figure 4-1).[2] In this process, two major sources of hydrogen (water and hydrocarbons) are reacted to produce a mixture of carbon monoxide and hydrogen (synthesis gas). Hydrogen can then be separated from the mixture after shift converting carbon monoxide to carbon dioxide. Carbon oxides are removed by passing the mixture through a pressure swing adsorption system. The shift conversion reaction is discussed in relation to ammonia synthesis in Chapter 5. The production of synthesis gas by steam reforming liquid hydrocarbons is noted later in this chapter.

Recently, a new process has been developed to manufacture hydrogen by steam reforming methanol. In this process, an active catalyst is used to decompose methanol and shift convert carbon monoxide to carbon dioxide. The produced gas is cooled, and carbon dioxide is removed:

$$CH_3OH(g) + H_2O(g) \rightarrow CO_2(g) + 3\ H_2(g)$$

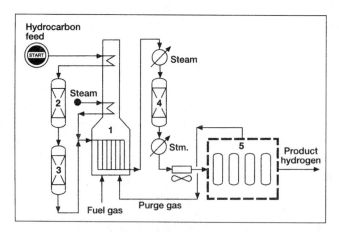

Figure 4.1. A process for producing hydrogen by steam reforming of hydrocarbons:[2] (1) reforming furnace (2,3) purification section, (4) shift converter, (5) pressure swing adsorption.

This process is used to produce relatively small quantities (0.18–1.8 MMscfd) of highly pure hydrogen when methanol is available at a reasonable price.

In the petroleum refining industry, hydrogen is essentially obtained from catalytic naphtha reforming, where it is a coproduct with reformed gasoline.

The use of hydrogen in the chemical and petroleum refining industries is of prime importance. Hydrogen is essentially a hydrogenating agent. For example, it is used with vegetable oils and fats to reduce unsaturated esters (triglycerides). It is also a reducing agent for sulfide ores such as zinc and iron sulfides (to get the metals from their ores).

Hydrogen use in the petroleum refining includes many processing schemes such as hydrocracking, hydrofinishing of lube oils, hydrodealkylation and hydrodesulfurization of petroleum fractions and residues. Hydrocracking of petroleum resids is becoming more important to produce lighter petroleum distillates of low sulfur and nitrogen content to meet stringent government-mandated product specifications to control pollution.

In the petrochemical field, hydrogen is used to hydrogenate benzene to cyclohexane and benzoic acid to cyclohexane carboxylic acid. These compounds are precursors for nylon production (Chapter 10). It is also used to selectively hydrogenate acetylene from C_4 olefin mixture.

As a constituent of synthesis gas, hydrogen is a precursor for ammonia, methanol, Oxo alcohols, and hydrocarbons from Fischer Tropsch processes. The direct use of hydrogen as a clean fuel for automobiles and buses is currently being evaluated compared to fuel cell vehicles that use hydrocarbon fuels which are converted through on-board reformers to a hydrogen-rich gas. Direct use of H_2 provides greater efficiency and environmental benefits.[3]

Due to the increasing demand for hydrogen, many separation techniques have been developed to recover it from purge streams vented from certain processing operations such as hydrocracking and hydrotreating. In addition to hydrogen, these streams contain methane and other light hydrocarbon gases. Physical separation techniques such as adsorption, diffusion, and cryogenic phase separation are used to achieve this goal.

Adsorption is accomplished using a special solid that preferentially adsorbs hydrocarbon gases, not hydrogen. The adsorbed hydrocarbons are released by reducing the pressure. Cryogenic phase separation on the other hand, depends on the difference between the volatilities of the components at the low temperatures and high pressures used. The vapor phase is rich in hydrogen, and the liquid phase contains the hydrocarbons. Hydrogen is separated from the vapor phase at high purity.

Diffusion separation processes depend on the permeation rate for gas mixtures passing through a special membrane. The permeation rate is a function of the type of gas feed, the membrane material, and the operating conditions. Gases with smaller molecular sizes such as helium and hydrogen permeate membranes more readily than larger molecules such as methane and ethane.[4] An example of membrane separator is the hollow fiber type shown in Figure 4-2. After the feed gas is preheated and filtered it enters the membrane separation section. This is made of a permeater vessel containing 12-inch diameter bundles (resemble filter cartridges) and consists of millions of hollow fibers. The gas mixture is distributed in the annulus between the fiber bundle and the vessel wall. Hydrogen, being more permeable, diffuses through the wall of the hollow fiber and exits at a lower pressure. The less permeable hydrocarbons flow around the fiber walls to a perforated center tube and exit at approximately feed pressure. It has been reported that this system can deliver a reliable supply of 95+% pure hydrogen from off-gas streams having as low as 15% H_2.[5]

SULFUR

Sulfur is a reactive, nonmetallic element naturally found in nature in a free or combined state. Large deposits of elemental sulfur are found in various parts of the world, with some of the largest being along the coastal plains of Louisiana. In its combined form, sulfur is naturally present in sulfide ores of metals such as iron, zinc, copper, and lead. It is also a constituent of natural gas and refinery gas streams in the form of hydrogen sulfide. Different processes have been developed for obtaining sulfur and sulfuric acid from these three sources.

The Frasch process, developed in 1894, produces sulfur from underground deposits.

Smelting iron ores produces large amounts of sulfur dioxide, which is catalytically oxidized to sulfur trioxide for sulfuric acid production. This process is declining due to pollution control measures and the presence of some impurities in the product acid.

Currently, sulfur is mainly produced by the partial oxidation of hydrogen sulfide through the Claus process. The major sources of hydrogen sulfide are natural gas and petroleum refinery streams treatment operations. It has been estimated that 90–95% of the world's recovered sulfur is produced through the Claus process.[6] Typical sulfur recovery ranges from 90% for a lean acid gas feed to 97% for a rich acid gas feed.[7]

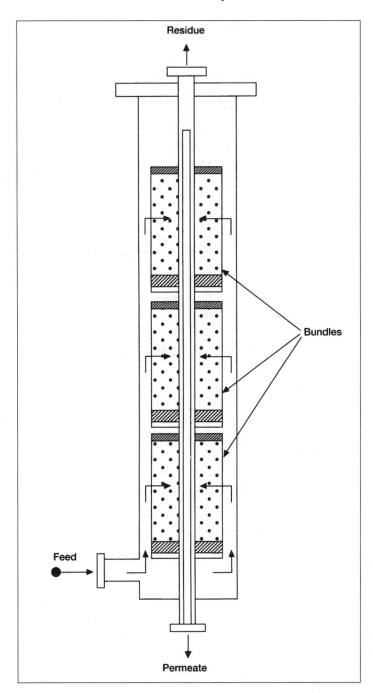

Figure 4-2. Permeator for gas separation.[5]

USES OF SULFUR

The most important use of sulfur is for sulfuric acid production. Other uses range from dusting powder for roses to rubber vulcanization to sulfur-asphalt pavements. Flower sulfur is used in match production and in certain pharmaceuticals. Sulfur is also an additive in high pressure lubricants.

Sulfur can replace 30–50% of the asphalt in the blends used for road construction. Road surfaces made from asphalt-sulfur blends have nearly double the strength of conventional pavement, and it has been claimed that such roads are more resistant to climatic conditions. The impregnation of concrete with molten sulfur is another potential large sulfur use. Concretes impregnated with sulfur have better tensile strength and corrosion resistance than conventional concretes. Sulfur is also used to produce phosphorous pentasulfide, a precursor for zinc dithiophosphates used as corrosion inhibitors.

Sulfur reacts with nitrogen to form polymeric sulfur nitrides (SN_x) or polythiazyls. These polymers were found to have the optical and electrical properties of metals.[8]

THE CLAUS PROCESS

This process includes two main sections: the burner section with a reaction chamber that does not have a catalyst, and a Claus reactor section. In the burner section, part of the feed containing hydrogen sulfide and some hydrocarbons is burned with a limited amount of air. The two main reactions that occur in this section are the complete oxidation of part of the hydrogen sulfide (feed) to sulfur dioxide and water and the partial oxidation of another part of the hydrogen sulfide to sulfur. The two reactions are exothermic:

$$H_2S + {}^3/_2O_2 \rightarrow SO_2 + H_2O \qquad \Delta H = -519 \text{ to } -577 \text{ KJ}$$

$$3H_2S + {}^3/_2O_2 \rightarrow 3/x\ S_x + 3H_2O \qquad \Delta H = -607 \text{ to } -724 \text{ KJ}$$

In the second section, unconverted hydrogen sulfide reacts with the produced sulfur dioxide over a bauxite catalyst in the Claus reactor. Normally more than one reactor is available. In the Super-Claus process (Figure 4-3), three reactors are used.[9] The last reactor contains a selective oxidation catalyst of high efficiency. The reaction is slightly exothermic:

$$2H_2S + SO_2 \rightarrow 3/x\ S_x + 2H_2O \qquad \Delta H = -88 \text{ to } -146 \text{ KJ}$$

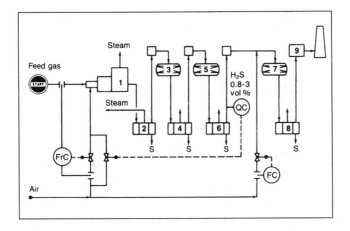

Figure 4–3. The Super Claus process for producing sulfur:[9] (1) main burner, (2,4, 6,8) condensers, (3,5) Claus reactors, (7) reactor with selective oxidation catalyst.

After each reaction stage, sulfur is removed by condensation so that it does not collect on the catalyst. The temperature in the catalytic converter should be kept over the dew point of sulfur to prevent condensation on the catalyst surface, which reduces activity.

Due to the presence of hydrocarbons in the gas feed to the burner section, some undesirable reactions occur, such as the formation of carbon disulfide (CS_2) and carbonyl sulfide (COS). A good catalyst has a high activity toward H_2S conversion to sulfur and a reconversion of COS and CS_2 to sulfur and carbon oxides. Mercaptans in the acid gas feed results in an increase in the air demand. For example, approximately 5–13% increase in the air required is anticipated if about 2 mol% mercaptans are present.[7] The increase in the air requirement is essentially a function of the type of mercaptans present. The oxidation of mercaptans could be represented as:

$$CH_3 SH + 3O_2 \rightarrow SO_2 + CO_2 + 2H_2O$$

$$C_2H_5SH + {}^9/_2O_2 \rightarrow SO_2 + 2CO_2 + 3H_2O$$

Sulfur dioxide is then reduced in the Claus reactor to elemental sulfur.

SULFURIC ACID (H_2SO_4)

Sulfuric acid is the most important and widely used inorganic chemical. The 1994 U.S. production of sulfuric acid was 89.2 billion pounds.

(most used industrial chemical).[10] Sulfuric acid is produced by the contact process where sulfur is burned in an air stream to sulfur dioxide, which is catalytically converted to sulfur trioxide. The catalyst of choice is solid vanadium pentoxide (V_2O_5). The oxidation reaction is exothermic, and the yield is favored at lower temperatures:

$$SO_2 (g) + \tfrac{1}{2}O_2 (g) \rightarrow SO_3 (g) \qquad \Delta H = -98.9 \text{ KJ}$$

The reaction occurs at about 450°C, increasing the rate at the expense of a higher conversion. To increase the yield of sulfur trioxide, more than one conversion stage (normally three stages) is used with cooling between the stages to offset the exothermic reaction heat. Absorption of SO_3 from the gas mixture exiting from the reactor favors the conversion of SO_2. The absorbers contain sulfuric acid of 98% concentration which dissolves sulfur trioxide. The unreacted sulfur dioxide and oxygen are recycled to the reactor. The absorption reaction is exothermic, and special coolers are used to cool the acid:

$$SO_3(g) + H_2O(l) \rightarrow H_2SO_4(l)$$

Uses of Sulfuric Acid

Sulfuric acid is primarily used to make fertilizers. It is also used in other major industries such as detergents, paints, pigments, and pharmaceuticals.

CARBON BLACK

Carbon black is an extremely fine powder of great commercial importance, especially for the synthetic rubber industry. The addition of carbon black to tires lengthens its life extensively by increasing the abrasion and oil resistance of rubber.

Carbon black consists of elemental carbon with variable amounts of volatile matter and ash. There are several types of carbon blacks, and their characteristics depend on the particle size, which is mainly a function of the production method.

Carbon black is produced by the partial combustion or the thermal decomposition of natural gas or petroleum distillates and residues. Petroleum products rich in aromatics such as tars produced from catalytic and thermal cracking units are more suitable feedstocks due to their high carbon/hydrogen ratios. These feeds produce blacks with a carbon content of approximately 92 wt%. Coke produced from delayed and fluid coking units with low sulfur and ash contents has been investigated as a possible substitute for carbon black.[11] Three processes are currently used for the manufacture of carbon blacks. These are the channel, the furnace, and the thermal processes.

THE CHANNEL PROCESS

This process is only of historical interest, because not more than 5% of the blacks are produced via this route. In this process, the feed (e.g., natural gas) is burned in small burners with a limited amount of air. Some methane is completely combusted to carbon dioxide and water, producing enough heat for the thermal decomposition of the remaining natural gas. The two main reactions could be represented as:

$$CH_4 + 2O_2 \rightarrow CO_2 + 2H_2O \qquad \Delta H = -799 \text{ KJ}$$

$$CH_4 \rightarrow C + H_2 \qquad \Delta H = +92 KJ$$

The formed soot collects on cooled iron channels from which the carbon black is scraped. Channel black is characterized by having a lower pH, higher volatile matter, and smaller average particle size than blacks from other processes.

THE FURNACE BLACK PROCESS

This is a more advanced partial combustion process. The feed is first preheated and then combusted in the reactor with a limited amount of air. The hot gases containing carbon particles from the reactor are quenched with a water spray and then further cooled by heat exchange with the air used for the partial combustion. The type of black produced depends on the feed type and the furnace temperature. The average particle diameter of the blacks from the oil furnace process ranges between 200–500 Å, while it ranges between 400–700 Å from the gas furnace process. Figure 4-4 shows the oil furnace black process.[12]

THE THERMAL PROCESS

In this process, the feed (natural gas) is pyrolyzed in preheated furnaces lined with a checker work of hot bricks. The pyrolysis reaction produces carbon, which collects on the bricks. The cooled bricks are then

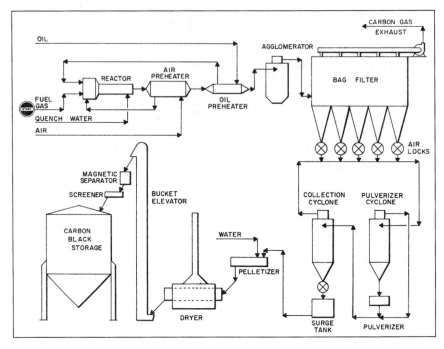

Figure 4-4. Carbon black (oil black) by furnace process of Ashland Chemical Co.[12]

reheated after carbon black is collected. The average particle diameter from this process is large and ranges between 1800 Å for the fine thermal and 5000 Å for medium thermal black.

PROPERTIES AND USES OF CARBON BLACK

The important properties of carbon black are particle size, surface area, and pH. These properties are functions of the production process and the feed properties. Channel blacks are generally acidic, while those produced by the Furnace and Thermal processes are slightly alkaline. The pH of the black has a pronounced influence on the vulcanization time of the rubber. (Vulcanization is a physicochemical reaction by which rubber changes to a thermosetting mass due to cross-linking of the polymer chains by adding certain agents such as sulfur.) The basic nature (higher pH) of furnace blacks is due to the presence of evaporation deposits from the water quench. Thermal blacks, due to their larger average particle size, are not suitable for tire bodies and tread bases, but they are used in inner tubes, footwear, and paint pigment. Gas and oil furnace blacks are the most important forms of carbon blacks and are generally used in tire treads and tire bodies. Table 4-1 shows a typical analysis of carbon black from an oil furnace process.

Carbon black is also used as a pigment for paints and printing inks, as a nucleation agent in weather modifications, and as a solar energy absorber. About 70% of the worlds' consumption of carbon black is used in the production of tires and tire products. Approximately 20% goes into other products such as footwear, belts, hoses, etc. and the rest is used in such items as paints, printing ink, etc. The world capacity of carbon black was approximately 17 billion pounds in 1998.[13] U.S. projected consumption for the year 2003 is approximately 3.9 billion pounds.

SYNTHESIS GAS

Synthesis gas generally refers to a mixture of carbon monoxide and hydrogen. The ratio of hydrogen to carbon monoxide varies according to the type of feed, the method of production, and the end use of the gas.

During World War II, the Germans obtained synthesis gas by gasifying coal. The mixture was used for producing a liquid hydrocarbon mixture in the gasoline range using Fischer-Tropsch technology. Although this route was abandoned after the war due to the high production cost of these hydrocarbons, it is currently being used in South Africa, where coal is inexpensive (SASOL, II, and III).

There are different sources for obtaining synthesis gas. It can be produced by steam reforming or partial oxidation of any hydrocarbon ranging from natural gas (methane) to heavy petroleum residues. It can also

Table 4-1
Selected properties of carbon black from an oil furnace process

Analysis	General purpose	High abrasion	Conductive
Volatile matter wt %	0.9	1.6	1.6
pH	9.1	9.0	8.0
Average particle diameter, Å	550	280	190
Surface area, m^2/g (electron microscope method)	40	75	120
Surface area, m^2/g (nitrogen adsorption method)	25	75	220

be obtained by gasifying coal to a medium Btu gas (medium Btu gas consists of variable amounts of CO, CO_2, and H_2 and is used principally as a fuel gas). Figure 4-5 shows the different sources of synthesis gas.

A major route for producing synthesis gas is the steam reforming of natural gas over a promoted nickel catalyst at about 800°C:

$$CH_4(g) + H_2O(g) \rightarrow CO(g) + 3H_2(g)$$

This route is used when natural gas is abundant and inexpensive, as it is in Saudi Arabia and the USA.

In Europe, synthesis gas is mainly produced by steam reforming naphtha. Because naphtha is a mixture of hydrocarbons ranging approximately from C_5-C_{10}, the steam reforming reaction may be represented using n-heptane:

$$CH_3(CH_2)_5CH_3 + 7H_2O(g) \rightarrow 7CO(g) + 15H_2(g)$$

As the molecular weight of the hydrocarbon increases (lower H/C feed ratio), the H_2/CO product ratio decreases. The H_2/CO product ratio is approximately 3 for methane, 2.5 for ethane, 2.1 for heptane, and less than 2 for heavier hydrocarbons. Noncatalytic partial oxidation of hydrocarbons is also used to produce synthesis gas, but the H_2/CO ratio is lower than from steam reforming:

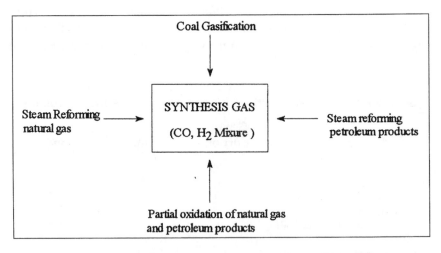

Figure 4-5. The different sources and routes to synthesis gas.

$$CH_4(g) + {}^1/_2O_2\ (g) \rightarrow CO\ (g) + 2H_2\ (g)$$

In practice, this ratio is even lower than what is shown by the stoichiometric equation because part of the methane is oxidized to carbon dioxide and water. When resids are partially oxidized by oxygen and steam at 1400–1450°C and 55–60 atmospheres, the gas consists of equal parts of hydrogen and carbon monoxide. Table 4-2 compares products from steam reforming natural gas with products from partial oxidation of heavy fuel oil.[14]

USES OF SYNTHESIS GAS

Synthesis gas is an important intermediate. The mixture of carbon monoxide and hydrogen is used for producing methanol. It is also used to synthesize a wide variety of hydrocarbons ranging from gases to naphtha to gas oil using Fischer Tropsch technology. This process may offer an alternative future route for obtaining olefins and chemicals. The hydroformylation reaction (Oxo synthesis) is based on the reaction of synthesis gas with olefins for the production of Oxo aldehydes and alcohols (Chapters 5, 7, and 8).

Synthesis gas is a major source of hydrogen, which is used for producing ammonia. Ammonia is the host of many chemicals such as urea, ammonium nitrate, and hydrazine. Carbon dioxide, a by-product from synthesis gas, reacts with ammonia to produce urea.

The production of synthesis gas from methane and the major chemicals based on it are noted in Chapter 5.

Hydrocarbons from Synthesis Gas (Fischer Tropsch Synthesis, FTS)

Most of the production of hydrocarbons by Fischer Tropsch method uses synthesis gas produced from sources that yield a relatively low

Table 4-2
Composition of synthesis gas from steam reforming natural gas and partial oxidation of fuel oil[14]

Process	Volume % dry sulfur free				
	CO	H_2	CO_2	N_2+A	CH_4
Steam reforming natural gas	15.5	75.7	8.1	0.2	0.5
Partial oxidation-heavy fuel oil	47.5	46.7	4.3	1.4	0.3

H_2/CO ratio, such as coal gasifiers. This, however, does not limit this process to low H_2/CO gas feeds. The only large-scale commercial process using this technology is in South Africa, where coal is an abundant energy source. The process of obtaining liquid hydrocarbons from coal through FTS is termed indirect coal liquefaction. It was originally intended for obtaining liquid hydrocarbons from solid fuels.[15] However, this method may well be applied in the future to the manufacture of chemicals through cracking the liquid products or by directing the reaction to produce more olefins.

The reactants in FTS are carbon monoxide and hydrogen. The reaction may be considered a hydrogenative oligomerization of carbon monoxide in presence of a heterogeneous catalyst.

The main reactions occurring in FTS are represented as:[16]

$$2nH_2 + nCO \rightarrow C_nH_{2n} + nH_2O \qquad \text{(olefins)}$$

$$(2n + 1)\, H_2 + nCO \rightarrow C_nH_{2n+2} + nH_2O \qquad \text{(paraffins)}$$

$$2nH_2 + nCO \rightarrow C_nH_{2n+2}\, O + (n\text{-}1)\, H_2O \qquad \text{(alcohols)}$$

The coproduct water reacts with carbon monoxide (the shift reaction), yielding hydrogen and carbon dioxide:

$$CO + H_2O \rightarrow CO_2 + H_2$$

The gained hydrogen from the water shift reaction reduces the hydrogen demand for FTS. Water gas shift proceeds at about the same rate as the FT reaction. Studies of the overall water shift reaction in FT synthesis have been reviewed by Rofer Deporter.[17] Another side reaction also occurring in FTS reactors is the disproportionation of carbon monoxide to carbon dioxide and carbon:

$$2CO \rightarrow CO_2 + C$$

This reaction is responsible for the deposition of carbon in the reactor tubes in fixed-bed reactors and reducing heat transfer efficiency.

Fischer Tropsch synthesis is catalyzed by a variety of transition metals such as iron, nickel, and cobalt. Iron is the preferred catalyst due to its higher activity and lower cost. Nickel produces large amounts of methane, while cobalt has a lower reaction rate and lower selectivity than iron. By comparing cobalt and iron catalysts, it was found that cobalt promotes more middle-distillate products. In FTS, cobalt produces hydrocarbons plus water while iron catalyst produces hydrocarbons and carbon dioxide.[18] It appears that the iron catalyst promotes the shift reaction more than the cobalt catalyst. Dry[19] reviewed types of catalysts used in FT processes and their preparation.

Two reactor types are used commercially in FTS, a fixed bed and a fluid-bed. The fixed-bed reactors usually run at lower temperatures to avoid carbon deposition on the reactor tubes. Products from fixed-bed reactors are characterized by low olefin content, and they are generally heavier than products from fluid-beds. Heat distribution in fluid-beds, however, is better than fixed-bed reactors, and fluid-beds are generally operated at higher temperatures. Figure 4-6 shows the Synthol fluid-bed reactor.[20] Products are characterized by having more olefins, a high percent of light hydrocarbon gases, and lower molecular weight product slate than from fixed bed types. Table 4-3 compares the feed, the reaction conditions, and the products from the two reactor systems.

Fischer Tropsch technology is best exemplified by the SASOL projects in South Africa. After coal is gasified to a synthesis gas mixture, it is purified in a rectisol unit. The purified gas mixture is reacted in a synthol unit over an iron-based catalyst. The main products are gasoline, diesel fuel, and jet fuels. By-products are ethylene, propylene, alpha olefins, sulfur, phenol, and ammonia which are used for the production of downstream chemicals.[21]

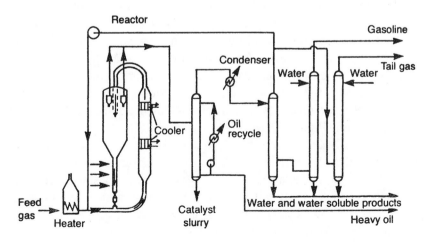

Figure 4-6. A flow chart of the Synthol process.[20]

Table 4-3
Typical analysis of products from Fischer-Tropsch fixed and fluid-bed reactors

Conditions	Fixed-Bed	Fluid-Bed
Temperature range °F	425–450	625–650
Conversion %	65	85
H_2/CO ratio	1.7	2.8
Products %		
Hydrocarbon Gases C_1-C_4	21.1	51.0
C_5-C_{12}	19.0	31.0
C_{13}-C_{18}	15.0	5.0
C_{19}-C_{31} (Heavy oil)	41.0	6.0
Oxygenates	3.9	7.0

A slurry bed reactor is in a pilot stage investigation. This type is characterized by having the catalyst in the form of a slurry. The feed gas mixture is bubbled through the catalyst suspension. Temperature control is easier than the other two reactor types. An added advantage to slurry-bed reactor is that it can accept a synthesis gas with a lower H_2/CO ratio than either the fixed-bed or the fluid-bed reactors.

Reactions occurring in FTS are essentially bond forming, and they release a large amount of heat. This requires an efficient heat removal system.

The FTS mechanism could be considered a simple polymerization reaction, the monomer being a C_1 species derived from carbon monoxide.[16] This polymerization follows an Anderson-Schulz-Flory distribution of molecular weights. This distribution gives a linear plot of the logarithm of yield of product (in moles) versus carbon number.[22] Under the assumptions of this model, the entire product distribution is determined by one parameter, α, the probability of the addition of a carbon atom to a chain (Figure 4-7).[16]

Much work has been undertaken to understand the steps and intermediates by which the reaction occurs on the heterogeneous catalyst surface. However, the exact mechanism is not fully established. One approach assumes a first-step adsorption of carbon monoxide on the catalyst surface followed by a transfer of an adsorbed hydrogen atom from an adjacent site to the metal carbonyl (M-CO):

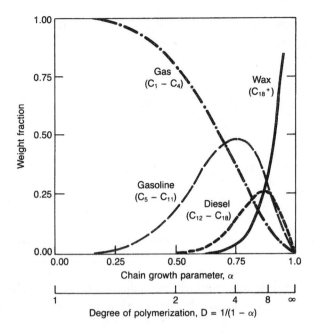

Figure 4-7. Yields of various products from FTS.[16]

$$M \ + \ CO \longrightarrow \begin{matrix} CO \\ | \\ M \end{matrix}$$

$$\begin{matrix} CO \\ | \\ M \end{matrix} + \begin{matrix} H \\ | \\ M \end{matrix} \longrightarrow \begin{matrix} H \quad CO \\ \diagdown \diagup \\ M \end{matrix} \ + \ M$$

Note: M represents a catalyst surface adsorption site.
Successive hydrogenation produces a metal-methyl species accompanied by the release of water:

$$\begin{matrix} H \qquad CO \ H \\ \diagdown \quad \diagup \quad | \\ M \ + \ 2M \end{matrix} \rightarrow \ \rightarrow \ M{-}CH_2OH \ + \ 2M$$

$$M{-}CH_2OH \ + \ \overset{\overset{\textstyle H}{|}}{2M} \ \rightarrow \ M{-}CH_3 \ + \ H_2O \ + \ 2M$$

In a subsequent step, the insertion of CO between the metal and the adsorbed methyl group occurs, followed by hydrogenation and elimination of water.

$$M-CH_3 + CO \longrightarrow M-COCH_3$$

$$M-COCH_3 + 2\overset{\overset{\displaystyle H}{|}}{M} \longrightarrow M-CHOHCH_3 + 2M$$

$$M-CHOHCH_3 + \overset{\overset{\displaystyle H}{|}}{M} \longrightarrow M-CH_2CH_3 + M + H_2O$$

The polymerization continues (as in the last three steps shown above) until termination occurs and the hydrocarbon is desorbed:

$$M-CH_2CH_2R + \overset{\overset{\displaystyle H}{|}}{M} \longrightarrow 2M + RCH_2CH_3$$
(paraffins)

$$M + RCH_2CH_3 \longrightarrow \overset{\overset{\displaystyle H}{|}}{M} + RCH{=}CH_2$$
(olefins)

$$M-CHOHR \longrightarrow \overset{\overset{\displaystyle H}{|}}{M} + RCHO$$
(aldehydes)

$$M-CHOHR + \overset{\overset{\displaystyle H}{|}}{M} \longrightarrow 2M + RCH_2OH$$
(alcohols)

The last two steps shown above explain the presence of oxygenates in FTS products.

Alternatively, an intermediate formation of an adsorbed methylene on the catalyst surface through the dissociative adsorption of carbon monoxide has been considered:

$$
M + CO \longrightarrow \overset{\displaystyle CO}{\underset{\displaystyle M}{|}}
$$

$$
M\text{—}CO + M \longrightarrow \overset{\displaystyle C}{\underset{\displaystyle M}{|}} + \overset{\displaystyle O}{\underset{\displaystyle M}{|}}
$$

The formed metal carbide (M-C) is then hydrogenated to a reactive methylene metal species.

$$
\overset{\displaystyle C}{\underset{\displaystyle M}{|}} + 2\overset{\displaystyle H}{\underset{\displaystyle M}{|}} \longrightarrow \overset{\displaystyle CH_2}{\underset{\displaystyle M}{||}} + 2M
$$

The methylene intermediate abstracts a hydrogen and is converted to an adsorbed methyl. Reaction of the methyl with the methylene produces an ethyl-metal species. Successive reactions of the methylene with the formed ethyl produces a long chain adsorbed alkyl.

$$
\overset{\displaystyle CH_2}{\underset{\displaystyle M}{||}} + \overset{\displaystyle H}{\underset{\displaystyle M}{|}} \rightarrow M\text{—}CH_3 + M
$$

$$
M{=}CH_2 + M\text{—}CH_3 \rightarrow M\text{—}CH_2\text{—}CH_3 + M
$$

$$
M\text{—}CH_2\text{—}CH_3 + nM{=}CH_2 \rightarrow M\text{—}(CH_2)_{\overline{n+1}}CH_3 + nM
$$

The adsorbed alkyl species can either terminate to a paraffin by a hydrogenation step or to an olefin by a dehydrogenation step:

$$
M\text{-}(CH_2)_{\overline{n+1}}CH_3
\begin{cases}
\xrightarrow{+M\text{-}H} CH_3(CH_2)_{\overline{n}}CH_3 + 2M \\
\xrightarrow{-H} CH_3(CH_2)_{\overline{n-1}}CH{=}CH_2 + M\text{—}H
\end{cases}
$$

The carbide mechanism, however, does not explain the formation of oxygenates in FTS products.[23]

NAPHTHENIC ACIDS

Naphthenic acids are a mixture of cyclo-paraffins with alkyl side chains ending with a carboxylic group. The low-molecular-weight naphthenic acids (8–12 carbons) are compounds having either a cyclopentane or a cyclohexane ring with a carboxyalkyl side chain. These compounds are normally found in middle distillates such as kerosine and gas oil. High boiling napthenic acids from the lube oils are monocarboxylic acids, (C_{14}-C_{19}) with an average of 2.6 rings.

Naphthenic acids constitute about 50 wt% of the total acidic compounds in crude oils. Naphthenic-based crudes contain a higher percentage of naphthenic acids. Consequently, it is more economical to isolate these acids from naphthenic-based crudes.[24]

The production of naphthenic acids from middle distillates occurs by extraction with 7–10% caustic solution.

The formed sodium salts, which are soluble in the lower aqueous layer, are separated from the hydrocarbon layer and treated with a mineral acid to spring out the acids. The free acids are then dried and distilled.

Using strong caustic solutions for the extraction may create separation problems because naphthenic acid salts are emulsifying agents. Properties of two naphthenic acid types are shown in Table 4-4.[25]

USES OF NAPHTHENIC ACIDS AND ITS SALTS

Free naphthenic acids are corrosive and are mainly used as their salts and esters. The sodium salts are emulsifying agents for preparing agricultural insecticides, additives for cutting oils, and emulsion breakers in the oil industry.

Other metal salts of naphthenic acids have many varied uses. For example, calcium naphthenate is a lubricating oil additive, and zinc naphthenate is an antioxidant. Lead, zinc, and barium naphthenates are wetting agents used as dispersion agents for paints. Some oil soluble metal naphthenates, such as those of zinc, cobalt, and lead, are used as

Table 4-4
Properties of two types of naphthenic acids[25]

Test	Type A*	Type B**
Density (d_4^{20})	0.972	0.987
Viscosity SU/210, °F	40.1	159.0
Pour point, °F	−30	40
Refractive index (d_4^{20})	1.476	1.503
Average molecular weight of deoiled acids	206	330
Unsaponifiable matter (wt%)	12.5	6.3
Acid number, mg KOH/g	235	

Used to produce driers
**Used to produce inhibitors and emulsifiers*

driers in oil-based paints. Among the diversified uses of naphthenates is the use of aluminum naphthenates as gelling agents for gasoline flame throwers (napalm). Manganese naphthenates are well-known oxidation catalysts.

CRESYLIC ACID

Cresylic acid is a commercial mixture of phenolic compounds including phenol, cresols, and xylenols. This mixture varies widely according to its source. Properties of phenol, cresols, and xylenols are shown in Table 4-5[26] Cresylic acid constitutes part of the oxygen compounds found in crudes that are concentrated in the naphtha fraction obtained principally from naphthenic and asphaltic-based crudes. Phenolic compounds, which are weak acids, are extracted with relatively strong aqueous caustic solutions.

Originally cresylic acid was obtained from caustic waste streams that resulted from treating light distillates with caustic solutions to reduce H_2S and mercaptans. Currently, most of these streams are hydrodesulfurized, and the product streams practically do not contain phenolic compounds.

However, cresylic acid is still obtained to a lesser extent from petroleum fractions, especially cracked gasolines, which contain higher percentages of phenols. It is also extracted from coal liquids.

Strong alkaline solutions are used to extract cresylic acid. The aqueous layer contains, in addition to sodium phenate and cresylate, a small amount of sodium naphthenates and sodium mercaptides. The reaction between cresols and sodium hydroxide gives sodium cresylate.

Table 4-5
Properties of Phenol, Cresols and Xylenols[26]

Name	Formula	MP(°C)	BP(°C)	20/4°C	pKa	Ka × 10⁻¹⁰
Phenol		42.5	182	1.0722	10.0	1.1
Cresols						
o-Cresol		31	191	1.02734	10.2	0.63
m-Cresol		11	202	1.0336	10.01	0.98
p-Cresol		35.5	202	1.0178	10.17	0.67
Xylenols						
2,4-Dimethylphenol		26	211	0.9650		
2,5-Dimethylphenol		75	212			
3,4-Dimethylphenol		62.5	225	0.9830		
3,5-Dimethylphenol		68	219.5	0.9680		

Mercaptans in the aqueous extract are oxidized to the disulfides, which are insoluble in water and can be separated from the cresylate solution by decantation:

$$2RSH + \tfrac{1}{2}O_2 \rightarrow R\text{-}S\text{-}S\text{-}R + H_2O$$

Free cresylic acid is obtained by treating the solution with a weak acid or dilute sulfuric acid. Refinery flue gases containing CO_2 are sometimes used to release cresylic acid. Aqueous streams with low cresylic acid concentrations are separated by adsorption by passing them through one or more beds containing a high adsorbent resin. The resin is regenerated with 1% sodium hydroxide solution.[27]

It should be noted that the extraction of cresylic acid does not create an isolation problem with naphthenic acids which are principally present in heavier fractions. Naphthenic acids, which are relatively stronger acids (lower pKa value), are extracted with less concentrated caustic solution.

Uses of Cresylic Acid

Cresylic acid is mainly used as degreasing agent and as a disinfectant of a stabilized emulsion in a soap solution. Cresols are used as flotation agents and as wire enamel solvents. Tricresyl phosphates are produced from a mixture of cresols and phosphorous oxychloride. The esters are plasticizers for vinyl chloride polymers. They are also gasoline additives for reducing carbon deposits in the combustion chamber.

REFERENCES

1. Ohta, T., *Solar Energy,* Pergamon Press, Oxford, England, 1979. p. 9.
2. "Gas Processing Handbook," *Hydrocarbon Processing,* Vol. 71, No. 4, 1992, p. 110.
3. Raman, V., *Oil and Gas Journal,* July 12, 1999, p. 5.
4. Chiu, C. H., "Advances in Gas Separation," *Hydrocarbon Processing,* Vol. 69, No. 1, 1990, pp. 69–72.
5. Shaver, K. G., Poffenbarger, G. L., and Grotewold, D. R., "Membranes Recover Hydrogen," *Hydrocarbon Processing,* Vol. 70, No. 6, 1991, pp. 77–79.
6. Chou, J. S. et al., "Mercaptans Affect Claus Units," *Hydrocarbon Processing,* Vol. 70, No. 4, 1991, pp. 39–42.
7. Yen, C., Chen, D. H., and Maddox, R. N., *Chemical Engineering Communications,* Vol. 52, 1987, p. 237.
8. *Chemical and Engineering News,* May 26, 1976, pp. 18–19.
9. Gas Processing Handbook," *Hydrocarbon Processing,* Vol. 69, No. 4, 1990, p. 97.
10. *Chemical and Engineering News,* April 10, 1995, p. 17.
11. Gotshall, W. W., Reprints Division of Petroleum Chemistry, ACS, Vol. 20, No. 2, 1975.
12. "Petrochemical Handbook," *Hydrocarbon Processing,* Vol. 58, No. 11, 1979, p. 162.
13. *Chemical Industries Newsletter,* Jan.–Mar. 1999, p. 5
14. Foo, K. W. and Shortland, I., *Hydrocarbon Processing,* Vol. 55, No. 5, 1976, pp. 149–152.
15. Bukur, D. B., Lang, X., Patel, S. A., Zimmerman, W. H., Rosynek, M. P., and Withers, H. P., Texas A & M Univ. (TAMU), Proc. 8th Indirect Liquefaction Contractors, Review Meeting, Pittsburgh, 1988.
16. Srivastava, R. D. et al., "Catalysts for Fischer Tropsch," *Hydrocarbon Processing,* Vol. 69, No. 2, 1990, pp. 59–68.

17. Rofer-Depoorter, C. K., "Water Gas Shift from Fischer Tropsch," in *Catalytic Conversions of Synthesis Gas and Alcohols to Chemicals,* edited by R. G. Herman, Plenum, New York, 1984.

18. Lingung Xu et al., "Don't Rule Out Iron Catalysts for Fischer-Tropsch Synthesis," *CHEMTECH,* Vol. 29, No. 1, 1998, pp. 47–53.

19. Dry, M. E., "The Fischer Tropsch Synthesis," in *Catalysis Science and Technology,* edited by J. R. Anderson and M. Boudart, Springer Verlag, 1981.

20. Deckwer, W. D., "FT Process Alternatives Hold Promise," *Oil and Gas Journal,* Vol. 78, 10 Nov., 1980, pp. 198–208.

21. Layman, P. L., *Chemical and Engineering News,* Aug. 8, 1994, pp. 12–24.

22. Anderson, R. B., *The Fischer Tropsch Synthesis,* Academic Press, Orlando, Fla., 1984.

23. Rober, M. "Fischer-Tropsch Synthesis" in *Catalysis in C_1 Chemistry,* edited by W. Keim, D. Reidel Publishing Company, Dordrecht, The Netherlands, 1983, pp. 41–87.

24. Lochte, H. L. and Littman, E. R., *Petroleum Acids and Bases,* Chemical Publishing Company, Inc., 1955, p. 124.

25. Matson, J. A., *Oil and Gas Journal,* March 24, 1980, pp. 93–94.

26. Hatch, L. F. and Matar, S., *From Hydrocarbons to Petrochemicals,* Gulf Publishing Co., 1981, p. 46.

27. Fox, C. R., *Hydrocarbon Processing,* Vol. 54, No. 7, 1975, pp. 109–111.

CHAPTER FIVE

Chemicals Based on Methane

INTRODUCTION

As mentioned in Chapter 2, methane is a one-carbon paraffinic hydrocarbon that is not very reactive under normal conditions. Only a few chemicals can be produced directly from methane under relatively severe conditions. Chlorination of methane is only possible by thermal or photochemical initiation. Methane can be partially oxidized with a limited amount of oxygen or in presence of steam to a synthesis gas mixture. Many chemicals can be produced from methane via the more reactive synthesis gas mixture. Synthesis gas is the precursor for two major chemicals, ammonia and methanol. Both compounds are the hosts for many important petrochemical products. Figure 5-1 shows the important chemicals based on methane, synthesis gas, methanol, and ammonia.[1]

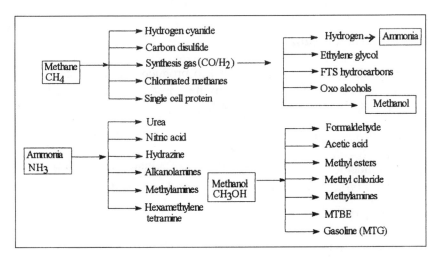

Figure 5-1. Important chemicals based on methane, synthesis gas, ammonia, and methanol.[1]

135

CHEMICALS BASED ON DIRECT REACTIONS OF METHANE

A few chemicals are based on the direct reaction of methane with other reagents. These are carbon disulfide, hydrogen cyanide chloromethanes, and synthesis gas mixture. Currently, a redox fuel cell based on methane is being developed.[2]

CARBON DISULFIDE (CS$_2$)

Methane reacts with sulfur (an active nonmetal element of group 6A) at high temperatures to produce carbon disulfide. The reaction is endothermic, and an activation energy of approximately 160 KJ is required.[3] Activated alumina or clay is used as the catalyst at approximately 675°C and 2 atmospheres. The process starts by vaporizing pure sulfur, mixing it with methane, and passing the mixture over the alumina catalyst. The reaction could be represented as:

$$CH_4(g) + 2S_2(g) \rightarrow CS_2(g) + 2H_2S(g) \qquad \Delta H^\circ_{298} = +150 \text{ KJ/mol}$$

Hydrogen sulfide, a coproduct, is used to recover sulfur by the Claus reaction. A CS$_2$ yield of 85–90% based on methane is anticipated. An alternative route for CS$_2$ is by the reaction of liquid sulfur with charcoal. However, this method is not used very much.

Uses of Carbon Disulfide

Carbon disulfide is primarily used to produce rayon and cellophane (regenerated cellulose). CS$_2$ is also used to produce carbon tetrachloride using iron powder as a catalyst at 30°C:

$$CS_2 + 3Cl_2 \rightarrow CCl_4 + S_2Cl_2$$

Sulfur monochloride is an intermediate that is then reacted with carbon disulfide to produce more carbon tetrachloride and sulfur:

$$2S_2Cl_2 + CS_2 \rightarrow CCl_4 + 6S$$

The net reaction is:

$$CS_2 + 2Cl_2 \rightarrow CCl_4 + 2S$$

Carbon disulfide is also used to produce xanthates ROC(S)SNa as an ore flotation agent and ammonium thiocyanate as a corrosion inhibitor in ammonia handling systems.

HYDROGEN CYANIDE (HCN)

Hydrogen cyanide (hydrocyanic acid) is a colorless liquid (b.p. 25.6°C) that is miscible with water, producing a weakly acidic solution. It is a highly toxic compound, but a very useful chemical intermediate with high reactivity. It is used in the synthesis of acrylonitrile and adiponitrile, which are important monomers for plastic and synthetic fiber production.

Hydrogen cyanide is produced via the Andrussaw process using ammonia and methane in presence of air. The reaction is exothermic, and the released heat is used to supplement the required catalyst-bed energy:

$$2CH_4 + 2NH_3 + 3O_2 \rightarrow 2HCN + 6H_2O$$

A platinum-rhodium alloy is used as a catalyst at 1100°C. Approximately equal amounts of ammonia and methane with 75 vol % air are introduced to the preheated reactor. The catalyst has several layers of wire gauze with a special mesh size (approximately 100 mesh).

The Degussa process, on the other hand, reacts ammonia with methane in absence of air using a platinum, aluminum-ruthenium alloy as a catalyst at approximately 1200°C. The reaction produces hydrogen cyanide and hydrogen, and the yield is over 90%. The reaction is endothermic and requires 251 KJ/mol.

$$CH_4 + NH_3 + 251\ KJ \rightarrow HCN + 3H_2$$

Hydrogen cyanide may also be produced by the reaction of ammonia and methanol in presence of oxygen:

$$NH_3 + CH_3OH + O_2 \rightarrow HCN + 3H_2O$$

Hydrogen cyanide is a reactant in the production of acrylonitrile, methyl methacrylates (from acetone), adiponitrile, and sodium cyanide. It is also used to make oxamide, a long-lived fertilizer that releases nitrogen steadily over the vegetation period. Oxamide is produced by the reaction of hydrogen cyanide with water and oxygen using a copper nitrate catalyst at about 70°C and atmospheric pressure:

$$4\,HCN + O_2 + 2\,H_2O \longrightarrow 2\,H_2N-\overset{\overset{\displaystyle O}{\|}}{C}-\overset{\overset{\displaystyle O}{\|}}{C}-NH_2$$

CHLOROMETHANES

The successive substitution of methane hydrogens with chlorine produces a mixture of four chloromethanes:

- Monochloromethane (methyl chloride, CH_3Cl)
- Dichloromethane (methylene chloride, CH_2Cl_2)
- Trichloromethane (chloroform, $CHCl_3$)
- Tetrachloromethane (carbon tetrachloride, CCl_4)

Each of these four compounds has many industrial applications that will be dealt with separately.

Production of Chloromethanes

Methane is the most difficult alkane to chlorinate. The reaction is initiated by chlorine free radicals obtained via the application of heat (thermal) or light (hv). Thermal chlorination (more widely used industrially) occurs at approximately 350–370°C and atmospheric pressure. A typical product distribution for a CH_4/Cl_2 feed ratio of 1.7 is: mono- (58.7%), di- (29.3%) tri- (9.7%) and tetra- (2.3%) chloromethanes.

The highly exothermic chlorination reaction produces approximately 95 KJ/mol of HCI. The first step is the breaking of the Cl–Cl bond (bond energy = + 584.2 KJ), which forms two chlorine free radicals (Cl atoms):

$$\overset{hv}{Cl_2 \rightarrow 2\dot{C}l}$$

The Cl atom attacks methane and forms a methyl free radical plus HCl. The methyl radical reacts in a subsequent step with a chlorine molecule, forming methyl chloride and a Cl atom:

$$\dot{C}l + CH_4 \rightarrow \dot{C}H_3 + HCl$$

$$\dot{C}H_3 + Cl_2 \rightarrow CH_3Cl + \dot{C}l$$

The new Cl atom either attacks another methane molecule and repeats the above reaction, or it reacts with a methyl chloride molecule to form a chloromethyl free radical CH_2Cl and HCl.

$$\dot{C}l + CH_3Cl \rightarrow \dot{C}H_2Cl + HCl$$

The chloromethyl free radical then attacks another chlorine molecule and produces dichloromethane along with a Cl atom:

$$\dot{C}H_2Cl + Cl_2 \rightarrow CH_2Cl_2 + \dot{C}l$$

This formation of Cl free radicals continues until all chlorine is consumed. Chloroform and carbon tetrachloride are formed in a similar way by reaction of $\dot{C}HCl_2$ and $\dot{C}Cl_3$ free radicals with chlorine.

Product distribution among the chloromethanes depends primarily on the mole ratio of the reactants. For example, the yield of monochloromethane could be increased to 80% by increasing the CH_4/Cl_2 mole ratio to 10:1 at 450°C. If dichloromethane is desired, the CH_4/Cl_2 ratio is lowered and the monochloromethane recycled. Decreasing the CH_4/Cl_2 ratio generally increases polysubstitution and the chloroform and carbon tetrachloride yield.

An alternative way to produce methyl chloride (monochloromethane) is the reaction of methanol with HCl (see later in this chapter, "Chemicals from Methanol"). Methyl chloride could be further chlorinated to give a mixture of chloromethanes (dichloromethane, chloroform, and carbon tetrachloride).

Uses of Chloromethanes

The major use of methyl chloride is to produce silicon polymers. Other uses include the synthesis of tetramethyl lead as a gasoline octane booster, a methylating agent in methyl cellulose production, a solvent, and a refrigerant.

Methylene chloride has a wide variety of markets. One major use is a paint remover. It is also used as a degreasing solvent, a blowing agent for polyurethane foams, and a solvent for cellulose acetate.

Chloroform is mainly used to produce chlorodifluoromethane (Fluorocarbon 22) by the reaction with hydrogen fluoride:

$$CHCl_3 + 2\,HF \rightarrow CHClF_2Cl + 2HCl$$

This compound is used as a refrigerant and as an aerosol propellant. It is also used to synthesize tetrafluoroethylene, which is polymerized to a heat resistant polymer (Teflon):

$$2CHClF_2 \rightarrow CF_2{=}CF_2 + 2HCl$$

Carbon tetrachloride is used to produce chlorofluorocarbons by the reaction with hydrogen fluoride using an antimony pentachloride ($SbCl_5$) catalyst:

$$CCl_4 + HF \rightarrow CCl_3F + HCl$$

$$CCl_4 + 2HF \rightarrow CCl_2F_2 + 2HCl$$

The formed mixture is composed of trichlorofluoromethane (Freon-11) and dichlorodifluoromethane (Freon-12). These compounds are used as aerosols and as refrigerants. Due to the depleting effect of chlorofluoro-carbons (CFCs) on the ozone layer, the production of these compounds may be reduced appreciably.

Much research is being conducted to find alternatives to CFCs with lit-tle or no effect on the ozone layer. Among these are HCFC-123 ($HCCl_2CF_3$) to replace Freon-11 and HCFC-22 ($CHClF_2$) to replace Freon-12 in such uses as air conditioning, refrigeration, aerosol, and foam. These compounds have a much lower ozone depletion value com-pared to Freon-11, which was assigned a value of 1. Ozone depletion values for HCFC-123 and HCFC-22 relative to Freon-11 equals 0.02 and 0.055, respectively.[4]

SYNTHESIS GAS (STEAM REFORMING OF NATURAL GAS)

As mentioned in Chapter 4, synthesis gas may be produced from a vari-ety of feedstocks. Natural gas is the preferred feedstock when it is avail-able from gas fields (nonassociated gas) or from oil wells (associated gas).

The first step in the production of synthesis gas is to treat natural gas to remove hydrogen sulfide. The purified gas is then mixed with steam and introduced to the first reactor (primary reformer). The reactor is con-structed from vertical stainless steel tubes lined in a refractory furnace. The steam to natural gas ratio varies from 4–5 depending on natural gas composition (natural gas may contain ethane and heavier hydrocarbons) and the pressure used.

A promoted nickel type catalyst contained in the reactor tubes is used at temperature and pressure ranges of 700–800°C and 30–50 atmos-pheres, respectively. The reforming reaction is equilibrium limited. It is favored at high temperatures, low pressures, and a high steam to carbon ratio. These conditions minimize methane slip at the reformer outlet and yield an equilibrium mixture that is rich in hydrogen.[5]

The product gas from the primary reformer is a mixture of H_2, CO, CO_2, unreacted CH_4, and steam.

The main steam reforming reactions are:

$$CH_4(g) + H_2O(g) \rightarrow CO\ (g) + 3H_2\ (g)$$ $\Delta H° = +206$ KJ

$$\Delta H°_{800°C} = +226 \text{ KJ}$$

$$CH_4(g) + 2H_2O(g) \rightarrow CO_2(g) + 4H_2(g)$$ $\Delta H° = +164.8$ KJ

For the production of methanol, this mixture could be used directly with no further treatment except adjusting the $H_2/(CO + CO_2)$ ratio to approximately 2:1.

For producing hydrogen for ammonia synthesis, however, further treatment steps are needed. First, the required amount of nitrogen for ammonia must be obtained from atmospheric air. This is done by partially oxidizing unreacted methane in the exit gas mixture from the first reactor in another reactor (secondary reforming).

The main reaction occurring in the secondary reformer is the partial oxidation of methane with a limited amount of air. The product is a mixture of hydrogen, carbon dioxide, carbon monoxide, plus nitrogen, which does not react under these conditions. The reaction is represented as follows:

$$CH_4 + {}^1/_2\ (O_2 + 3.76\ N_2) \rightarrow CO + 2H_2 + 1.88\ N_2 \qquad \Delta H° = -32.1 \text{ KJ}$$

The reactor temperature can reach over 900°C in the secondary reformer due to the exothermic reaction heat. Typical analysis of the exit gas from the primary and the secondary reformers is shown in Table 5-1.

The second step after secondary reforming is removing carbon monoxide, which poisons the catalyst used for ammonia synthesis. This is done in three further steps, shift conversion, carbon dioxide removal, and methanation of the remaining CO and CO_2.

Table 5-1
Typical analysis of effluent from primary and secondary reformers

Constituent	Primary reformer	Secondary reformer
H_2	47	39.0
CO	10.2	12.2
CO_2	6.3	4.2
CH_4	7.0	0.6
H_2O	29.4	27.0
N_2	0.02	17.0

Shift Conversion

The product gas mixture from the secondary reformer is cooled then subjected to shift conversion.

In the shift converter, carbon monoxide is reacted with steam to give carbon dioxide and hydrogen. The reaction is exothermic and independent of pressure:

$$CO(g) + H_2O (g) \rightarrow CO_2(g) + H_2(g) \qquad \Delta H° = -41 \text{ KJ}$$

The feed to the shift converter contains large amounts of carbon monoxide which should be oxidized. An iron catalyst promoted with chromium oxide is used at a temperature range of 425–500°C to enhance the oxidation.

Exit gases from the shift conversion are treated to remove carbon dioxide. This may be done by absorbing carbon dioxide in a physical or chemical absorption solvent or by adsorbing it using a special type of molecular sieves. Carbon dioxide, recovered from the treatment agent as a byproduct, is mainly used with ammonia to produce urea. The product is a pure hydrogen gas containing small amounts of carbon monoxide and carbon dioxide, which are further removed by methanation.

Methanation

Catalytic methanation is the reverse of the steam reforming reaction. Hydrogen reacts with carbon monoxide and carbon dioxide, converting them to methane. Methanation reactions are exothermic, and methane yield is favored at lower temperatures:

$$3H_2(g) + CO(g) \rightarrow CH_4(g) + H_2O(g) \qquad \Delta H° = -206 \text{ KJ}$$

$$4H_2(g) + CO_2 (g) \rightarrow CH_4(g) + 2H_2O(g) \qquad \Delta H° = -164.8 \text{ KJ}$$

The forward reactions are also favored at higher pressures. However, the space velocity becomes high with increased pressures, and contact time becomes shorter, decreasing the yield. The actual process conditions of pressure, temperature, and space velocity are practically a compromise of several factors. Rany nickel is the preferred catalyst. Typical methanation reactor operating conditions are 200–300°C and approximately 10 atmospheres. The product is a gas mixture of hydrogen and nitrogen having an approximate ratio of 3:1 for ammonia production. Figure 5-2

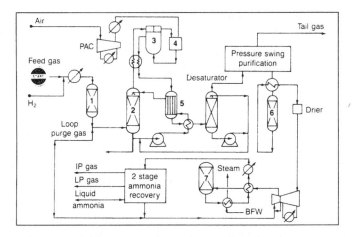

Figure 5-2. The ICI process for producing synthesis gas and ammonia:[6] (1) desul-furization, (2) feed gas saturator, (3) primary reformer, (4) secondary reformer, (5) shift converter, (6) methanator, (7) ammonia reactor.

shows the ICI process for the production of synthesis gas for the manu-facture of ammonia.[6]

CHEMICALS BASED ON SYNTHESIS GAS

Many chemicals are produced from synthesis gas. This is a conse-quence of the high reactivity associated with hydrogen and carbon monoxide gases, the two constituents of synthesis gas. The reactivity of this mixture was demonstrated during World War II, when it was used to produce alternative hydrocarbon fuels using Fischer Tropsch technology. The synthesis gas mixture was produced then by gasifying coal. Fischer Tropsch synthesis of hydrocarbons is discussed in Chapter 4.

Synthesis gas is also an important building block for aldehydes from olefins. The catalytic hydroformylation reaction (Oxo reaction) is used with many olefins to produce aldehydes and alcohols of commercial importance.

The two major chemicals based on synthesis gas are ammonia and methanol. Each compound is a precursor for many other chemicals. From ammonia, urea, nitric acid, hydrazine, acrylonitrile, methylamines and many other minor chemicals are produced (see Figure 5-1). Each of these chemicals is also a precursor of more chemicals.

Methanol, the second major product from synthesis gas, is a unique compound of high chemical reactivity as well as good fuel properties. It is a building block for many reactive compounds such as formaldehyde, acetic acid, and methylamine. It also offers an alternative way to produce hydrocarbons in the gasoline range (Mobil to gasoline MTG process). It may prove to be a competitive source for producing light olefins in the future.

AMMONIA (NH$_3$)

Ammonia is one of the most important inorganic chemicals, exceeded only by sulfuric acid and lime. This colorless gas has an irritating odor, and is very soluble in water, forming a weakly basic solution. Ammonia could be easily liquefied under pressure (liquid ammonia), and it is an important refrigerant. Anhydrous ammonia is a fertilizer by direct application to the soil. Ammonia is obtained by the reaction of hydrogen and atmospheric nitrogen, the synthesis gas for ammonia. The 1994 U.S. ammonia production was approximately 40 billion pounds (sixth highest volume chemical).

Ammonia Production (Haber Process)

The production of ammonia is of historical interest because it represents the first important application of thermodynamics to an industrial process. Considering the synthesis reaction of ammonia from its elements, the calculated reaction heat (ΔH) and free energy change (ΔG) at room temperature are approximately –46 and –16.5 KJ/mol, respectively. Although the calculated equilibrium constant $K_c = 3.6 \times 108$ at room temperature is substantially high, no reaction occurs under these conditions, and the rate is practically zero. The ammonia synthesis reaction could be represented as follows:

$$N_2 \text{ (g)} + 3H_2 \text{ (g)} \rightarrow 2NH_3 \text{ (g)} \qquad \Delta \overset{\circ}{H} = -46.1 \text{ KJ/mol}$$

Increasing the temperature increases the reaction rate, but decreases the equilibrium (K_c @ 500°C = 0.08). According to LeChatlier's principle, the equilibrium is favored at high pressures and at lower temperatures. Much of Haber's research was to find a catalyst that favored the formation of ammonia at a reasonable rate at lower temperatures. Iron oxide promoted with other oxides such as potassium and aluminum oxides is currently used to produce ammonia in good yield at relatively low temperatures.

In a commercial process, a mixture of hydrogen and nitrogen (exit gas from the methanator) in a ratio of 3:1 is compressed to the desired pressure (150–1,000 atmospheres). The compressed mixture is then preheated by heat exchange with the product stream before entering the ammonia reactor. The reaction occurs over the catalyst bed at about 450°C. The exit gas containing ammonia is passed through a cooling chamber where ammonia is condensed to a liquid, while unreacted hydrogen and nitrogen are recycled (see Figure 5-2). Usually, a conversion of approximately 15% per pass is obtained under these conditions.

Uses of Ammonia

The major end use of ammonia is the fertilizer field for the production of urea, ammonium nitrate and ammonium phosphate, and sulfate. Anhydrous ammonia could be directly applied to the soil as a fertilizer. Urea is gaining wide acceptance as a slow-acting fertilizer.

Ammonia is the precursor for many other chemicals such as nitric acid, hydrazine, acrylonitrile, and hexamethylenediamine. Ammonia, having three hydrogen atoms per molecule, may be viewed as an energy source. It has been proposed that anhydrous liquid ammonia may be used as a clean fuel for the automotive industry. Compared with hydrogen, anhydrous ammonia is more manageable. It is stored in iron or steel containers and could be transported commercially via pipeline, railroad tanker cars, and highway tanker trucks.[7] The oxidation reaction could be represented as:

$$4NH_3 + 3O_2 \rightarrow 2N_2 + 6H_2O \qquad \Delta H = -316.9 \text{ KJ/mol}$$

Only nitrogen and water are produced. However, many factors must be considered such as the coproduction of nitrogen oxides, the economics related to retrofitting of auto engines, etc. The following describes the important chemicals based on ammonia.

$$
\overset{\textstyle O}{\overset{\textstyle \|}{}}
$$

Urea (H2N–C–NH2)

The highest fixed nitrogen-containing fertilizer 46.7 wt %, urea is a white solid that is soluble in water and alcohol. It is usually sold in the form of crystals, prills, flakes, or granules. Urea is an active compound that reacts with many reagents. It forms adducts and clathrates with many substances such as phenol and salicylic acid. By reacting with formalde-hyde, it produces an important commercial polymer (urea formaldehyde resins) that is used as glue for particle board and plywood.

Production. The technical production of urea is based on the reaction of ammonia with carbon dioxide:

$$2NH_3 \text{ (g)} + CO_2 \text{ (g)} \longrightarrow H_2N\text{-}COONH_4 \text{ (s)} \qquad \Delta H° = -126 \text{ KJ/mol}$$

$$H_2N\text{—}COONH_{4(s)} \longrightarrow H_2N\text{—}\overset{\overset{\displaystyle O}{\|}}{C}\text{—}NH_{2(aq)} + H_2O_{(1)}$$
$$\Delta H° = +29 \text{ KJ/mol}$$

The reaction occurs in two steps: ammonium carbamate is formed first, followed by a decomposition step of the carbamate to urea and water. The first reaction is exothermic, and the equilibrium is favored at lower temperatures and higher pressures. Higher operating pressures are also desirable for the separation absorption step that results in a higher carbamate solution concentration. A higher ammonia ratio than stoichio-metric is used to compensate for the ammonia that dissolves in the melt. The reactor temperature ranges between 170–220°C at a pressure of about 200 atmospheres.

The second reaction represents the decomposition of the carbamate. The reaction conditions are 200°C and 30 atmospheres. Decomposition in presence of excess ammonia limits corrosion problems and inhibits the decomposition of the carbamate to ammonia and carbon dioxide. The urea solution leaving the carbamate decomposer is expanded by heating at low pressures and ammonia recycled. The resultant solution is further concentrated to a melt, which is then prilled by passing it through special sprays in an air stream. Figure 5-3 shows the Snamprogetti process for urea production.[8]

Uses of Urea. The major use of urea is the fertilizer field, which accounts for approximately 80% of its production (about 16.2 billion pounds were produced during 1994 in U.S.). About 10% of urea is used for the production of adhesives and plastics (urea formaldehyde and melamine formaldehyde resins). Animal feed accounts for about 5% of the urea produced.

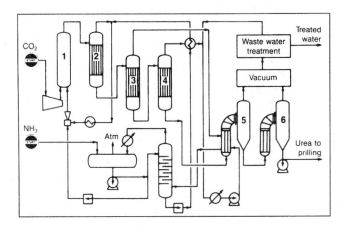

Figure 5-3. The Snamprogetti process for producing urea[8]: (1) reactor, (2,3,4) carbonate decomposers, (5,6) crystallizing and prilling.

Urea possesses a unique property of forming adducts with n-paraffins. This is used in separating C_{12}-C_{14} n-paraffins from kerosines for detergent production (Chapter 6).

Nitric Acid (HNO₃)

Nitric acid is one of the most used chemicals. The 1994 U.S. production was approximately 17.65 billion pounds. It is a colorless to a yellow liquid, which is very corrosive. It is a strong oxidizing acid that can attack almost any metal. The most important use of nitric acid is to produce ammonium nitrate fertilizer.

Nitric acid is commercially produced by oxidizing ammonia with air over a platinum-rhodium wire gauze. The following sequence represents the reactions occurring over the heterogeneous catalyst:

$$4NH_3(g) + 5O_2(g) \rightarrow 4NO(g) + 6H_2O(g) \qquad \Delta H° = -226.4 \text{ KJ/mol}$$

$$2NO(g) + O_2(g) \rightarrow 2NO_2(g) \qquad \Delta H° = -56.5 \text{ KJ/mol}$$

$$3NO_2(g) + H_2O(1) \rightarrow 2HNO_3(aq) + NO(g) \qquad \Delta H° = -33.4 \text{ KJ/mol}$$

The three reactions are exothermic, and the equilibrium constants for the first two reactions fall rapidly with increase of temperature. Increasing pressure favors the second reaction but adversely affects the first reaction.

For this reason, operation around atmospheric pressures is typical. Space velocity should be high to avoid the reaction of ammonia with oxygen on the reactor walls, which produces nitrogen and water, and results in lower conversions. The concentration of ammonia must be kept below the inflammability limit of the feed gas mixture to avoid explosion. Optimum nitric acid production was found to be obtained at approximately 900°C and atmospheric pressure.

Uses of Nitric Acid. The primary use of nitric acid is for the production of ammonium nitrate for fertilizers. A second major use of nitric acid is in the field of explosives. It is also a nitrating agent for aromatic and paraffinic compounds, which are useful intermediates in the dye and explosive industries. It is also used in steel refining and in uranium extraction.

Hydrazine (H_2N-NH_2).

A colorless, fuming liquid miscible with water, hydrazine (diazine) is a weak base but a strong reducing agent. Hydrazine is used as a rocket fuel because its combustion is highly exothermic and produces 620 KJ/mol:

$$H_2N\text{-}NH_2 + O_2 \rightarrow N_2 + 2H_2O + 620 \text{ KJ}$$

Hydrazine is produced by the oxidation of ammonia using the Rashig process. Sodium hypochlorite is the oxidizing agent and yields chloramine NH_2Cl as an intermediate. Chloramine further reacts with ammonia producing hydrazine:

$$2NH_3 + NaOCl \rightarrow H_2N\text{-}NH_2 + NaCl + H_2O$$

Hydrazine is then evaporated from the sodium chloride solution.

Hydrazine can also be produced by the Puck process. The oxidizing agent is hydrogen peroxide:

$$2NH_3 + H_2O_2 \rightarrow H_2N\text{-}NH_2 + 2H_2O$$

Uses of Hydrazine. In addition to rocket fuel, hydrazine is used as a blowing agent and in the pharmaceutical and fertilizer industries. Due to the weak N-N bond, it is used as a polymerization initiator. As a reducing agent, hydrazine is used as an oxygen scavenger for steam boilers. It is also a selective reducing agent for nitro compounds. Hydrazine is a good building block for many chemicals, especially agricultural products, which dominates its use.

METHYL ALCOHOL (CH₃OH)

Methyl alcohol (methanol) is the first member of the aliphatic alcohol family. It ranks among the top twenty organic chemicals consumed in the U.S. The current world demand for methanol is approximately 25.5 million tons/year (1998) and is expected to reach 30 million tons by the year 2002.[9] The 1994 U.S. production was 10.8 billion pounds.

Methanol was originally produced by the destructive distillation of wood (wood alcohol) for charcoal production. Currently, it is mainly produced from synthesis gas.

As a chemical compound, methanol is highly polar, and hydrogen bonding is evidenced by its relatively high boiling temperature (65°C), its high heat of vaporization, and its low volatility. Due to the high oxygen content of methanol (50 wt%), it is being considered as a gasoline blending compound to reduce carbon monoxide and hydrocarbon emissions in automobile exhaust gases. It was also tested for blending with gasolines due to its high octane (RON = 112). During the late seventies and early eighties, many experiments tested the possible use of pure (straight) methanol as an alternative fuel for gasoline cars. Several problems were encountered, however, in its use as a fuel, such as the cold engine startability due to its high vaporization heat (heat of vaporization is 3.7 times that for gasoline), its lower heating value, which is approximately half that of gasoline, and its corrosive properties. The subject has been reviewed by Keller.[10] However, methanol is a potential fuel for gas turbines because it burns smoothly and has exceptionally low nitrogen oxide emission levels.

Due to the high reactivity of methanol, many chemicals could be derived from it. For example, it could be oxidized to formaldehyde, an important chemical building block, carbonylated to acetic acid, and dehydrated and polymerized to hydrocarbons in the gasoline range (MTG process). Methanol reacts almost quantitatively with isobutene and isoamylenes, producing methyl t-butylether (MTBE) and tertiary amyl methyl ether (TAME), respectively. Both are important gasoline additives for raising the octane number and reducing carbon monoxide and hydrocarbon exhaust emissions. Additionally, much of the current work is centered on the use of shape-selective catalysts to convert methanol to light olefins as a possible future source of ethylene and propylene. The subject has been reviewed by Chang.[11]

Production of Methanol

Methanol is produced by the catalytic reaction of carbon monoxide and hydrogen (synthesis gas). Because the ratio of $CO:H_2$ in synthesis gas from natural gas is approximately 1:3, and the stoichiometric ratio required for methanol synthesis is 1:2, carbon dioxide is added to reduce the surplus hydrogen. An energy-efficient alternative to adjusting the $CO:H_2$ ratio is to combine the steam reforming process with autothermal reforming (combined reforming) so that the amount of natural gas fed is that required to produce a synthesis gas with a stoichiometric ratio of approximately 1:2.05. Figure 5-4 is a combined reforming diagram.[12] If an autothermal reforming step is added, pure oxygen should be used. (This is a major difference between secondary reforming in case of ammonia production, where air is used to supply the needed nitrogen).

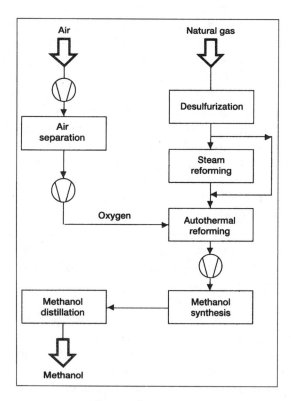

Figure 5-4. A block flow diagram showing the combined reforming for methanol synthesis.[12]

An added advantage of combined reforming is the decrease in NO_x emission. However, a capital cost increase (for air separation unit) of roughly 15% is anticipated when using combined reforming in comparison to plants using a single train steam reformer. The process scheme is viable and is commercially proven.[13] The following reactions are representative for methanol synthesis.

$$CO_{(g)} + 2H_{2(g)} \rightarrow CH_3OH_{(1)} \qquad \Delta H° = -128 \text{ KJ/mol}$$

$$CO_2 + 3H_2 \rightarrow CH_3OH_{(\ell)} + H_2O$$

Old processes use a zinc-chromium oxide catalyst at a high-pressure range of approximately 270–420 atmospheres for methanol production.

A low-pressure process has been developed by ICI operating at about 50 atm (700 psi) using a new active copper-based catalyst at 240°C. The synthesis reaction occurs over a bed of heterogeneous catalyst arranged in either sequential adiabatic beds or placed within heat transfer tubes. The reaction is limited by equilibrium, and methanol concentration at the converter's exit rarely exceeds 7%. The converter effluent is cooled to 40°C to condense product methanol, and the unreacted gases are recycled. Crude methanol from the separator contains water and low levels of by-products, which are removed using a two-column distillation system. Figure 5-5 shows the ICI methanol synthesis process.[14]

Methanol synthesis over the heterogeneous catalyst is thought to occur by a successive hydrogenation of chemisorbed carbon monoxide.

$$\equiv\!\!-M + CO \longrightarrow \equiv\!\!-M\text{-}CO \longrightarrow \equiv\!\!-M = C{=}O$$

$$\equiv\!\!-M{=}C{=}O + H_2 \longrightarrow \equiv\!\!-M{=}C\!\!\begin{smallmatrix} H \\ \\ OH \end{smallmatrix}$$

$$\equiv\!\!-M{=}C\!\!\begin{smallmatrix} H \\ \\ OH \end{smallmatrix} + H_2 \longrightarrow \equiv\!\!-M \ C(OH)H_2 + H$$

$$\equiv\!\!-M\text{-}C(OH)H_2 + H \longrightarrow CH_3OH + \equiv\!\!-M$$

Other mechanisms have been also proposed.[15]

Uses of Methanol

Methanol has many important uses as a chemical, a fuel, and a building block. Approximately 50% of methanol production is oxidized to

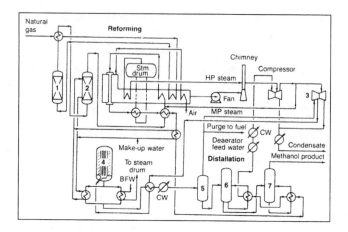

Figure 5-5. The ICI low-pressure process for producing methanol:[14] (1) desulfurization, (2) saturator (for producing process steam), (3) synthesis loop circulator, (4) reactor, (5) heat exchanger and separator, (6) column for light ends recovery, (7) column for water removal.

formaldehyde. As a methylating agent, it is used with many organic acids to produce the methyl esters such as methyl acrylate, methylmethacrylate, methyl acetate, and methyl terephthalate. Methanol is also used to produce dimethyl carbonate and methyl-t-butyl ether, an important gasoline additive. It is also used to produce synthetic gasoline using a shape selective catalyst (MTG process). Olefins from methanol may be a future route for ethylene and propylene in competition with steam cracking of hydrocarbons. The use of methanol in fuel cells is being investigated. Fuel cells are theoretically capable of converting the free energy of oxidation of a fuel into electrical work. In one type of fuel cells, the cathode is made of vanadium which catalyzes the reduction of oxygen, while the anode is iron (III) which oxidizes methane to CO_2 and iron (II) is formed in aqueous H_2SO_4.[16] The benefits of low emission may be offest by the high cost. The following describes the major chemicals based on methanol.

$$O$$
$$\|$$
Formaldehyde (H-C-H)

The main industrial route for producing formaldehyde is the catalyzed air oxidation of methanol.

$$CH_3OH_{(g)} + \tfrac{1}{2} O_{2(g)} \quad \longrightarrow \quad H-\overset{\overset{\displaystyle O}{\|}}{C}-H_{(g)} + H_2O_{(g)} \;\; \Delta H^\circ = -148.9 \text{ KJ/mol}$$

A silver-gauze catalyst is still used in some older processes that operate at a relatively higher temperature (about 500°C). New processes use an iron-molybdenum oxide catalyst. Chromium or cobalt oxides are sometimes used to dope the catalyst. The oxidation reaction is exothermic and occurs at approximately 400–425°C and atmospheric pressure. Excess air is used to keep the methanol air ratio below the explosion limits. Figure 5-6 shows the Haldor Topsoe iron-molybdenum oxide catalyzed process.[17]

Uses of Formaldehyde. Formaldehyde is the simplest and most reactive aldehyde. Condensation polymerization of formaldehyde with phenol, urea, or melamine produces phenol-formaldehyde, urea formaldehyde, and melamine formaldehyde resins, respectively. These are important glues used in producing particle board and plywood.

Condensation of formaldehyde with acetaldehyde in presence of a strong alkali produces pentaerythritol, a polyhydric alcohol for alkyd resin production:

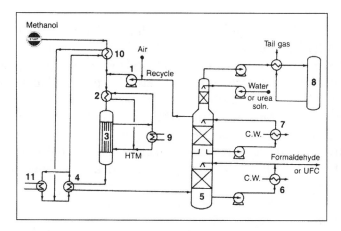

Figure 5-6. The Haldor Topsoe and Nippon Kasei process for producing formaldehyde:[17] (1) blower, (2) heat exchanger, (3) reactor, (4) steam boiler, (5) absorber, (6,7) coolers, (8) incinerator, (9) heat recovery, (10) methanol evaporator, (11) boiler feed water.

$$4 \; H-\overset{O}{\overset{\|}{C}}-H + CH_3\overset{O}{\overset{\|}{C}}-H + NaOH \longrightarrow C(CH_2OH)_4 + HCOONa$$

$$Pentaerythritol$$

Formaldehyde reacts with ammonia and produces hexamethylenete-tramine (hexamine):

$$6 \; H-\overset{O}{\overset{\|}{C}}-H + 4\,NH_3 \longrightarrow (CH_2)_6N_4 + 6\,H_2O$$

Hexamine is a cross-linking agent for phenolic resins.

Methyl Chloride (CH₃CI)

Methyl chloride is produced by the vapor phase reaction of methanol and hydrogen chloride:

$$CH_3OH + HCI \rightarrow CH_3CI + H_2O$$

Many catalysts are used to effect the reaction, such as zinc chloride on pumice, cuprous chloride, and ignited alumina gel. The reaction conditions are 350°C at nearly atmospheric pressure. The yield is approximately 95%.

Zinc chloride is also a catalyst for a liquid-phase process using concentrated hydrochloric acid at 100–150°C. Hydrochloric acid may be generated in situ by reacting sodium chloride with sulfuric acid. As mentioned earlier, methyl chloride may also be produced directly from methane with other chloromethanes. However, methyl chloride from methanol may be further chlorinated to produce dichloromethane, chloroform, and carbon tetrachloride.

Methyl chloride is primarily an intermediate for the production of other chemicals. Other uses of methyl chloride have been mentioned with chloromethanes.

Acetic Acid (CH₃COOH)

The carbonylation of methanol is currently one of the major routes for acetic acid production. The basic liquid-phase process developed by BASF uses a cobalt catalyst at 250°C and a high pressure of about 70 atmospheres. The newer process uses a rhodium complex catalyst in presence of CH_3I, which acts as a promoter. The reaction occurs at 150°C and atmospheric pressure. A 99% selectivity is claimed with this catalyst:

$$CH_3OH + CO \rightarrow CH_3COOH$$

The mechanism of the carbonylation reaction is thought to involve a first-step oxidative addition of the methyl iodide promotor to the Rh(I) complex, followed by a carbonyl cis insersion step:

$$[Rh(CO)_2I_2]^{1-} + CH_3I \longrightarrow [CH_3Rh(CO)_2I_3]^{1-}$$

$$[CH_3Rh(CO)_2I_3]^{1-} \longrightarrow [CH_3\overset{\overset{O}{\|}}{C}Rh(CO)I_3]^{1-}$$

Carbonylation followed by reductive elimination produces back the Rh(I) catalyst:

$$[CH_3\overset{\overset{O}{\|}}{C}Rh(CO)I_3]^{1-} + CO \longrightarrow [Rh(CO)_2I_2]^{1-} + CH_3\overset{\overset{O}{\|}}{C}-I$$

The final step is the reaction between acetyl iodide and methyl alcohol, yielding acetic acid and the promotor:

$$CH_3\overset{\overset{O}{\|}}{C}-I + CH_3OH \longrightarrow CH_3COOH + CH_3I$$

Figure 5-7 is a flow diagram showing the Monsanto carbonylation process.[18]

Acetic acid is also produced by the oxidation of acetaldehyde and the oxidation of n-butane. However, acetic acid from the carbonylation route has an advantage over the other commercial processes because both methanol and carbon monoxide come from synthesis gas, and the process conditions are quite mild.

Uses of Acetic Acid. The main use of acetic acid is to produce vinyl acetate (44%), followed by acetic acid esters (13%) and acetic anhydride (12%). Vinyl acetate is used for the production of adhesives, film, paper

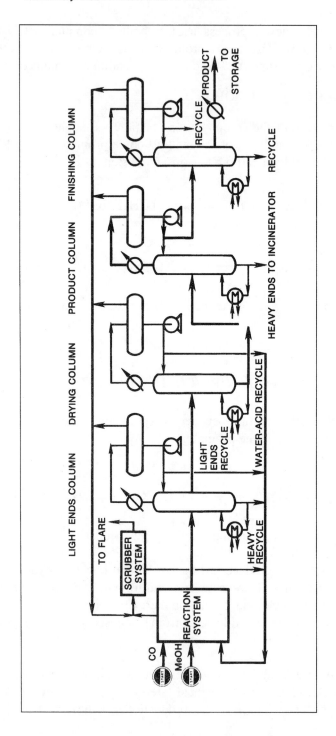

Figure 5-7. The Monsanto methanol carbonylation process for producing acetic acid.[18]

and textiles. Terephthalic acid consumes 12% of acetic acid demand.[19] Acetic acid is also used to produce pharmaceuticals, dyes, and insecticides. Chloroacetic acid (from acetic acid) is a reactive intermediate used to manufacture many chemicals such as glycine and carboxymethyl cellulose.

Methyl Tertiary Butyl Ether ((CH₃)₃C-O-CH₃)

MTBE is produced by the reaction of methanol and isobutene:

$$CH_3OH + CH_3-\underset{\underset{CH_3}{|}}{C}=CH_2 \longrightarrow CH_3-O-\underset{\underset{CH_3}{|}}{\overset{\overset{CH_3}{|}}{C}}-CH_3$$

The reaction occurs in the liquid phase at relatively low temperatures (about 50°C) in the presence of a solid acid catalyst. Few side reactions occur such as the hydration of isobutene to tertiary butyl alcohol, and methanol dehydration and formation of dimethyl ether and water. However, only small amounts of these compounds are produced. Figure 5-8 is a simplified flow diagram of the BP Etherol process.[20]

The MTBE reaction is equilibrium limited. Higher temperatures increase the reaction rate, but the conversion level is lower. Lower temperatures shift the equilibrium toward ether production, but more catalyst

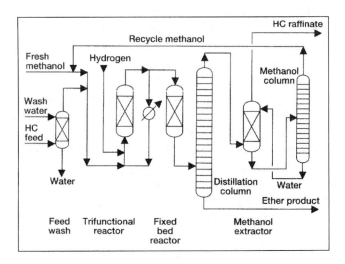

Figure 5-8. Simplified flow diagram of the British Petroleum Etherol process.[20]

inventory is required. Therefore, conventional MTBE units are designed with two reactors in series. Most of the etherification reaction is achieved at an elevated temperature in the first reactor and then finished at a thermodynamically favorable lower temperature in the second reactor.[21]

An alternative way for the production of MTBE is by using isobutane, propene, and methanol. This process coproduces propylene oxide. In this process, isobutane reacts with oxygen giving t-butyl hydroperoxide. The epoxide reacts with propene yielding t-butyl alcohol and propylene oxide. t-Butyl alcohol loses water giving isobutene which reacts with methanol yielding MTBE.[22] The following shows the sequence of the reactions:

$$
\begin{array}{ccc}
& CH_3 & & CH_3 \\
& | & & | \\
CH_3\!-\!C\!-\!H + {}^1\!/_2O_2 & \rightarrow & CH_3\!-\!C\!-\!OOH \\
& | & & | \\
& CH_3 & & CH_3
\end{array}
$$

$$
(CH_3)_3COOH + CH_2\!=\!CH\!-\!CH_3 \rightarrow CH_2\!-\!CH\!-\!CH_3 + (CH_3)_3COH
$$
$$
\diagdown\!\!O\!\!\diagup
$$

$$
(CH_3)_3COH + CH_3OH \rightarrow (CH_3)_3COCH_3 + H_2O
$$

MTBE is an important gasoline additive because of its high octane rating. Currently, it is gaining more importance for the production of lead-free gasolines. It reduces carbon monoxide and hydrocarbon exhaust emissions probably (the exact means is not known) by reducing the aromatics in gasolines. In the past few years, many arguments existed regarding the use of MTBE as a gasoline additive. It was found that leakage from old gasoline storage tanks pollutes underground water. Compared to other constituents of gasoline, MTBE is up to 10 times more soluble in water. It also has little affinity for soil, and unlike other gasoline components, it passes through the soil and is carried by the water.[23]

Many recommendations are being thought to reduce pollution effects of MTBE. One way is to use alternative oxygenates which are not as soluble in water as MTBE. Another way is by phasing out the 2% oxygen by weight required in reformulated gasoline. These changes will affect the future demand for MTBE. Currently, the worldwide consumption of MTBE reached 6.6 billion gallons of which 65% is consumed in the U.S.[23]

Tertiary Amyl Methyl Ether (CH₃CH₂C(CH₃)₂-O-CH₃)

TAME can also be produced by the reaction of methanol with iso-amylenes. The reaction conditions are similar to those used with MTBE, except the temperature is a little higher:

$$CH_3OH + CH_3CH=\overset{\overset{\displaystyle CH_3}{|}}{C}-CH_3 \longrightarrow CH_3CH_2-\overset{\overset{\displaystyle CH_3}{|}}{\underset{\underset{\displaystyle CH_3}{|}}{C}}-OCH_3$$

Similar to MTBE, TAME is used as gasoline additive for its high octane rating and its ability to reduce carbon monoxide and hydrocarbon exhaust emissions. Properties of oxygenates used as gasoline additives are shown in Table 5-2.[20]

Dimethyl Carbonate (CO(OCH₃)₂)

Dimethyl carbonate (DMC) is a colorless liquid with a pleasant odor. It is soluble in most organic solvents but insoluble in water. The classical synthesis of DMC is the reaction of methanol with phosgene. Because phosgene is toxic, a non-phosgene-route may be preferred. The new route reacts methanol with urea over a tin catalyst. However, the yield is low. Using electron donor solvents such as trimethylene glycol dimethyl ether and continually distilling off the product increases the yield.[24]

Dimethyl carbonate is used as a specialty solvent. It could be used as an oxygenate to replace MTBE. It has almost three times the oxygen content as MTBE. It has also a high octane rating. However, it must be evaluated in regard to economics and toxicity.

$$H_2N-\overset{\overset{\displaystyle O}{\|}}{C}-NH_2 + 2CH_3OH \rightarrow CH_3O-\overset{\overset{\displaystyle O}{\|}}{C}-OCH_3 + 2NH_3$$

Methylamines

Methylamines can be synthesized by alkylating ammonia with methyl halides or with methyl alcohol. The reaction with methanol usually occurs at approximately 500°C and 20 atmospheres in the presence of an

Table 5-2
Properties of oxygenates (MTBE, TAME, and ETBE)[20]

Property	MTBE	ETBE	TAME
Blending octane (R + M/2)	110	111	105
Blending octane (RON)	112–130	120	105–115
Blending octane (MON)	97–115	102	95–105
Reid vapor pressure (psi)	7.8	4.0	2.5
Boiling point			
(°C)	55	72	88
(°F)	131	161	187
Density			
(kg/l)	.742	.743	.788
(lb/gal)	6.19	6.20	8.41
Energy density			
(kcal/l)	89.3	92.5	98.0
(kBtu/gal)	93.5	96.9	100.8
Heat of vaporization			
(kcal/l)	0.82	0.79	0.86
(kBtu/gal) @ nbp	0.86	0.83	0.90
Oxygenate requirement (vol% @ 2.7 wt% ox.)	15.0	17.2	16.7
Solubility in water (wt%)	4.3	1.2	1.2
Water pickup (wt%)	1.4	0.5	0.6
Heat of reaction			
(kcal/mol)	9.4	6.6	11
(kBtu/lb mol)	17	12	20

aluminum silicate or phosphate catalyst. The alkylation does not stop at the monomethylamine stage, because the produced amine is a better nucleophile than ammonia. The product distribution at equilibrium is: monomethylamine MMA (43%), dimethylamine DMA (24%), and trimethylamine TMA (33%):

$$CH_3OH + NH_3 \rightarrow CH_3NH_2 + H_2O$$

$$CH_3OH + CH_3NH_2 \rightarrow (CH_3)_2NH + H_2O$$

$$CH_3OH + (CH_3)_2NH \rightarrow (CH_3)_3N + H_2O$$

To improve the yield of mono- and dimethylamines, a shape selective catalyst has been tried. Carbogenic sieves are microporous materials (similar to zeolites), which have catalytic as well as shape selective properties. Combining the amorphous aluminum silicate catalyst (used for producing the amines) with carbogenic sieves gave higher yeilds of the more valuable MMA and DMA.[25]

Uses of Methylamines. Dimethylamine is the most widely used of the three amines. Excess methanol and recycling monomethylamine increases the yield of dimethylamine. The main use of dimethylamine is the synthesis of dimethylformamide and dimethylacetamide, which are solvents for acrylic and polyurethane fibers.

Monoethylamine is used in the synthesis of Sevin, an important insecticide. Trimethylamine has only one major use, the synthesis of choline, a high-energy additive for poultry feed.

Hydrocarbons from Methanol (Methanol to Gasoline MTG Process)

Methanol may have a more important role as a basic building block in the future because of the multisources of synthesis gas. When oil and gas are depleted, coal and other fossil energy sources could be converted to synthesis gas, then to methanol, from which hydrocarbon fuels and chemicals could be obtained. During the early seventies, oil prices escalated (as a result of 1973 Arab-Israeli War), and much research was directed toward alternative energy sources. In 1975, a Mobil research group discovered that methanol could be converted to hydrocarbons in the gasoline range with a special type of zeolite (ZSM-5) catalyst.[26]

The reaction of methanol over a ZSM-5 catalyst could be considered a dehydration, oligomerization reaction. It may be simply represented as:

$$nCH_3OH \rightarrow (CH_2)_n + nH_2O$$

where $(CH_2)_n$ represents the hydrocarbons (paraffins + olefins + aromatics). The hydrocarbons obtained are in the gasoline range. Table 5-3 shows the analysis of hydrocarbons obtained from the conversion of methanol to gasoline (MTG Process).[27] The MTG process has been operating in New Zealand since 1985. The story of the discovery of the MTG process has been reviewed by Meisel.[28]

Converting methanol to hydrocarbons is not as simple as it looks from the previous equation. Many reaction mechanisms have been proposed,

Table 5-3
Analysis of gasoline from MTG process[27]

Components, wt%		
Butanes	3.2	
Alkylates	28.6	
C_5 gasoline	68.2	
	100.0	
Components, wt%		
Paraffins	56	
Olefins	7	
Naphthenes	4	
Aromatics	33	
	100	
Octane	Research	Motor
Clear	96.8	87.4
Leaded (3 cc TEL/U.S. gal)	102.6	95.8
Reid vapor pressure		
psi	9	
kPa	62	
Specific gravity	0.730	
Sulfur, wt%	Nil	
Nitrogen, wt%	Nil	
Durene, wt%	3.8	
Corrosion, copper strip	1A	
ASTM distillation, °C.		
10%	47	
30%	70	
50%	103	
90%	169	

and most of them are centered around the intermediate formation of dimethyl ether followed by olefin formation. Olefins are thought to be the precursors for paraffins and aromatics:

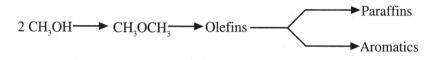

$$2 \; CH_3OH \longrightarrow CH_3OCH_3 \longrightarrow Olefins$$
Paraffins
Aromatics

The product distribution is influenced by the catalyst properties as well as the various reaction parameters. The catalyst activity and selectivity are functions of acidity, crystalline size, silica/alumina ratio, and even the synthetic procedure. Since the discovery of the MTG process, much work has been done on other catalyst types to maximize light olefins production.

The important property of ZSM-5 and similar zeolites is the intercrystalline catalyst sites, which allow one type of reactant molecule to diffuse, while denying diffusion to others. This property, which is based on the shape and size of the reactant molecules as well as the pore sizes of the catalyst, is called shape selectivity. Chen and Garwood document investigations regarding the various aspects of ZSM-5 shape selectivity in relation to its intercrystalline and pore structure.[29]

In general, two approaches have been found that enhance selectivity toward light olefin formation. One approach is to use catalysts with smaller pore sizes such as crionite, chabazite, and zeolite T. The other approach is to modify ZSM-5 and similar catalysts by reducing the pore size of the catalyst through incorporation of various substances in the zeolite channels and/or by lowering its acidity by decreasing the Al_2O/SiO_3 ratio. This latter approach is used to stop the reaction at the olefin stage, thus limiting the steps up to the formation of olefins and suppressing the formation of higher hydrocarbons. Methanol conversion to light olefins has been reviewed by Chang.[30]

Table 5-4 shows the product distribution, when methanol was reacted over different catalysts for maximizing olefin yield.[11]

Table 5-4
Methanol conversion to hydrocarbons over various zeolites[11]
(370°C, 1 atm, 1 LHSV)

| | Hydrocarbon distribution (wt%) in | | | | |
	Erionite	ZSM-5	ZSM-11	ZSM-4	Mordenite
C_1	5.5	1.0	0.1	8.5	4.5
C_2	0.4	0.6	0.1	1.8	0.3
C_2^-	36.3	0.5	0.4	11.2	11.0
C_3	1.8	16.2	6.0	19.1	5.9
C_3^-	39.1	1.0	2.4	8.7	15.7
C_4	5.7	24.2	25.0	8.8	13.8
C_4^-	9.0	1.3	5.0	3.2	9.8
C_5^+ aliphatic	2.2	14.0	32.7	4.8	18.6
A_6	–	1.7	0.8	0.1	0.4
A_7	–	10.5	5.3	0.5	0.9
A_8	–	18.0	12.4	1.3	1.0
A_9	–	7.5	8.4	2.2	1.0
A_{10}	–	3.3	1.5	3.2	2.0
A_{11}^+	–	0.2	–	26.6	15.1

OXO ALDEHYDES AND ALCOHOLS
(Hydroformylation Reaction)

Hydroformylation of olefins (Oxo reaction) produces aldehydes with one more carbon than the reacting olefin. For example, when ethylene is used, propionaldehyde is produced. This reaction is especially important for the production of higher aldehydes that are further hydrogenated to the corresponding alcohols. The reaction is catalyzed with cobalt or rhodium complexes. Olefins with terminal double bonds are more reactive and produce aldehydes which are hydrogenated to the corresponding primary alcohols. With olefins other than ethylene, the hydroformylation reaction mainly produces a straight chain aldehyde with variable amounts of branched chain aldehydes. The reaction could be generally represented as:

$$2RCH=CH_2 + 2H_2 + 2CO \longrightarrow RCH_2CH_2CHO + \overset{\overset{\textstyle CH_3}{\textstyle |}}{R}CHCHO$$

The largest commercial process is the hydroformylation of propene, which yields n-butyraldehyde and isobutyraldehyde. n-Butyraldehyde (n-butanal) is either hydrogenated to n-butanol or transformed to 2-ethyl-hexanol via aldol condensation and subsequent hydrogenation. 2-Ethylhexanol is an important plasticizer for polyvinyl chloride. This reaction is noted in Chapter 8.

Other olefins applied in the hydroformylation process with subsequent hydrogenation are propylene trimer and tetramer for the production of decyl and tridecyl alcohols, respectively, and C_7 olefins (from copolymers of C_3 and C_4 olefins) for isodecyl alcohol production.

Several commercial processes are currently operative. Some use a rhodium catalyst complex incorporating phosphine ligands $HRhCO(PPh_3)_2$ at relatively lower temperatures and pressures and produce less branched aldehydes. Older processes use a cobalt carbonyl complex $HCo(CO)_4$ at higher pressures and temperatures and produce a higher ratio of the branched aldehydes. The hydroformylation reaction using phosphine ligands occurs in an aqueous medium. A higher catalyst activity is anticipated in aqueous media than in hydrocarbons. Selectivity is also higher. Having more than one phase allows for complete separation of the catalyst and the products.

In order to make the catalysts soluble in water, ionic ligands are attached to the catalyst. The Rhurchemie/Rhone-Poulenc process for the production of butyraldehyde from propylene is based on this technology.[31] Hydroformylation of higher olefins using ionic phosphine catalysts that are solubilized in both reactants and products was investigated by Union Carbide researchers. This yields a one-phase homogeneous system. The catalyst is recovered outside the reaction zone. Although this is a single-phase system, these catalysts could be induced to separate into a nonpolar product and polar catalyst phases. This technology provides an effective means of catalyst recovery.[32] Cobalt catalysts have also been investigated. Hoechest researchers have developed a water soluble cobalt cluster compound that can hydroformylate olefins in a two-phase system. Hydroformylation of higher olefins is possible when polyethylene glycol is used as a solvent. Higher olefins have greater affinity for ethylene glycol than for water, therefore allowing greater reaction rates. To facilitate the separation of the products, pentane is added to the system. The reaction takes place at 120°C and 70 KPa. When 1-hexene is used, the ratio of n-heptanal to the iso- was 0.73–3.75.[33] Table 5-5 shows the hydroformylation conditions of some commercial processes.

A simplified mechanism for the hydroformylation reaction using the rhodium complex starts by the addition of the olefin to the catalyst (A) to form complex (B). The latter rearranges, probably through a four-centered intermediate, to the alkyl complex (C). A carbon monoxide insertion gives the square-planar complex (D). Successive H_2 and CO addition produces the original catalyst and the product:[34]

Table 5-5
Catalysts used in some commerical oxo processes and approximate conditions for propylene hydroformylation

Process	Catalyst	Conditions	% Normal
Ruhrchemie	Co^{2+}, Co^{0}	150°C, 300 atm.	70
BASF	$HCO(CO)_4$	150°C, 30 MPa	70
ICI	Co^{2+}	high pressure	70
Shell	CO/PR_3	180, 50 atm	88
UCC	$HRh(CO)(PPh_3)_3$	100, 30 atm	94

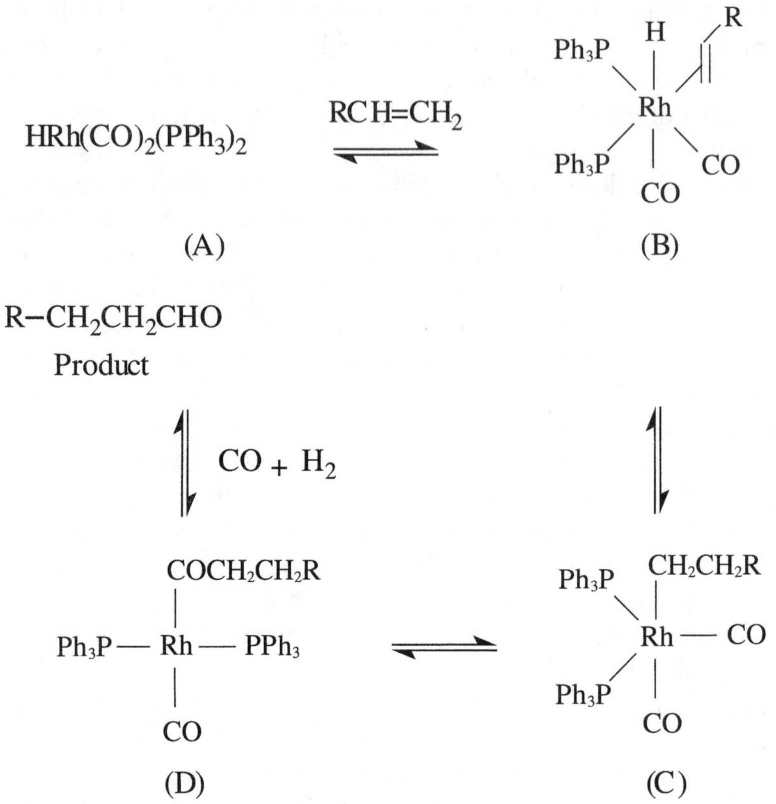

$$HRh(CO)_2(PPh_3)_2 \xrightarrow{\ RCH=CH_2\ } \rightleftharpoons$$

(A)

(B)

R—CH$_2$CH$_2$CHO
Product

$$\big\updownarrow\ CO + H_2$$

COCH$_2$CH$_2$R
|
Ph$_3$P — Rh — PPh$_3$ $\rightleftharpoons$
|
CO

(D)

Ph$_3$P   CH$_2$CH$_2$R
  \ |
   Rh — CO
  / |
Ph$_3$P   CO

(C)

PPh$_3$ is triphenyl phosphine.

ETHYLENE GLYCOL

Ethylene glycol could be produced directly from synthesis gas using an Rh catalyst at 230°C at very high pressure (3,400 atm). In theory, five moles synthesis gas mixture are needed to produce one mole ethylene glycol:[35]

$$3H_2 + 2CO \rightarrow HOCH_2—CH_2OH$$

Other routes have been tried starting from formaldehyde or paraformaldehyde. One process reacts formaldehyde with carbon monoxide and H$_2$ (hydroformylation) at approximately 4,000 psi and 110°C using a rhodium triphenyl phosphine catalyst with the intermediate formation of glycolaldehyde. Glycolaldehyde is then reduced to ethylene glycol:

$$\underset{\text{H}}{\overset{\text{O}}{\underset{\|}{\text{C}}}}\text{—H} + \text{CO} + \text{H}_2 \ \longrightarrow \ \underset{\text{H}_2\text{C}}{\overset{\text{OH}}{\underset{|}{\text{C}}}}\overset{\text{O}}{\underset{\|}{\text{C}}}\text{—H} \ \overset{\text{H}_2}{\longrightarrow} \ \text{HOCH}_2\text{CH}_2\text{OH}$$

The DuPont process (the oldest syngas process to produce ethylene glycol) reacts formaldehyde with CO in the presence of a strong mineral acid. The intermediate is glycolic acid, which is esterified with methanol. The ester is then hydrogenated to ethylene glycol and methanol, which is recovered. The net reaction from either process could be represented as:

$$\text{H}\overset{\text{O}}{\underset{\|}{\text{—C}}}\text{—H} + \text{CO} + 2\text{H}_2 \ \longrightarrow \ \text{HOCH}_2\text{CH}_2\text{OH}$$

REFERENCES

1. Hatch, L. F. and Matar S., "Petrochemicals from Methane" *From Hydrocarbons to Petrochemicals,* Gulf Publishing Co., Houston, 1981, p. 49.
2. *Chemical and Engineering News,* Aug. 16, 1999, p. 7.
3. Stevenson, R. M., *Introduction to the Chemical Process Industries,* Reinhold Publishing Corporation, 1966, p. 293.
4. Al-Najjar, I. M., CFC's Symposium: Phase out Chlorofluorocarbons Chamber of Commerce and Industry, Dammam, Saudi Arabia. No. 24, 1992, pp. 398–441.
5. Shahani, G. H., et al., "Hydrogen and Utility Optimization," *Hydrocarbon Processing,* Vol. 77, No. 9, 1998, pp. 143–150.
6. "Petrochemicals Handbook," *Hydrocarbon Processing,* Vol. 70, No. 3, 1991, p. 134.
7. Steele, R. B., "A Proposal for an Ammonia Economy," *CHEMTECH,* Vol. 29, No. 8, 1999, p. 28.
8. "Petrochemicals Handbook," *Hydrocarbon Processing,* Vol. 70, No. 3, 1991, p. 191.
9. *Hydrocarbon Processing,* Vol. 78, No. 1, 1999, p. 29.
10. Keller, J. L. "Alcohols as Motor Fuel," *Hydrocarbon Processing,* Vol. 58, No. 5, 1979, pp. 127–137.
11. Chang, C. D., "Hydrocarbons from Methanol," *Catal. Rev. Sci. Eng.* Vol. 25, No. 1, 1983 pp. 1–118, and Chang, C. D., Lang, W. H. and Bell, K., *Catalysis of Organic Reactions,* Dekker, New York, 1981.
12. Farina, G. L. and Supp, E., "Produce Syngas from Methanol" *Hydrocarbon Processing,* Vol. 71, No. 3, 1992, pp. 77–79.

13. Schneider, R. V. and LeBlanc, J. R., Jr., "Choose Optional Syngas Route," *Hydrocarbon Processing,* Vol. 71, No. 3, 1992, pp. 51–57.
14. "Petrochemical Handbook," *Hydrocarbon Processing,* Vol. 70, No. 3, 1991, p. 164.
15. Matar, S., Mirbach, M. and Tayim, H., *Catalysis in Petrochemical Processes,* Kluwer Publishing Company, 1989, p. 158.
16. *Chemical and Engineering News,* September 5, 1994, p. 21.
17. "Petrochemical Handbook," *Hydrocarbon Processing,* Vol. 69, No. 3, 1991, p. 158.
18. Grove, H. D., *Hydrocarbon Processing,* Vol. 51, No. 11, 1972, pp. 76–78.
19. *Hydrocarbon Processing,* Vol. 76, No. 2, 1997, p. 29.
20. Rock, K., "TAME: Technology Merits," *Hydrocarbon Processing,* Vol. 71, No. 5, 1992, p. 87.
21. Chang, E. J. and Leiby, S. M., "Ethers Help Gasoline Quality," *Hydrocarbon Processing,* Vol. 71, No. 2, 1992, pp. 41–44.
22. Morse, P. M. "Producers brace for MTBE Phaseout," *Chemical and Engineering News,* April 12, 1999, p. 26.
23. Nakamura, D. N., "HP in Processing," *Hydrocarbon Processing,* Vol. 77, No. 1, 1998, p. 15.
24. *CHEMTECH,* Vol. 29, No. 8, 1999, p. 26, US patent 5902894, 11 May, 1999.
25. Haggin, J., "Carbogenic Sieves", *Chemical and Engineering News,* Dec. 19, 1994, pp. 36–37.
26. Chang, C. D. and Silverstri, A. J., "MTG: Origin, Evolution, Operation," *CHEMTECH,* Oct. 1987, pp. 624–631.
27. *Oil and Gas Journal,* "New Zealand Methanol to Gasoline," Jan. 14, 1980, pp. 95–96.
28. Meisel, S. L., "Catalysis Research Bears Fruit," *CHEMTECH,* Vol. 18, No. 1, 1988, pp. 32–37.
29. Chen, N. Y., and Garwood, W. E., "Some Catalytic Properties of ZSM-5, a New Shape-Selective Zeolite," *J. Cat.,* Vol. 52, 1978, pp. 453–458.
30. Chang, C. D., "Methanol Conversion To Light Olefins," *Catal. Rev. Sci. Eng.,* 26, No. 344, 1984, pp. 323–345.
31. *Chemical and Engineering News,* October 10, 1994, p. 28.
32. *Chemical and Engineering News,* April 17, 1995, pp. 25–26.
33. *CHEMTECH,* Vol. 29, No. 3, 1999, p. 32
34. Gates, B., Katzer, J. and Schuit, G. C., "Chemistry of Catalytic Processes," McGraw-Hill Book Company, 1979, p. 144.
35. Kollar, J., "Ethylene Glycol From Syngas," *CHEMTECH,* August 1984, pp. 504–510.

CHAPTER SIX

Ethane and Higher Paraffins-Based Chemicals

INTRODUCTION

As discussed in Chapter 2, paraffinic hydrocarbons are less reactive than olefins; only a few chemicals are directly based on them. Nevertheless, paraffinic hydrocarbons are the starting materials for the production of olefins. Methane's relation with petrochemicals is primarily through synthesis gas (Chapter 5). Ethane, on the other hand, is a major feedstock for steam crackers for the production of ethylene. Few chemicals could be obtained from the direct reaction of ethane with other reagents. The higher paraffins—propane, butanes, pentanes, and heavier—also have limited direct use in the chemical industry except for the production of light olefins through steam cracking. This chapter reviews the petrochemicals directly produced from ethane and higher paraffins.

ETHANE CHEMICALS

The main source for ethane is natural gas liquids. Approximately 40% of the available ethane is recovered for chemical use. The only large consumer of ethane is the steam cracking process for ethylene production.

A minor use of ethane is its chlorination to ethyl chloride:

$$CH_3CH_3 + Cl_2 \rightarrow CH_3CH_2Cl + HCl$$

By-product HCl may be used for the hydrochlorination of ethylene to produce more ethyl chloride. Hydrochlorination of ethylene, however, is the main route for the production of ethyl chloride:

$$CH_2 = CH_2 + HCl \rightarrow CH_3CH_2Cl$$

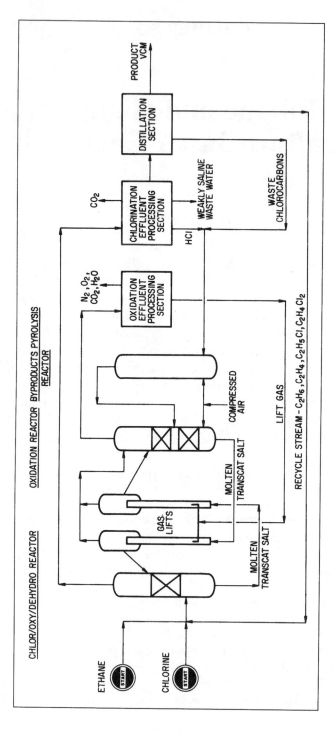

Figure 6-1. The Transcat process for producing vinyl chloride from ethane.[1]

Major uses of ethyl chloride are the manufacture of tetraethyl lead and the synthesis of insecticides. It is also used as an alkylating agent and as a solvent for fats and wax.

A small portion of vinyl chloride is produced from ethane via the Transcat process. In this process a combination of chlorination, oxychlorination, and dehydrochlorination reactions occur in a molten salt reactor. The reaction occurs over a copper oxychloride catalyst at a wide temperature range of 310–640°C. During the reaction, the copper oxychloride is converted to copper(I) and copper(II) chlorides, which are air oxidized to regenerate the catalyst. Figure 6-1 is a flow diagram of the Transcat process for producing vinyl chloride from ethane.[1]

Vinyl chloride is an important monomer for polyvinyl chloride (PVC). The main route for obtaining this monomer, however, is via ethylene (Chapter 7). A new approach to utilize ethane as an inexpensive chemical intermediate is to ammoxidize it to acetonitrile. The reaction takes place in presence of a cobalt-B-zeolite.

$$CH_3\text{–}CH_3 + NH_3 + {}^3/_2O_2 \rightarrow CH_3CN + 3H_2O$$

However, the process is not yet commercial.[2]

PROPANE CHEMICALS

A major use of propane recovered from natural gas is the production of light olefins by steam cracking processes. However, more chemicals can be obtained directly from propane by reaction with other reagents than from ethane. This may be attributed to the relatively higher reactivity of propane than ethane due to presence of two secondary hydrogens, which are easily substituted.

The following reviews some of the important reactions and chemicals based on propane.

OXIDATION OF PROPANE

The noncatalytic oxidation of propane in the vapor phase is nonselective and produces a mixture of oxygenated products. Oxidation at temperatures below 400°C produces a mixture of aldehydes (acetaldehyde and formaldehyde) and alcohols (methyl and ethyl alcohols). At higher temperatures, propylene and ethylene are obtained in addition to hydrogen peroxide. Due to the nonselectivity of this reaction, separation of the products is complex, and the process is not industrially attractive.

CHLORINATION OF PROPANE
(Production of Perchloroethylene)

Chlorination of propane with chlorine at 480–640°C yields a mixture of perchloroethylene (Perchlor) and carbon tetrachloride:

$$CH_3CH_2CH_3 + 8Cl_2 \rightarrow CCl_2=CCl_2 + CCl_4 + 8HCl$$
$$\text{Perchlor}$$

Carbon tetrachloride is usually recycled to produce more perchloroethylene:

$$2CCl_4 \rightarrow CCl_2=CCl_2 + 2Cl_2$$

Perchlor may also be produced from ethylene dichloride (1,2-dichloroethane) through an oxychlorination-oxyhydrochlorination process. Trichloroethylene (trichlor) is coproduced (Chapter 7).

Perchlor and trichlor are used as metal degreasing agents and as solvents in dry cleaning. Perchlor is also used as a cleaning and drying agent for electronic equipment and as a fumigant.

DEHYDROGENATION OF PROPANE (Propene Production)

The catalytic dehydrogenation of propane is a selective reaction that produces mainly propene:

$$CH_3CH_2CH_3 \rightarrow CH_2=CH-CH_3 + H_2 \qquad \Delta H = + ve$$

The process could also be used to dehydrogenate butane, isobutane, or mixed LPG feeds. It is a single-stage system operating at a temperature range of 540–680°C and 5–20 absolute pressures. Conversions in the range of 55–65% are attainable, and selectivities may reach up to 95%. Figure 6-2 shows the Lummus-Crest Catofin dehydrogenation process.[3]

For a given dehydrogenation system, i.e., operating temperature and pressure, thermodynamic theory provides a limit to the per pass conversion that can be achieved.[4] A general formula is

$$Kp = X^2P/ (I-X^2)$$
Kp = equilibrium constant at a given temperature
 X = fraction paraffin converted to mono-olefins
 P = reaction pressure in atmospheres

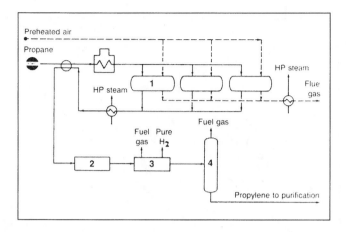

Figure 6-2. The Lummus Crest Catofin dehydrogenation process:[3] (1) reactor, (2) compressor, (3) liquid product recovery, (4) product purification.

According to Le Chatelier's principle, conversion is increased by increasing the temperature and decreasing the pressure. Figure 6-3 shows the effect of temperature on the dehydrogenation of different light paraffins.[4]

NITRATION OF PROPANE (Production of Nitroparaffins)

Nitrating propane produces a complex mixture of nitro compounds ranging from nitromethane to nitropropanes. The presence of lower nitroparaffins is attributed to carbon-carbon bond fission occurring at the temperature used. Temperatures and pressures are in the range of 390°–440°C and 100–125 psig, respectively. Increasing the mole ratio of propane to nitric acid increases the yield of nitropropanes. Typical product composition for 25:1 propane/acid ratio is:[5]

$$CH_3CH_2CH_3 + HNO_3 \longrightarrow \begin{array}{ll} CH_3CHNO_2CH_3 \\ CH_3CH_2CH_2NO_2 \end{array} \Bigg\} \; 55\text{–}65 \; wt\% \\ \qquad\qquad\qquad\qquad CH_3CH_2NO_2 \qquad 20\text{–}25 \; wt\% \\ \qquad\qquad\qquad\qquad CH_3NO_2 \qquad\quad 10\text{–}30 \; wt\%$$

Nitropropanes are good solvents for vinyl and epoxy resins. They are also used to manufacture rocket propellants. Nitromethane is a fuel additive for racing cars.

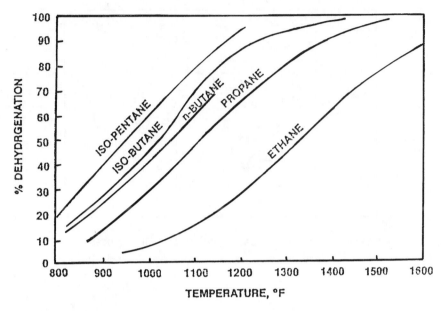

Figure 6-3. Effect of temperature on the dehydrogenation of light paraffins at one atmosphere.[4]

Nitropropane reacts with formaldehyde producing nitroalcohols:

$$CH_3CH_2CH_2NO_2 + HCHO \rightarrow CH_3CH_2CH(NO_2)CH_2OH$$

These difunctional compounds are versatile solvents, but they are expensive.

n-BUTANE CHEMICALS

Like propane, n-butane is mainly obtained from natural gas liquids. It is also a by-product from different refinery operations. Currently, the major use of n-butane is to control the vapor pressure of product gasoline. Due to new regulations restricting the vapor pressure of gasolines, this use is expected to be substantially reduced. Surplus n-butane could be isomerized to isobutane, which is currently in high demand for producing isobutene. Isobutene is a precursor for methyl and ethyl tertiary butyl ethers, which are important octane number boosters.[6] Another alternative outlet for surplus n-butane is its oxidation to maleic anhydride. Almost all new maleic anhydride processes are based on butane oxidation.

n-Butane has been the main feedstock for the production of butadiene. However, this process has been replaced by steam cracking hydrocarbons, which produce considerable amounts of by-product butadiene.

The chemistry of n-butane is more varied than that of propane, partly because n-butane has four secondary hydrogen atoms available for substitution and three carbon-carbon bonds that can be cracked at high temperatures:

$$CH_3 - \overset{\overset{\displaystyle H}{|}}{\underset{\underset{\displaystyle H}{|}}{C}} - \overset{\overset{\displaystyle H}{|}}{\underset{\underset{\displaystyle H}{|}}{C}} - CH_3$$

Like propane, the noncatalytic oxidation of butane yields a variety of products including organic acids, alcohols, aldehydes, ketones, and olefins. Although the noncatalytic oxidation of butane produces mainly aldehydes and alcohols, the catalyzed oxidation yields predominantly acids.

OXIDATION OF n-BUTANE (Acetic Acid and Acetaldehyde)

The oxidation of n-butane represents a good example illustrating the effect of a catalyst on the selectivity for a certain product. The noncatalytic oxidation of n-butane is nonselective and produces a mixture of oxygenated compounds including formaldehyde, acetic acid, acetone, and alcohols. Typical weight % yields when n-butane is oxidized in the vapor phase at a temperature range of 360–450°C and approximately 7 atmospheres are: formaldehyde 33%, acetaldehyde 31%, methanol 20%, acetone 4%, and mixed solvents 12%.

On the other hand, the catalytic oxidation of a n-butane, using either cobalt or manganese acetate, produces acetic acid at 75–80% yield. Byproducts of commercial value are obtained in variable amounts. In the Celanese process, the oxidation reaction is performed at a temperature range of 150–225°C and a pressure of approximately 55 atmospheres.[7]

$$CH_3CH_2CH_2CH_3 + O_2 \rightarrow CH_3COOH + \text{by-products} + H_2O$$

The main by-products are formic acid, ethanol, methanol, acetaldehyde, acetone, and methylethyl ketone (MEK). When manganese acetate is used as a catalyst, more formic acid ($\approx 25\%$) is obtained at the expense of acetic acid.

Maleic Anhydride:

Catalytic oxidation of n-butane at 490° over a cerium chloride, Co-Mo oxide catalyst produces maleic anyhydride:

$$2 \ CH_3CH_2CH_2CH_3 + 7 \ O_2 \ \rightarrow \ 2 \quad \text{[structure]} \quad + 8H_2O$$

Other catalyst systems such as iron V_2O_5-P_2O_5 over silica alumina are used for the oxidation. In the Monsanto process (Figure 6-4), n-butane and air are fed to a multitube fixed-bed reactor, which is cooled with molten salt. The catalyst used is a proprietary modified vanadium oxide. The exit gas stream is cooled, and crude maleic anhydride is absorbed then recovered from the solvent in the stripper. Maleic anhydride is further purified using a proprietary solvent purification system.[8]

A new process for the partial oxidation of n-butane to maleic anhydride was developed by DuPont. The important feature of this process is the use of a circulating fluidized bed-reactor. Solids flux in the rizer-reactor is high and the superficial gas velocities are also high, which encounters short residence times usually in seconds. The developed catalyst for this process is based on vanadium phosphorous oxides

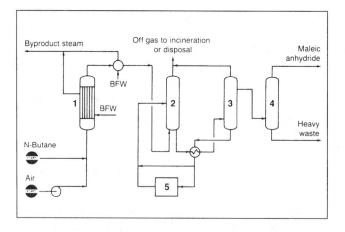

Figure 6-4. The Monsanto process for producing maleic anhydride from butane:[8] (1) reactor, (2) absorber (3) stripper, (4) fractionator, (5) solvent purification.

$(VO)_2P_2O_7$ type, which provides the oxygen needed for oxidation. The selective oxidation of n-butane to maleic anhydride involves a redox mechanism where the removal of eight hydrogen atoms as water and the insertion of three oxygen atoms into the butane molecule occurs. The reaction temperature is approximately 500°C. Subsequent hydrogenation of maleic anhydride produces tetrahydrofuran.[9] Figure 6-5 shows the DuPont butane to maleic anhydride process.

Oxidation of n-butane to maleic anhydride is becoming a major source for this important chemical. Maleic anhydride could also be produced by the catalytic oxidation of n-butenes (Chapter 9) and benzene (Chapter 10). The principal use of maleic anhydride is in the synthesis of unsaturated polyester resins. These resins are used to fabricate glass-fiber reinforced materials. Other uses include fumaric acid, alkyd resins, and pesticides. Maleic acid esters are important plasticizers and lubricants. Maleic anhydride could also be a precursor for 1,4-butanediol (Chapter 9).

Aromatics Production

Liquefied petroleum gas (LPG), a mixture of propane and butanes, is catalytically reacted to produce an aromatic-rich product. The first step is

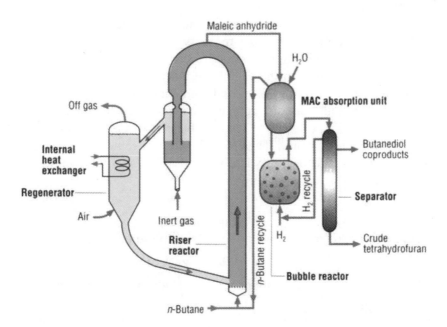

Figure 6-5. The DuPont butane to maleic anhydride process.[9]

assumed to be the dehydrogenation of propane and butane to the corresponding olefins followed by oligomerization to C_6, C_7, and C_8 olefins. These compounds then dehydrocyclize to BTX aromatics. The following reaction sequence illustrates the formation of benzene from 2 propane molecules:

$$2CH_3CH_2CH_3 \rightarrow CH_3CH_2CH_2CH_2CH=CH_2 + 2H_2$$
1-Hexene

Although olefins are intermediates in this reaction, the final product contains a very low olefin concentration. The overall reaction is endothermic due to the predominance of dehydrogenation and cracking. Methane and ethane are by-products from the cracking reaction. Table 6-1 shows the product yields obtained from the Cyclar process developed jointly by British Petroleum and UOP.[10] A simplified flow scheme for the Cyclar process is shown in Figure 6-6.

The process consists of a reactor section, continuous catalyst regeneration unit (CCR), and product recovery section. Stacked radial-flow reactors are used to minimize pressure drop and to facilitate catalyst recirculation to and from the CCR. The reactor feed consists solely of LPG plus the recycle of unconverted feed components; no hydrogen is recycled. The liquid product contains about 92 wt% benzene, toluene, and xylenes (BTX) (Figure 6-7), with a balance of C_9^+ aromatics and a low nonaromatic content.[10] Therefore, the product could be used directly for the recovery of benzene by fractional distillation (without the extraction step needed in catalytic reforming).

Table 6-1
Product yield from saturated LPG feed to the cyclar process[10]

	Yields, wt% of fresh feed		
Feedstock	Aromatics	Hydrogen	Fuel gas
Propane (100%)	63.1	5.9	31.0
Butanes (100%)	65.9	5.2	28.9

Basis: High-yield mode. Lower cost Cyclar units can be designed, but for lower overall yields.

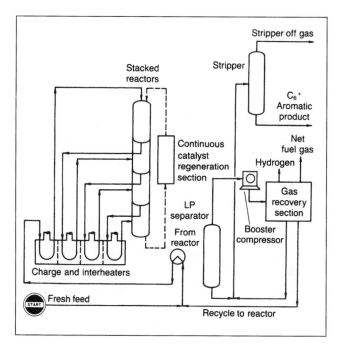

Figure 6-6. A flow diagram showing the Cyclar process for aromatization of LPG.[10]

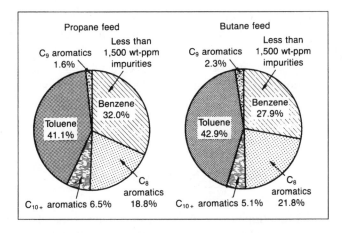

Figure 6-7. The liquid (C_6^+) product breakdown in weight units obtained from the Cyclar process.[10]

Interest in the use of lower-value light paraffins for the production of aromatics led to the introduction of two new processes similar to the Cyclar process, the Z-forming and the Aroformer processes, which were developed in Japan and Australia, respectively.[12,13]

Research is also being conducted in Japan to aromatize propane in presence of carbon dioxide using a Zn-loaded HZSM-5 catalyst.[14] The effect of CO_2 is thought to improve the equilibrium formation of aromatics by the consumption of product hydrogen (from dehydrogenation of propane) through the reverse water gas shift reaction.

$$CO_2 + H_2 \rightleftharpoons CO + H_2O$$

However, it was found that the effect on the equilibrium formation of aromatics is not substantial due to thermodynamic considerations. A more favorable effect was found for the reaction between ethylene (formed via cracking during aromatization of propane) and hydrogen. The reverse shift reaction consumes hydrogen and decreases the chances for the reduction of ethylene to ethane byproduct.

$$CH_2=CH_2 + H_2 \rightarrow CH_3-CH_3$$

ISOMERIZATION OF n-BUTANE (Isobutane Production)

Because of the increasing demand for isobutylene for the production of oxygenates as gasoline additives, a substantial amount of n-butane is isomerized to isobutane, which is further dehydrogenated to isobutene. The Butamer process (Figure 6-8) has a fixed-bed reactor containing a highly selective catalyst that promotes the conversion of n-butane to isobutane equilibrium mixture.[15] Isobutane is then separated in a deisobutanizer tower. The n-butane is recycled with make-up hydrogen. The isomerization reaction occurs at a relatively low temperature:

$$CH_3CH_2CH_2CH_3 \rightarrow CH_3CH(CH_3)_2$$
$$\text{Isobutane}$$

ISOBUTANE CHEMICALS

As has been mentioned in Chapter 3, isobutane is mainly used as an alkylating agent to produce different compounds (alkylates) with a high octane number to supplement the gasoline pool. Isobutane is in high

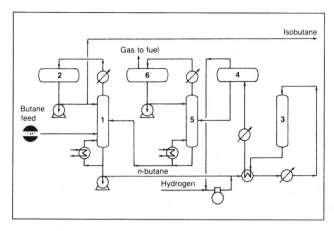

Figure 6-8. The UOP Butamer process for isomerization of n-butane to isobutane:[15] (1,2) deisobutanizer, (3) reactor, (4) separator (for separation and recycling H_2), (5,6) stabilizer.

demand as an isobutene precursor for producing oxygenates such as methyl and ethyl tertiary butyl ethers (MTBE and ETBE). The production and use of MTBE are discussed in Chapter 5. Accordingly, greater amounts of isobutane are produced from n-butane through isomerization followed by dehydrogenation to isobutene. The Catofin process is currently used to dehydrogenate isobutane to isobutene. Alternatively, isobutane could be thermally cracked to yield predominantly isobutene plus propane. Other by-products are fuel gas and C_5^+ liquid. The steam cracking process is made of three sections: a cracking furnace, a vapor recovery section, and a product fractionation section. The Coastal isobutane cracking process is reviewed by Soudek and Lacatena.[16]

NAPHTHA-BASED CHEMICALS

Light naphtha containing hydrocarbons in the C_5-C_7 range is the preferred feedstock in Europe for producing acetic acid by oxidation. Similar to the catalytic oxidation of n-butane, the oxidation of light naphtha is performed at approximately the same temperature and pressure ranges (170–200°C and ≈50 atmospheres) in the presence of manganese acetate catalyst. The yield of acetic acid is approximately 40 wt%.

Light naphtha + O_2 → CH_3COOH + by-products + H_2O

The product mixture contains essentially oxygenated compounds (acids, alcohols, esters, aldehydes, ketones, etc.). As many as 13 distillation columns are used to separate the complex mixture. The number of products could be reduced by recycling most of them to extinction.

Manganese naphthenate may be used as an oxidation catalyst. Rouchaud and Lutete have made an in-depth study of the liquid phase oxidation of n-hexane using manganese naphthenate. A yield of 83% of C_1-C_5 acids relative to n-hexane was reported. The highest yield of these acids was for acetic acid followed by formic acid. The lowest yield was observed for pentanoic acid.[17]

In Europe naphtha is the preferred feedstock for the production of synthesis gas, which is used to synthesize methanol and ammonia (Chapter 4). Another important role for naphtha is its use as a feedstock for steam cracking units for light olefins production (Chapter 3). Heavy naphtha, on the other hand, is a major feedstock for catalytic reforming. The product reformate containing a high percentage of C_6-C_8 aromatic hydrocarbons is used to make gasoline. Reformates are also extracted to separate the aromatics as intermediates for petrochemicals.

CHEMICALS FROM HIGH MOLECULAR WEIGHT n-PARAFFINS

High molecular weight n-paraffins are obtained from different petroleum fractions through physical separation processes. Those in the range of C_8-C_{14} are usually recovered from kerosines having a high ratio of these compounds. Vapor phase adsorption using molecular sieve 5A is used to achieve the separation. The n-paraffins are then desorbed by the action of ammonia. Continuous operation is possible by using two adsorption sieve columns, one bed on stream while the other bed is being desorbed. n-Paraffins could also be separated by forming an adduct with urea. For a paraffinic hydrocarbon to form an adduct under ambient temperature and atmospheric pressure, the compound must contain a long unbranched chain of at least six carbon atoms. Ease of adduct formation and adduct stability increases with increase of chain length.[18] Table 6-2 shows some physical properties of C_5-C_{16} n-paraffins. As with shorter-chain n-paraffins, the longer chain compounds are not highly reactive. However, they may be oxidized, chlorinated, dehydrogenated, sulfonated, and fermented under special conditions. The C_9-C_{17} paraffins are used to produce olefins or monochlorinated paraffins for the production of detergents. The 1996 capacity for the U.S., Europe, and Japan was 3.0 billion pounds.[19]

Table 6-2
Selected properties of n-paraffins from C_5-C_{16}

Name	Formula	Density	B.P.°C	M.P.°C
Pentane	$CH_3(CH_2)_3CH_3$	0.626	36.0	−130.0
Hexane	$CH_3(CH_2)_4CH_3$	0.695	69.0	−95.0
Heptane	$CH_3(CH_2)_5CH_3$	0.684	98.0	−90.5
Octane	$CH_3(CH_2)_6CH_3$	0.703	126.0	−57.0
Nonane	$CH_3(CH_2)_7CH_3$	0.718	151.0	−54.0
Decane	$CH_3(CH_2)_8CH_3$	0.730	174.0	−30.0
Undecane	$CH_3(CH_2)_9CH_3$	0.740	196.0	−26.0
Dodecane	$CH_3(CH_2)_{10}CH_3$	0.749	216.0	−10.0
Tridecane	$CH_3(CH_2)_{11}CH_3$	0.757	234.0	−6.0
Tetradecane	$CH_3(CH_2)_{12}CH_3$	0.764	252.0	5.5
Pentadecane	$CH_3(CH_2)_{13}CH_3$	0.769	266.0	10.0
Hexadecane	$CH_3(CH_2)_{14}CH_3$	0.775	280.0	18.0

OXIDATION OF PARAFFINS (Fatty Acids and Fatty Alcohols)

The catalytic oxidation of long-chain paraffins (C_{18}-C_{30}) over manganese salts produces a mixture of fatty acids with different chain lengths. Temperature and pressure ranges of 105–120°C and 15–60 atmospheres are used. About 60 wt% yield of fatty acids in the range of C_{12}-C_{14} is obtained. These acids are used for making soaps. The main source for fatty acids for soap manufacture, however, is the hydrolysis of fats and oils (a nonpetroleum source). Oxidation of paraffins to fatty acids may be illustrated as:

$$RCH_2(CH_2)_nCH_2CH_2R + {}^5/_2O_2 \rightarrow R(CH_2)_nCOOH + RCH_2COOH + H_2O$$

Oxidation of C_{12}-C_{14} n-paraffins using boron trioxide catalysts was extensively studied for the production of fatty alcohols.[20] Typical reaction conditions are 120–130°C at atmospheric pressure. ter-Butyl hydroperoxide (0.5%) was used to initiate the reaction. The yield of the alcohols was 76.2 wt% at 30.5% conversion. Fatty acids (8.9 wt%) were also obtained. Product alcohols were essentially secondary with the same number of carbons and the same structure per molecule as the parent paraffin hydrocarbon. This shows that no cracking has occurred under the conditions used. The oxidation reaction could be represented as:

$$RCH_2CH_2 R' + {}^1/_2O_2 \rightarrow R\text{-}CH_2CHOHR'$$

n-Paraffins can also be oxidized to alcohols by a dilute oxygen stream (3–4%) in the presence of a mineral acid. The acid converts the alcohols to esters, which prohibit further oxidation of the alcohols to fatty acids. The obtained alcohols are also secondary. These alcohols are of commercial importance for the production of nonionic detergents (ethyoxylates):

$$
\overset{\displaystyle O}{\overset{\displaystyle /\,\backslash}{RCH_2CHOHR' + nCH_2{-}CH_2}} \longrightarrow RCH_2CHO(CH_2CH_2O)_{\overline{n}}{-}
$$
$$
\underset{\displaystyle R'}{|}
$$

Nonionic detergents are discussed in Chapter 7. Other uses of these alcohols are in the plasticizer field and in monoolefin production.

CHLORINATION OF n-PARAFFINS (Chloroparaffins)

Chlorination of n-paraffins (C_{10}-C_{14}) in the liquid phase produces a mixture of chloroparaffins. Selectivity to monochlorination could be increased by limiting the reaction to a low conversion and by decreasing the chlorine to hydrocarbon ratio. Substitution of secondary hydrogen predominates. The reaction may be represented as:

$$R\ CH_2\ CH_2R' + Cl_2 \rightarrow R\ CHCl\ CH_2R' + HCl$$

Monochloroparaffins in this range may be dehydrochlorinated to the corresponding monoolefins and used as alkylating agents for the production of biodegradable detergents. Alternatively, the monochloroparaffins are used directly to alkylate benzene in presence of a Lewis acid catalyst to produce alkylates for the detergent production. These reactions could be illustrated as follows:

$$RCH_2CHClR' \longrightarrow RCH{=}CHR' + HCl$$

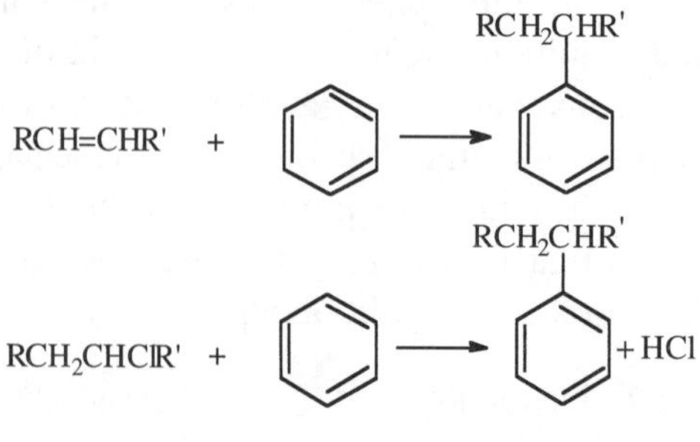

An alkylate

Detergent production is further discussed in Chapter 10.

Polychlorination, on the other hand, can be carried out on the whole range of n-paraffins from C_{10}-C_{30} at a temperature range of 80–120°C (using a high Cl_2/paraffin ratio). The product has a chlorine content of approximately 70%. Polychloroparaffins are used as cutting oil additives, plasticizers, and retardant chemicals.

SULFONATION OF n-PARAFFINS
(Secondary Alkane Sulfonates SAS)

Linear secondary alkane sulfonates are produced by the reaction between sulfur dioxide and n-paraffins in the range of C_{15}-C_{17}.

$$R\text{-}H + 2SO_2 + 2O_2 + H_2O \rightarrow RSO_3H + H_2SO_4$$

The reaction is catalyzed by ultraviolet light with a wave-length between 3,300–3,600Å.[21] The sulfonates are nearly 100% biodegradable, soft and stable in hard water, and have good washing properties.

Sodium alkanesulfonates for detergent manufacture can also be produced from the free-radical addition of sodium bisulfite and alpha olefins:

$$RCH=CH_2 + NaHSO_3 \rightarrow RCH_2CH_2SO_3Na$$

FERMENTATION USING n-PARAFFINS (Single Cell Protein SCP)

The term single cell protein is used to represent a group of microbial cells such as algae and yeast that have high protein content. The production of these cells is not generally considered a synthetic process but microbial farming via fermentation in which n-paraffins serve as the substrate. Substantial research efforts were invested in the past two decades to grow algae, fungi, and yeast on different substrates such as n-paraffins, methane, methanol, and even carbon dioxide. The product SCP is constituted mainly of protein and variable amounts of lipids, carbohydrates, vitamins, and minerals. Some of the constituents of SCP limit its usefulness for use as food for human beings but can be used for animal feed. A commercial process using methanol as the substrate was developed by ICI. The product Pruteen is an energy-rich material containing over 70% protein.[22]

One of the problems facing the use of n-paraffins as a substrate for Candida yeast is the presence of residual hydrocarbons in the product.[23]

The reliability and economics of producing high-quality n-paraffins is a critical factor in the use of n-paraffins for the production of SCP.

REFERENCES

1. "Petrochemical Handbook," *Hydrocarbon Processing,* Vol. 52, No. 11, 1973, p.92.
2. *CHEMTECH,* March, 1998, p. 3.
3. "Petrochemical Handbook," *Hydrocarbon Processing,* Vol. 70, No. 3, 1991, p. 185.
4. Tucci, E., Dufallo, J. M. and Feldman, R. J., "Commercial Performance of the Houdry CATOFIN Process for Isobutylene Production for MTBE, Catalysts, and Catalytic Processes Used in Saudi Arabia Workshop," KFUPM, Nov. 6, 1991.
5. Hatch, L. F. and Matar, S., "Petrochemicals from n-Paraffins," *Hydrocarbon Processing,* Vol. 56, No. 11, 1977, pp. 349–357.
6. Iborra, M., Izquierdo, J. F., Tejero, J. and Cunill, F., "Getting the Lead Out of t-Butyl Ether," *CHEMTECH,* Feb. 1988, pp. 120–122.
7. Saunby, J. B. and Kiff, B. W., *Hydrocarbon Processing,* Vol. 55, No. 11, 1974, pp. 247–252.
8. "Petrochemical Handbook," *Hydrocarbon Processing,* Vol. 70, No. 3, 1991, p. 164.
9. Haggin, J. "Innovation in Catalysis Create Environmentally Friendly THF Process" *Chemical and Engineering News,* April 3, 1995, pp. 20–23.
10. Doolan, P. C. and Pujado, P. R., "Make Aromatics from LPG," *Hydrocarbon Processing,* Vol. 68, No. 9, 1989, pp. 72–76.
11. "Petrochemical Handbook," *Hydrocarbon Processing,* Vol. 78, No. 3, p. 100.
12. Kondoh, T., et al., *Zeoraito,* Vol. 9, 1992, p. 20.
13. Babier, J. C. and Minkkinen, A., JPI Petroleum Refining Conference, Tokyo, 1990.
14. Syoichi, Y. et al., "Aromatization of Propane in CO_2 Atmosphere," Second Joint Saudi Japanese Workshop on Recent Developments in Selected Petroleum Refining and Petrochemical Processes, KFUPM, Dhahran, Saudi Arabia, 12–13, Dec. 1992.
15. "Gas Processing Handbook," *Hydrocarbon Processing,* Vol. 69, No. 4, 1990, pp. 73–76.
16. Saudek, M. and Lacatena, J. J., "Crack Isobutane for Isobutylene," *Hydrocarbon Processing,* Vol. 69, No. 5, 1990, pp. 73–76.

17. Rouchaud, J. and Lutete, B., *Industrial and Engineering Chemistry, Product Research Division,* Vol. 7, No. 4, 1968, pp. 266–270.
18. Speight, J. G., *The Chemistry and Technology of Petroleum,* 2nd Ed., Marcel Dekker, Inc. New York, 1991, p. 344.
19. *Chemical Industries News Letter,* April–June, 1998, p.8.
20. Marer, A. and Hussain, M. M., Second Arab Conference on Petrochemicals, United Arab Emirates, paper No. 6 (p. 3) March 15–23, 1976.
21. "Petrochemical Handbook," *Hydrocarbon Processing,* Vol. 58, No. 11, 1979, p. 186.
22. "Petrochemical Handbook," *Hydrocarbon Processing,* Vol. 64, No. 11, 1985, p. 167.
23. Kent, J. A. (ed.) *Riegel's Handbook of Industrial Chemistry,* 8th Ed., Van Nostrand Reinhold Co. New York, 1983, p. 685.

CHAPTER SEVEN

Chemicals Based on Ethylene

INTRODUCTION

Ethylene is sometimes known as the "king of petrochemicals" because more commercial chemicals are produced from ethylene than from any other intermediate. This unique position of ethylene among other hydrocarbon intermediates is due to some favorable properties inherent in the ethylene molecule as well as to technical and economical factors. These could be summarized in the following:

- Simple structure with high reactivity.
- Relatively inexpensive compound.
- Easily produced from any hydrocarbon source through steam cracking and in high yields.
- Less by-products generated from ethylene reactions with other compounds than from other olefins.

Ethylene reacts by addition to many inexpensive reagents such as water, chlorine, hydrogen chloride, and oxygen to produce valuable chemicals. It can be initiated by free radicals or by coordination catalysts to produce polyethylene, the largest-volume thermoplastic polymer. It can also be copolymerized with other olefins producing polymers with improved properties. For example, when ethylene is polymerized with propylene, a thermoplastic elastomer is obtained. Figure 7-1 illustrates the most important chemicals based on ethylene.

Global demand for ethylene is expected to increase from 79 million tons in 1997 to 114 million tons in 2005.[1] In 1998, the U.S. consumption of ethylene was approximately 52 billion pounds. Figure 7-2 shows the breakdown of the 1998 U.S. ethylene consumption.[2]

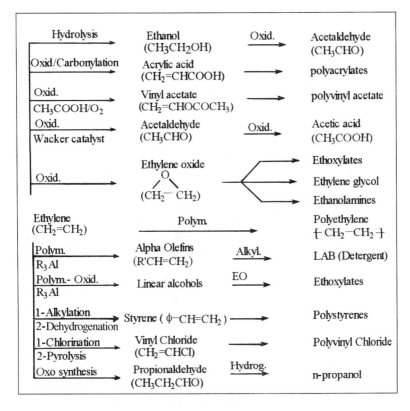

Figure 7-1. Major chemicals based on ethylene.

OXIDATION OF ETHYLENE

Ethylene can be oxidized to a variety of useful chemicals. The oxidation products depend primarily on the catalyst used and the reaction conditions. Ethylene oxide is the most important oxidation product of ethylene. Acetaldehyde and vinyl acetate are also oxidation products obtained from ethylene under special catalytic conditions.

$$\overset{\displaystyle O}{\overset{\displaystyle /\diagdown}{}}$$

Ethylene Oxide $(CH_2\!-\!CH_2)$

Ethylene oxide (EO) is a colorless gas that liquefies when cooled below 12°C. It is highly soluble in water and in organic solvents.

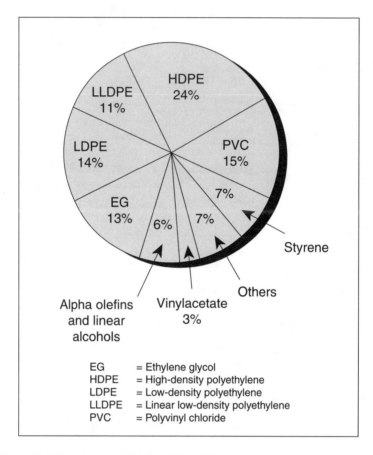

Figure 7-2. Breakdown of U.S. 1998 ethylene consumption of 52 billion lb.[2]

Ethylene oxide is a precursor for many chemicals of great commercial importance, including ethylene glycols, ethanolamines, and alcohol ethoxylates. Ethylene glycol is one of the monomers for polyesters, the most widely-used synthetic fiber polymers. The current US production of EO is approximately 8.1 billion pounds.

Production

The main route to ethylene oxide is oxygen or air oxidation of ethylene over a silver catalyst. The reaction is exothermic; heat control is important:

$$CH_2{=}CH_2 + 1/2O_2 \longrightarrow \overset{\displaystyle O}{\overset{\displaystyle /\diagdown}{CH_2{-}CH_2}} \qquad \Delta H = -147 \text{ KJ/mol}$$

A concomitant reaction is the complete oxidation of ethylene to carbon dioxide and water:

$$CH_2{=}CH_2 + 3O_2 \longrightarrow 2\,CO_2 + 2H_2O \qquad \Delta H = -1{,}421 \text{ KJ/mol}$$

This reaction is highly exothermic; the excessive temperature increase reduces ethylene oxide yield and causes catalyst deterioration. Overoxidation can be minimized by using modifiers such as organic chlorides.

It seems that silver is a unique epoxidation catalyst for ethylene. All other catalysts are relatively ineffective, and the reaction to ethylene is limited among lower olefins. Propylene and butylenes do not form epoxides through this route.[3]

Using oxygen as the oxidant versus air is currently favored because it is more economical.[4]

In the process (Figure 7-3), compressed oxygen, ethylene, and recycled gas are fed to a multitubular reactor.[5] The temperature of oxidation

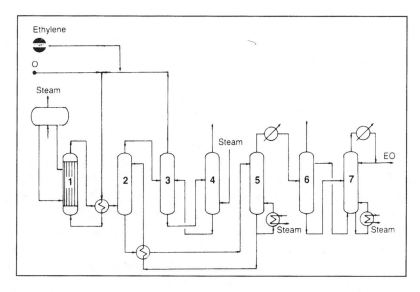

Figure 7-3. The Scientific Design Co. Ethylene Oxide process:[5] (1) reactor, (2) scrubber, (3,4) CO2 removal, (5) stripper, (6,7) fractionators.

is controlled by boiling water in the shell side of the reactor. Effluent gases are cooled and passed to the scrubber where ethylene oxide is absorbed as a dilute aqueous solution. Unreacted gases are recycled. Epoxidation reaction occurs at approximately 200–300°C with a short residence time of one second. A selectivity of 70–75% can be reached for the oxygen based process. Selectivity is the ratio of moles of ethylene oxide produced per mole of ethylene reacted. Ethylene oxide selectivity can be improved when the reaction temperature is lowered and the conversion of ethylene is decreased (higher recycle of unreacted gases).

Derivatives of Ethylene Oxide

Ethylene oxide is a highly active intermediate. It reacts with all compounds that have a labile hydrogen such as water, alcohols, organic acids, and amines. The epoxide ring opens, and a new compound with a hydroxyethyl group is produced. The addition of a hydroxyethyl group increases the water solubility of the resulting compound. Further reaction of ethylene oxide produces polyethylene oxide derivatives with increased water solubility.

Many commercial products are derived from ethylene oxide by reacting with different reagents. The following reviews the production and the utility of these chemicals.

Ethylene Glycol (CH_2OHCH_2OH)

Ethylene glycol (EG) is colorless syrupy liquid, and is very soluble in water. The boiling and the freezing points of ethylene glycol are 197.2° and –13.2°C, respectively.

Current world production of ethylene glycol is approximately 15 billion pounds. Most of that is used for producing polyethylene terephthalate (PET) resins (for fiber, film, bottles), antifreeze, and other products. Approximately 50% of the world EG was consumed in the manufacture of polyester fibers and another 25% went into the antifreeze.

EG consumption in the US was nearly 1/3 of the world's. The use pattern, however, is different; about 50% of EG is consumed in antifreeze. The US production of ethylene glycol was 5.55 billion pounds in 1994, the 30th largest volume chemical.

The main route for producing ethylene glycol is the hydration of ethylene oxide in presence of dilute sulfuric acid (Figure 7-4):[6]

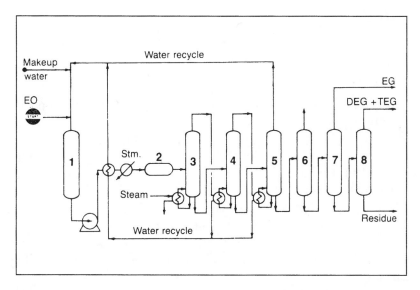

Figure 7-4. The Scientific Design Co. process for producing ethylene glycols from ethylene oxide:[5] (1) feed tank, (2) reactor, (3,4,5) multiple stage evaporators, #4 operates at lower pressure than #3, while #5 operates under vacuum, evaporated water is recycled to feed tank, (6) light ends stripper, (7,8) vacuum distillation columns.

$$CH_2\!\!-\!\!CH_2 + H_2O \xrightarrow{\ H^+\ } HO-CH_2-CH_2OH$$

The hydrolysis reaction occurs at a temperature range of 50–100°C. Contact time is approximately 30 minutes. Di- and triethylene glycols are coproducts with the monoglycol. Increasing the water/ethylene oxide ratio and decreasing the contact time decreases the formation of higher glycols. A water/ethylene oxide ratio of 10 is normally used to get approximately 90% yield of the monoglycol. However, the di- and the triglycols are not an economic burden, because of their commercial uses.

A new route to ethylene glycol from ethylene oxide via the intermediate formation of ethylene carbonate has recently been developed by Texaco. Ethylene carbonate may be formed by the reaction of carbon monoxide, ethylene oxide, and oxygen. Alternatively, it could be obtained by the reaction of phosgene and methanol.

Ethylene carbonate is a reactive chemical. It reacts smoothly with methanol and produces ethylene glycol in addition to dimethyl carbonate:

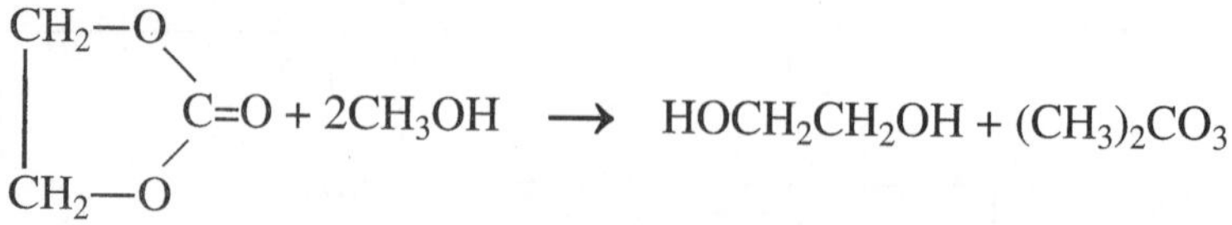

$$\begin{array}{c} CH_2-O \\ | \quad\quad\quad C=O + 2CH_3OH \quad \longrightarrow \quad HOCH_2CH_2OH + (CH_3)_2CO_3 \\ CH_2-O \end{array}$$

The reaction occurs at approximately 80–130°C using the proper catalyst. Many catalysts have been tried for this reaction, and there is an indication that the best catalyst types are those of the tertiary amine and quaternary ammonium functionalized resins.[7] This route produces ethylene glycol of a high purity and avoids selectivity problems associated with the hydrolysis of ethylene oxide.

The coproduct dimethyl carbonate is a liquid soluble in organic solvents. It is used as a specialty solvent, a methylating agent in organic synthesis, and a monomer for polycarbonate resins. It may also be considered as a gasoline additive due to its high oxygen content and its high octane rating.

Alternative Routes to Producing Ethylene Glycol

Ethylene glycol could also be obtained directly from ethylene by two methods, the Oxirane acetoxylation and the Teijin oxychlorination processes. The production of ethylene glycol from formaldehyde and carbon monoxide is noted in Chapter 5.

In the Oxirane process, ethylene is reacted in the liquid phase with acetic acid in the presence of a TeO_2 catalyst at approximately 160° and 28 atmospheres.[8] The product is a mixture of mono- and diacetates of ethylene glycol:

$$2CH_2=CH_2 + 3CH_3COOH + O_2 \longrightarrow \overset{\displaystyle O}{\overset{\displaystyle \|}{CH_3C}}OCH_2CH_2OH +$$

$$\overset{\displaystyle O}{\overset{\displaystyle \|}{CH_3C}}OCH_2CH_2\overset{\displaystyle O}{\overset{\displaystyle \|}{OC}}CH_3 + H_2O$$

The acetates are then hydrolyzed to ethylene glycol and acetic acid. The hydrolysis reaction occurs at approximately 107–130°C and 1.2 atmospheres. Acetic acid is then recovered for further use:

$$CH_3\overset{\text{O}}{\overset{\|}{C}}OCH_2CH_2OH + CH_3\overset{\text{O}}{\overset{\|}{C}}OCH_2CH_2O\overset{\text{O}}{\overset{\|}{C}}CH_3 + 3\,H_2O \longrightarrow$$

$$2HOCH_2CH_2OH + 3CH_3COOH$$

A higher glycol yield (approximately 94%) than from the ethylene oxide process is anticipated. However, there are certain problems inherent in the Oxirane process such as corrosion caused by acetic acid and the incomplete hydrolysis of the acetates. Also, the separation of the glycol from unhydrolyzed monoacetate is hard to accomplish.

The Teijin oxychlorination, on the other hand, is considered a modern version of the obsolete chlorohydrin process for the production of ethylene oxide. In this process, ethylene chlorohydrin is obtained by the catalytic reaction of ethylene with hydrochloric acid in presence of thallium(III) chloride catalyst:

$$CH_2=CH_2 + TlCl_3 + H_2O \rightarrow ClCH_2CH_2OH + TlCl + HCl$$

Ethylene chlorohydrin is then hydrolyzed in situ to ethylene glycol.

Catalyst regeneration occurs by the reaction of thallium(I) chloride with copper(II) chloride in the presence of oxygen or air. The formed Cu(I) chloride is reoxidized by the action of oxygen in the presence of HCl:

$$TlCl + 2CuCl_2 \rightarrow TlCl_3 + Cu_2Cl_2$$

$$Cu_2Cl_2 + 2HCl + {}^1\!/_2O_2 \rightarrow 2CuCl_2 + H_2O$$

The overall reaction is represented as:

$$CH_2=CH_2 + H_2O + {}^1\!/_2O_2 \rightarrow HOCH_2CH_2OH$$

Ethoxylates

The reaction between ethylene oxide and long-chain fatty alcohols or fatty acids is called ethoxylation. Ethoxylation of C_{10}-C_{14} linear alcohols and linear alkylphenols produces nonionic detergents. The reaction with alcohols could be represented as:

$$ROH + n\overset{\text{O}}{\overset{\diagup\diagdown}{CH_2-CH_2}} \longrightarrow RO(CH_2CH_2O)_{\overline{n-1}}H$$

The solubility of the product ethoxylates can be varied according to the number of ethylene oxide units in the molecule. The solubility is also a function of the chain-length of the alkyl group in the alcohol or in the phenol. Longer-chain alkyl groups reduce water solubility. In practice, the number of ethylene oxide units and the chain-length of the alkyl group are varied to either produce water-soluble or oil-soluble surface active agents. Surfactants properties and micelle formation in polar and nonpolar solvents have been reviewed by Rosen.[9]

Linear alcohols used for the production of ethoxylates are produced by the oligomerization of ethylene using Ziegler catalysts or by the Oxo reaction using alpha olefins.

Similarly, esters of fatty acids and polyethylene glycols are produced by the reaction of long-chain fatty acids and ethylene oxide:

$$\underset{\substack{\| \\ R\overset{\text{O}}{C}OH}}{} + n\,CH_2\text{---}CH_2 \longrightarrow R\overset{\text{O}}{\underset{\|}{C}}O(CH_2CH_2O)_{\overline{n-1}}H$$

The C_{12}-C_{18} fatty acids such as oleic, palmitic, and stearic are usually ethoxylated with EO for the production of nonionic detergents and emulsifiers.

Ethanolamines

A mixture of mono-, di-, and triethanolamines is obtained by the reaction between ethylene oxide (EO) and aqueous ammonia. The reaction conditions are approximately 30–40°C and atmospheric pressure:

$$6\,CH_2\text{---}CH_2 + 3\,NH_3 \longrightarrow H_2NCH_2CH_2OH + HN(CH_2CH_2OH)_2$$
$$+ N(CH_2CH_2OH)_3$$

The relative ratios of the ethanolamines produced depend principally on the ethylene oxide/ammonia ratio. A low EO/NH_3 ratio increases monoethanolamine yield. Increasing this ratio increases the yield of di-and triethanolamines. Table 7-1 shows the weight ratios of ethanolamines as a function of the mole ratios of the reactants.[10]

Ethanolamines are important absorbents of acid gases in natural gas treatment processes. Another major use of ethanolamines is the production of surfactants. The reaction between ethanolamines and fatty acids

Table 7-1
Weight ratios of ethanolamines as a function
of the mole ratios of the reactants[10]

	Moles of ethylene oxide/moles of ammonia		
	0.1	0.5	1.0
Monoethanolamine	75–61	25–31	12–15
Diethanolamine	21–27	28–32	23–26
Triethanolamine	4–12	37	65–59

produces ethanolamides. For example, when lauric acid and mono-ethanolamine are used, N-(2hydroxyethyl)-lauramide is obtained:

$$CH_3(CH_2)_{10}\overset{\overset{\displaystyle O}{\|}}{C}OH + HOCH_2CH_2NH_2 \longrightarrow CH_3(CH_2)_{10}CONH(CH_2)_2OH + H_2O$$

Lauric acid is the main fatty acid used for producing ethanolamides. Monoethanolamides are used primarily in heavy-duty powder detergents as foam stabilizers and rinse improvers.

1,3-Propanediol

1,3-Propanediol is a colorless liquid that boils at 210–211°C. It is soluble in water, alcohol, and ether. It is an intermediate for polyester production. It could be produced via the hydroformylation of ethylene oxide which yields 3-hydroxypropionaldehyde. Hydrogenation of the product produces 1,3-propanediol.

$$\overset{\displaystyle O}{\overset{\displaystyle /\quad\backslash}{CH_2 - CH_2}} + CO + H_2 \rightarrow HO–(C_2H_4)CHO$$

$$HO\text{-}(C_2H_4)CHO + H_2 \rightarrow \underset{\underset{\displaystyle OH}{|}}{CH_2} - CH_2 - \underset{\underset{\displaystyle OH}{|}}{CH_2}$$

The catalyst is a cobalt carbonyl that is prepared in situ from cobaltous hydroxide, and nonylpyridine is the promotor. Oxidation of the aldehyde produces 3-hydroxypropionic acid. 1,3-Propanediol and 3-hydroxypropionic acid could also be produced from acrolein (Chaper 8).[11]

ACETALDEHYDE (CH₃CHO)

Acetaldehyde is a colorless liquid with a pungent odor. It is a reactive compound with no direct use except for the synthesis of other compounds. For example, it is oxidized to acetic acid and acetic anhydride. It is a reactant in the production of 2-ethylhexanol for the synthesis of plasticizers and also in the production of pentaerithritol, a polyhydric compound used in alkyd resins.

There are many ways to produce acetaldehyde. Historically, it was produced either by the silver-catalyzed oxidation or by the chromium activated copper-catalyzed dehydrogenation of ethanol. Currently, acetaldehyde is obtained from ethylene by using a homogeneous catalyst (Wacker catalyst). The catalyst allows the reaction to occur at much lower temperatures (typically 130°) than those used for the oxidation or the dehydrogenation of ethanol (approximately 500°C for the oxidation and 250°C for the dehydrogenation).

Ethylene oxidation is carried out through oxidation-reduction (redox). The overall reaction is the oxidation of ethylene by oxygen as represented by:

$$CH_2=CH_2 + \tfrac{1}{2}O_2 \longrightarrow CH_3-\overset{\displaystyle O}{\overset{\|}{C}}-H \qquad \Delta H = -218.6 \text{ KJ/mol}$$

The Wacker process uses an aqueous solution of palladium(II) chloride, copper(II) chloride catalyst system.

In the course of the reaction, the Pd^{2+} ions are reduced to Pd metal, and ethylene is oxidized to acetaldehyde:

$$CH_2=CH_2 + PdCl_2 + H_2O \rightarrow CH_3CHO + 2HCl + Pd°$$

The formed Pd° is then reoxidized by the action of Cu(II) ions, which are reduced to Cu(I) ions:

$$Pd° + 2CuCl_2 \rightarrow PdCl_2 + 2CuCl$$

The reduced Cu(I) ions are reoxidized to Cu(II) ions by reaction with oxygen and HCl:

$$2CuCl + {}^1/_2O_2 + 2HCl \rightarrow 2CuCl_2 + H_2O$$

The oxidation reaction may be carried out in a single-stage or a two-stage process. In the single-stage, ethylene, oxygen, and recycled gas are fed into a vertical reactor containing the catalyst solution. Heat is controlled by boiling off some of the water. The reaction conditions are approximately 130°C and 3 atmospheres. In the two-stage process, the reaction occurs under relatively higher pressure (approximately 8 atmospheres) to ensure higher ethylene conversion. The reaction temperature is approximately 130°C. The catalyst solution is then withdrawn from the reactor to a tube-oxidizer to effect the oxidation of the catalyst at approximately 10 atmospheres. The yield of acetaldehyde from either process is about 95%. By-products from this reaction include acetic acid, ethyl chloride, chloroacetaldehyde, and carbon dioxide.

The Wacker reaction can also be carried out for other olefins with terminal double bonds. With propene, for example, approximately 90% yield of acetone is obtained. l-Butene gave approximately 80% yield of methyl ethyl ketone.[12]

Acetaldehyde is an intermediate for many chemicals such as acetic acid, n-butanol, pentaerithritol, and polyacetaldehyde.

Important Chemicals from Acetaldehyde

Acetic Acid

Acetic acid is obtained from different sources. Carbonylation of methanol is currently the major route. Oxidation of butanes and butenes is an important source of acetic acid, especially in the U.S. (Chapter 6). It is also produced by the catalyzed oxidation of acetaldehyde:

$$\underset{CH_3CH}{\overset{\overset{\displaystyle O}{\|}}{}} + \; \tfrac{1}{2}O_2 \longrightarrow \underset{CH_3COH}{\overset{\overset{\displaystyle O}{\|}}{}}$$

The reaction occurs in the liquid phase at approximately 65°C using manganese acetate as a catalyst. Uses of acetic acid have been noted in Chapter 5.

n-Butanol

n-Butanol is normally produced from propylene by the Oxo reaction (Chapter 8). It may also be obtained from the aldol condensation of acetaldehyde in presence of a base.

$$2\underset{CH_3CH}{\overset{\overset{\displaystyle O}{\|}}{}} \; \xrightarrow{\;OH^-\;} \; \underset{CH_3CHOHCH_2CH}{\overset{\overset{\displaystyle \qquad\qquad O}{\qquad\qquad\|}}{}}$$

The formed 3-hydroxybutanal eliminates one mole of water in the presence of an acid producing crotonaldehyde. Hydrogenation of crotonaldehyde produces n-butanol:

$$CH_3CHOHCH_2\overset{\overset{\displaystyle O}{\|}}{C}H \quad \xrightarrow{\text{H}^+} \quad CH_3CH=CH\overset{\overset{\displaystyle O}{\|}}{C}H + H_2O$$

Crotonaldehyde

$$CH_3CH=CH\overset{\overset{\displaystyle O}{\|}}{C}H + H_2 \longrightarrow CH_3CH_2CH_2CH_2OH$$

n-Butanol

The uses of n-butanol are noted in Chapter 8.

Vinyl Acetate ($CH_2=CHO\overset{\overset{\displaystyle O}{\|}}{C}CH_3$)

Vinyl acetate is a reactive colorless liquid that polymerizes easily if not stabilized. It is an important monomer for the production of polyvinyl acetate, polyvinyl alcohol, and vinyl acetate copolymers. The U.S. production of vinyl acetate, the 40th highest-volume chemical, was approximately 3 billion pounds in 1994.

Vinyl acetate was originally produced by the reaction of acetylene and acetic acid in the presence of mercury(II) acetate. Currently, it is produced by the catalytic oxidation of ethylene with oxygen, with acetic acid as a reactant and palladium as the catalyst:

$$CH_2=CH_2 + CH_3\overset{\overset{\displaystyle O}{\|}}{C}OH + \tfrac{1}{2}O_2 \quad \longrightarrow \quad CH_2=CHO\overset{\overset{\displaystyle O}{\|}}{C}CH_3 + H_2O$$

The process is similar to the catalytic liquid-phase oxidation of ethylene to acetaldehyde. The difference between the two processes is the presence of acetic acid. In practice, acetaldehyde is a major coproduct. The mole ratio of acetaldehyde to vinyl acetate can be varied from 0.3:1 to 2.5:1.[13] The liquid-phase process is not used extensively due to corrosion problems and the formation of a fairly wide variety of by-products.

In the vapor-phase process, oxyacylation of ethylene is carried out in a tubular reactor at approximately 117°C and 5 atmospheres. The palladium acetate is supported on carriers resistant to attack by acetic acid. Conversions of about 10–15% based on ethylene are normally used to operate safely outside the explosion limits (approximately 10% O_2). Selectivities of 91–94% based on ethylene are attainable.

OXIDATIVE CARBONYLATION OF ETHYLENE

Acrylic acid: $CH_2{=}CH\overset{\displaystyle O}{\overset{\displaystyle \|}{C}}OH$

The liquid phase reaction of ethylene with carbon monoxide and oxygen over a Pd^{2+}/Cu^{2+} catalyst system produces acrylic acid. The yield based on ethylene is about 85%. Reaction conditions are approximately 140°C and 75 atmospheres:

$$CH_2{=}CH_2 + CO + \tfrac{1}{2}O_2 \longrightarrow CH_2{=}CH\overset{\displaystyle O}{\overset{\displaystyle \|}{C}}OH$$

The catalyst is similar to that of the Wacker reaction for ethylene oxidation to acetaldehyde, however, this reaction occurs in presence of carbon monoxide.

Currently, the main route to acrylic acid is the oxidation of propene (Chapter 8).

CHLORINATION OF ETHYLENE

The direct addition of chlorine to ethylene produces ethylene dichloride (1,2-dichloroethane). Ethylene dichloride is the main precursor for vinyl chloride, which is an important monomer for polyvinyl chloride plastics and resins.

Other uses of ethylene dichloride include its formulation with tetraethyl and tetramethyl lead solutions as a lead scavenger, as a degreasing agent, and as an intermediate in the synthesis of many ethylene derivatives.

The reaction of ethylene with hydrogen chloride, on the other hand, produces ethyl chloride. This compound is a small-volume chemical with diversified uses (alkylating agent, refrigerant, solvent).

Ethylene reacts also with hypochlorous acid, yielding ethylene chlorohydrin:

$$CH_2{=}CH_2 + HOCl \rightarrow ClCH_2CH_2OH$$

Ethylene chlorohydrin via this route was previously used for producing ethylene oxide through an epoxidation step. Currently, the catalytic oxy-chlorination route (the Teijin process discussed earlier in this chapter) is an alternative for producing ethylene glycol where ethylene chlorohydrin is an intermediate. In organic synthesis, ethylene chlorohydrin is a useful agent for introducing the ethylhydroxy group. It is also used as a solvent for cellulose acetate.

Vinyl Chloride ($CH_2{=}CHCl$)

Vinyl chloride is a reactive gas soluble in alcohol but slightly soluble in water. It is the most important vinyl monomer in the polymer industry. The U.S. production of vinyl chloride, the 16th highest-volume chemical, was approximately 14.8 billion pounds in 1994.

Vinyl chloride monomer (VCM) was originally produced by the reaction of hydrochloric acid and acetylene in the presence of $HgCl_2$ catalyst. The reaction is straightforward and proceeds with high conversion (96% on acetylene):

$$HC{\equiv}CH + HCl \rightarrow CH_2{=}CHCl$$

However, ethylene as a cheap raw material has replaced acetylene for obtaining vinyl chloride. The production of vinyl chloride via ethylene is a three-step process. The first step is the direct chlorination of ethylene to produce ethylene dichloride. Either a liquid- or a vapor-phase process is used:

$$CH_2{=}CH_2 + Cl_2 \rightarrow ClCH_2CH_2Cl$$

The exothermic reaction occurs at approximately 4 atmospheres and 40–50°C in the presence of $FeCl_3$, $CuCl_2$ or $SbCl_3$ catalysts. Ethylene bromide may also be used as a catalyst.

The second step is the dehydrochlorination of ethylene dichloride (EDC) to vinyl chloride and HCl. The pyrolysis reaction occurs at approximately 500°C and 25 atmospheres in the presence of pumice on charcoal:

$$ClCH_2CH_2Cl \rightarrow CH_2{=}CHCl + HCl$$

The third step, the oxychlorination of ethylene, uses by-product HCl from the previous step to produce more ethylene dichloride:

$$CH_2=CH_2 + 2HCl + {}^1\!/_2O_2 \rightarrow ClCH_2\text{-}CH_2Cl + H_2O$$

Ethylene dichloride from this step is combined with that produced from the chlorination of ethylene and introduced to the pyrolysis furnace.

The reaction conditions are approximately 225°C and 2–4 atmospheres.

In practice the three steps, chlorination, oxychlorination, and dehydrochlorination, are integrated in one process so that no chlorine is lost. Figure 7-5 illustrates the process.[14]

PERCHLORO- AND TRICHLOROETHYLENE

Perchloro- and trichloroethylenes could be produced from ethylene dichloride by an oxychlorination/oxyhydrochlorination process without by-product hydrogen chloride. A special catalyst is used:

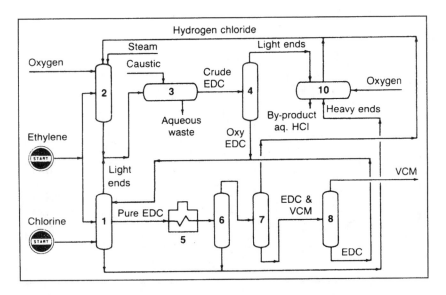

Figure 7-5. The European Vinyls Corporation process for producing vinyl chloride:[14] (1) chlorination section, (2) oxychlorination reactor, (3) steam stripping and caustic treatment of water effluent, (4) EDC distillation, (5) pyrolysis furnace, (6,7,8) VCM and EDC separation, (10) by-product reactor.

$$2ClCH_2\text{-}CH_2Cl + 1^1/_2Cl_2 + 7/4O_2 \rightarrow ClCH=CCl_2 + Cl_2C = CCl_2 + 3^1/_2H_2O$$

A fluid-bed reactor is used at moderate pressures at approximately 450°C. The reactor effluent, containing chlorinated organics, water, a small amount of HCl, carbon dioxide, and other impurities, is condensed in a water-cooled graphite exchanger, cooled in a refrigerated condenser, and then scrubbed. Separation of perchlor from the trichlor occurs by successive distillation. Figure 7-6 shows the PPG process.[15]

Perchloro- and trichloroethylene may also be produced from chlorination of propane (Chapter 6).

HYDRATION OF ETHYLENE
(Ethanol Production)

Ethyl alcohol (CH_3CH_2OH) production is considered by many to be the world's oldest profession. Fermenting carbohydrates is still the

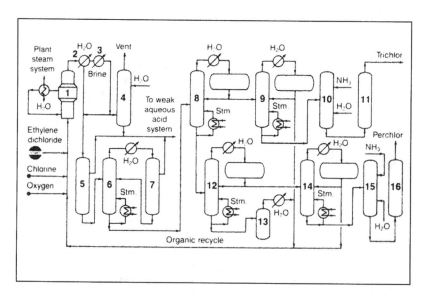

Figure 7-6. The PPG Industries Inc. Chloroethylene process for producing per-chloro- and trichloroethylene:[15] (1) reactor, (2) graphite exchanger, (3) refrigerated condenser, (4) scrubber, (5) phase separation of perchlor from trichlor, (6, 7) azeotropic distillation, (8) distillation train, (9–11) crude trichlor separation—purification, (10–16) crude perchlor separation—purification.

main route to ethyl alcohol in many countries with abundant sugar and grain sources.

Synthetic ethyl alcohol (known as ethanol to differentiate it from fermentation alcohol) was originally produced by the indirect hydration of ethylene in the presence of concentrated sulfuric acid. The formed mono- and diethyl sulfates are hydrolyzed with water to ethanol and sulfuric acid, which is regenerated:

$$3\ CH_2{=}CH_2 + 2H_2SO_4 \rightarrow CH_3CH_2OSO_3H + (CH_3CH_2O)_2SO_2$$

$$CH_3CH_2OSO_3H + (CH_3CH_2O)_2SO_2 + 3H_2O \rightarrow 3CH_3CH_2OH$$
$$+ 2H_2SO_4$$

The direct hydration of ethylene with water is the process currently used:

$$CH_2{=}CH_2 + H_2O \rightarrow CH_3CH_2OH \qquad \Delta H = -40\ KJ/mol$$

The hydration reaction is carried out in a reactor at approximately 300°C and 70 atmospheres. The reaction is favored at relatively lower temperatures and higher pressures. Phosphoric acid on diatomaceous earth is the catalyst. To avoid catalyst losses, a water/ethylene mole ratio less than one is used. Conversion of ethylene is limited to 4–5% under these conditions, and unreacted ethylene is recycled. A high selectivity to ethanol is obtained (95–97%).

Uses of Ethanol

Ethanol's many uses can be conveniently divided into solvent and chemical uses. As a solvent, ethanol dissolves many organic-based materials such as fats, oils, and hydrocarbons. As a chemical intermediate, ethanol is a precursor for acetaldehyde, acetic acid, and diethyl ether, and it is used in the manufacture of glycol ethyl ethers, ethylamines, and many ethyl esters.

OLIGOMERIZATION OF ETHYLENE

The addition of one olefin molecule to a second and to a third, etc. to form a dimer, a trimer, etc. is termed oligomerization. The reaction is normally acid-catalyzed. When propene or butenes are used, the formed compounds are branched because an intermediate carbocation is formed. These compounds were used as alkylating agents for producing benzene alkylates, but the products were nonbiodegradable.

Oligomerization of ethylene using a Ziegler catalyst produces unbranched alpha olefins in the C_{12}-C_{16} range by an insertion mechanism. A similar reaction using triethylaluminum produces linear alcohols for the production of biodegradable detergents.

Dimerization of ethylene to butene-1 has been developed recently by using a selective titanium-based catalyst. Butene-1 is finding new markets as a comonomer with ethylene in the manufacture of linear low-density polyethylene (LLDPE).

ALPHA OLEFINS PRODUCTION

The C_{12}-C_{16} alpha olefins are produced by dehydrogenation of n-paraffins, dehydrochlorination of monochloroparaffins, or by oligomerization of ethylene using trialkyl aluminum (Ziegler catalyst). Recently, it was found that iridium complexes catalyze the dehydrogenation of n-paraffins to α-olefins. The reaction uses a soluble iridium catalyst to transfer hydrogen to the olefinic acceptor.[16] The following shows the oligomerization of ethylene using triethylaluminum:

$$(CH_3CH_2)_3Al + 1\tfrac{1}{2}\, n\, CH_2{=}CH_2 \rightarrow [CH_3(CH_2)_{n+1}]_3Al$$

$$[CH_3(CH_2)_{n+1}]_3Al + 3CH_3CH_2CH{=}CH_2$$
$$\rightarrow 3CH_3(CH_2)_{\overline{n-1}}CH{=}CH_2 + (CH_3CH_2CH_2CH_2)_3Al$$
$$n = 4,6,8 \text{ etc.}$$

The triethylaluminum and 1-butene are recovered by the reaction between tributylaluminum and ethylene:

$$(CH_3CH_2CH_2CH_2)_3Al + 3CH_2{=}CH_2 \rightarrow (CH_3CH_2)_3Al$$
$$+ 3CH_3CH_2CH{=}CH_2$$

Alpha olefins are important compounds for producing biodegradable detergents. They are sulfonated and neutralized to alpha olefin sulfonates (AOS):

$$RCH{=}CH_2 + SO_3 \rightarrow RCH{=}CHSO_3H$$
$$RCH{=}CHSO_3H + NaOH \rightarrow RCH{=}CHSO_3Na + H_2O$$

Alkylation of benzene using alpha olefins produces linear alkylbenzenes, which are further sulfonated and neutralized to linear alkylbenzene sulfonates (LABS). These compounds constitute, with alcohol ethoxysulfates and ethoxylates, the basic active ingredients for household detergents. Production of LABS is discussed in Chapter 10.

Alpha olefins could also be carbonylated in presence of an alcohol using a cobalt catalyst to produce esters:

$$RCH=CH_2 + CO + R'OH \rightarrow RCH_2CH_2COOR'$$

Transesterification with penterithritol produces penterithritol esters and releases the alcohol.[17]

LINEAR ALCOHOLS

Linear alcohols (C_{12}-C_{26}) are important chemicals for producing various compounds such as plasticizers, detergents, and solvents. The production of linear alcohols by the hydroformylation (Oxo reaction) of alpha olefins followed by hydrogenation is discussed in Chapter 5. They are also produced by the oligomerization of ethylene using aluminum alkyls (Ziegler catalysts).

The Alfol process (Figure 7-7) for producing linear primary alcohols is a four-step process.[18] In the first step, triethylaluminum is produced by the reaction of ethylene with hydrogen and aluminum metal:

$$3\ CH_2=CH_2 + 1\tfrac{1}{2}\ H_2 + Al \rightarrow (CH_3CH_2)_3Al$$

In the next step, ethylene is polymerized by the action of triethylaluminum at approximately 120°C and 130 atmospheres to trialkylaluminum. Typical reaction time is approximately 140 minutes for an average C_{12} alcohol production:

$$n\ CH_2=CH_2 + (CH_3CH_2)_3Al \rightarrow \begin{array}{c} CH_3(CH_2)_x\text{-}CH_2 \\ \diagdown \\ CH_3(CH_2)_yCH_2\text{-}Al \\ \diagup \\ CH_3(CH_2)_zCH_2 \end{array}$$

Poisson distribution, x,y,z, 2,4,6,8.....26

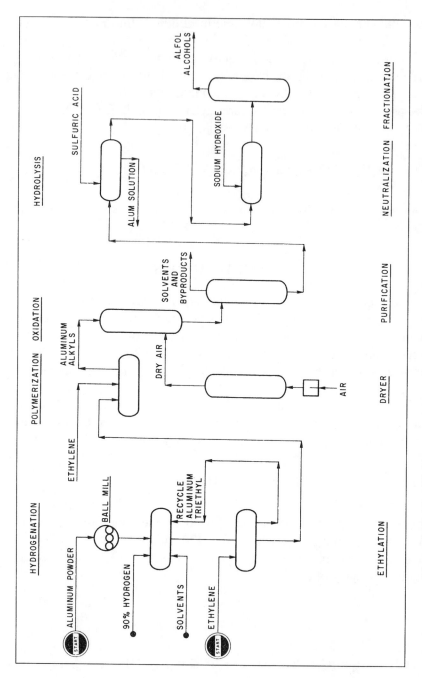

Figure 7-7. The Alfol process for making even-numbered straight-chain alpha alcohols.[18]

The oxidation of triethylaluminum is carried out between 20–50°C with "bone dry" air to aluminum trialkoxides.

$$CH_3(CH_2)_x\text{-}CH_2$$
$$CH_3(CH_2)_x\text{-}CH_2O$$

$$CH_3(CH_2)_y\text{-}CH_2\text{-}Al + 1\tfrac{1}{2}O_2 \longrightarrow CH_3(CH_2)_y\text{-}CH_2O\text{-}Al$$

$$CH_3(CH_2)_z\text{-}CH_2$$
$$CH_3(CH_2)_z\text{-}CH_2O$$

The final step is the hydrolysis of the trialkoxides with water to the corresponding even-numbered primary alcohols. Alumina is coproduced and is characterized by its high activity and purity:[19]

$$CH_3(CH_2)_x\text{-}CH_2O$$
$$CH_3(CH_2)_xCH_2OH$$

$$CH_3(CH_2)_y\text{-}CH_2O\text{-}Al + 3H_2O \longrightarrow CH_3(CH_2)_y\text{-}CH_2OH + Al(OH)_3$$

$$CH_3(CH_2)_z\text{-}CH_2O$$
$$CH_3(CH_2)_z\text{-}CH_2OH$$

Linear alcohols in the range of C_{10}–C_{12} are used to make plasticizers. Those in the range of C_{12}–C_{16} are used for making biodegradable detergents. They are either sulfated to linear alkylsulfates (ionic detergents) or reacted with ethylene oxide to the ethoxylated linear alcohols (nonionic detergents). The C_{16}–C_{18} alcohols are modifiers for wash and wear polymers. The higher alcohols, C_{20}–C_{26}, are synthetic lubricants and mold release agents.

BUTENE-1

A new process developed by Institut Francais du Petrole produces butene-1 (1-butene) by dimerizing ethylene.[20] A homogeneous catalyst system based on a titanium complex is used. The reaction is a concerted coupling of two molecules on a titanium atom, affording a titanium (IV) cyclic compound, which then decomposes to butene-1 by an intramolecular β-hydrogen transfer reaction.[21]

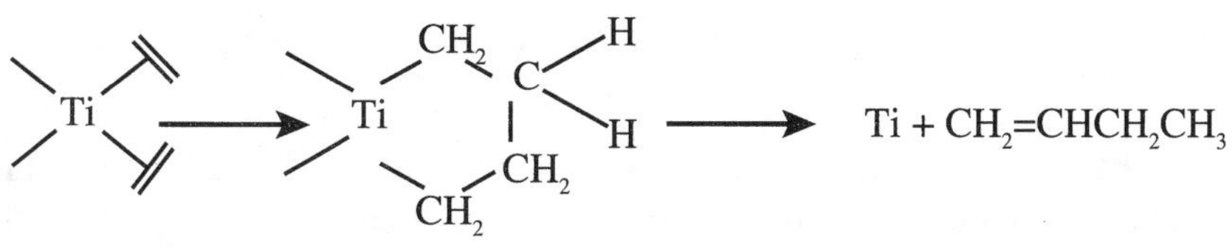

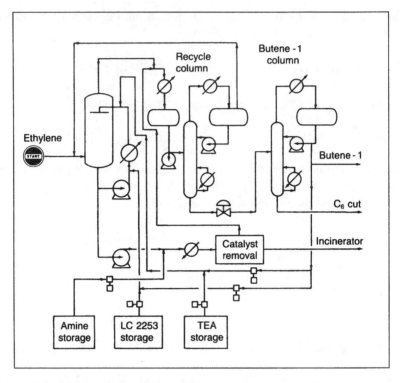

Figure 7-8. A flow diagram of the Institute Francais du Petrole process for producing 1-butene from ethylene.[21]

The Alphabutol process (Figure 7–8) operates at low temperatures (50–55°C) and relatively low pressures (22–27 atm). The reaction occurs in the liquid phase without a solvent. The process scheme includes four sections: the reactor, the co-catalyst injection, catalyst removal, and distillation. The continuous co-catalyst injection of an organo-basic compound deactivates the catalyst downstream of the reactor withdrawal valve to limit isomerization of l-butene to 2-butene. Table 7-2 shows the feed and product quality from the dimerization process.[21]

ALKYLATION USING ETHYLENE

Ethylene is an active alkylating agent. It can be used to alkylate aromatic compounds using Friedel-Crafts type catalysts. Commercially,

Table 7-2
Feed and product quality from dimerization
of ethylene to 1-butene[21]

Feed, polymer grade ethylene:	
Ethylene, vol%	99.90 min
Ethane + methane, vol%	0.10 max
Impurities, max.	
Methane, ppmv	250
C_3 and heavier, ppmv	10
Acetylene, H_2, H_2O, methanol, ppmv	5 each
CO, CO_2, O_2, ppmv	1 each
Sulfur, chlorine, ppmw	1 each
Product, polymerization grade butene-1:	
Butene-1, wt %	99.50 min
Other C_4s, wt %	0.30 max
Ethane, wt %	0.15 max
Ethylene, wt %	0.05 max
Impurities, max.	
C_6 olefins, ppmw	50
Ethers (as DME), ppmw	2
Sulfur, chlorine, ppmw	1
Dienes, acetylenics, ppmw	5 each
CO, CO_2, O_2, H_2O, methanol, ppmw	5 each
By-product, C_6 cut:	
3-Methyl 1-pentene, wt %	23.0
1-Hexene, wt %	5.8
2-Ethyl 1-butene, wt %	57.7
Hexadienes, wt %	1.3
Other C_6s, wt %	2 5
C_8^+, wt %	9.7
Properties	
Specific gravity, g/cm^3	0.68
Octane number, RON	95
MON	82
Distillation end point, °C	less than 200

ethylene is used to alkylate benzene for the production of ethyl benzene, a precursor for styrene. The subject is noted in Chapter 10.

REFERENCES

1. *Hydrocarbon Processing,* Vol. 78, No. 3, 1999, p. 29.
2. *Chemical and Engineering News,* July 5, 1999, p. 20.

3. Matar, S., Mirbach, M. and Tayim, H., *Catalysis in Petrochemical Processes,* Kluwer Academic Publishers, Dordrecht, 1989, p. 85.
4. DeMaglie, B. *Hydrocarbon Processing,* Vol. 55, No. 3, 1976, pp. 78–80.
5. "Petrochemical Handbook,"*Hydrocarbon Processing,* Vol. 70, No. 3, 1991, p. 156.
6. "Olefins Industrial Outlook II," *Chemical Industries Newsletter,* SRI International, Menlo Park, California, July–August 1989, p. 5.
7. Hajjin, J., "Catalytic Cosynthesis Method Developed," *Chemical and Engineering News,* Vol. 70, No. 18, May 4, 1992, pp. 24–25.
8. Brownstein, A. M., *Trends in Petrochemical Technology,* Tulsa, Petroleum Publishing Co., 1976, pp. 153–154.
9. Rosen, M. J. "Surfactants: Designing Structure for Performance," *CHEMTECH,* May, 1985, pp. 292–298.
10. *Petroleum Refiner,* Nov. 1957, pp. 36, 231.
11. Piccolinie R. and Plotkin, J. "Patent Watch" *CHEMTECH,* April 1999, p. 19
12. Stern, E. W., *Catal. Rev.,* Vol. 73, No. 1, 1967.
13. Hatch, L. F. and Matar, S., "Chemicals from Ethylene," *Hydrocarbon Processing,* Vol. 57, No. 4, 1978, pp. 155–166.
14. "Petrochemical Handbook," *Hydrocarbon Processing,* Vol. 70, No. 3, 1991 p. 192.
15. "Petrochemical Handbook," *Hydrocarbon Processing,* Vol. 70, No. 3, 1991 p. 150.
16. *Chemical and Engineering News,* July 5, 1999, p. 38.
17. Herron, S., *Chemical and Engineering News,* July 18, 1994, p. 156.
18. "Petrochemical Handbook," *Hydrocarbon Processing,* Vol. 54, No. 11, 1975, p. 110.
19. *Oil And Gas Journal,* May 26, 1975, pp. 103–108.
20. Commereuc, D. et al., "Dimerize Ethylene to Butene-1," *Hydrocarbon Processing,* Vol. 63, No. 11, 1984, p. 118.
21. Hennico, A. et al., "Butene-1 Is Made from Ethylene," *Hydrocarbon Processing,* Vol. 69, No. 3, 1990, pp. 73–75.

CHAPTER EIGHT

Chemicals Based on Propylene

INTRODUCTION

Propylene, "the crown prince of petrochemicals," is second to ethylene as the largest-volume hydrocarbon intermediate for the production of chemicals.

As an olefin, propylene is a reactive compound that can react with many common reagents used with ethylene such as water, chlorine, and oxygen. However, structural differences between these two olefins result in different reactivities toward these reagents. For example, direct oxidation of propylene using oxygen does not produce propylene oxide as in the case of ethylene. Instead, an unsaturated aldehyde, acrolein, is obtained. This could be attributed to the ease of oxidation of allylic hydrogens in propylene. Similar to the oxidation reaction, the direct catalyzed chlorination of propylene produces allyl chloride through substitution of allylic hydrogens by chlorine. Substitution of vinyl hydrogens in ethylene by chlorine, however, does not occur under normal conditions.

The current chemical demand for propylene is a little over one half that for ethylene. This is somewhat surprising because the added complexity of the propylene molecule (due to presence of a methyl group) should permit a wider spectrum of end products and markets. However, such a difference can lead to the production of undesirable by-products, and it frequently does. This may explain the relatively limited use of propylene in comparison to ethylene. Nevertheless, many important chemicals are produced from propylene.

The 1997 U.S. propylene demand ws 31 billion pounds and most of it was used to produce polypropylene polymers and copolymers (about 46%). Other large volume uses are acrylonitrile for synthetic fibers (Ca 13%), propylene oxide (Ca 10%), cumene (Ca 8%) and oxo alcohols (Ca 7%).[1]

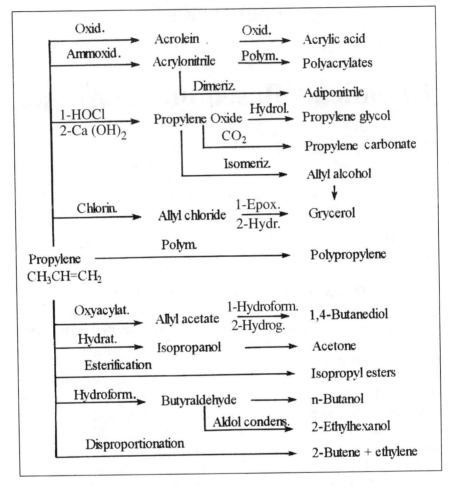

Figure 8-1. Important chemicals based on propylene.

Figure 8-1 shows the important chemicals based on propylene. The following discusses the chemistry of the production of these chemicals.

OXIDATION OF PROPYLENE

The direct oxidation of propylene using air or oxygen produces acrolein. Acrolein may further be oxidized to acrylic acid, which is a monomer for polyacrylic resins.

Ammoxidation of propylene is considered under oxidation reactions because it is thought that a common allylic intermediate is formed in both the oxidation and ammoxidation of propylene to acrolein and to acrylonitrile, respectively.

The use of peroxides for the oxidation of propylene produces propylene oxide. This compound is also obtained via a chlorohydrination of propylene followed by epoxidation.

ACROLEIN (CH_2=CHCHO)

Acrolein (2-propenal) is an unsaturated aldehyde with a disagreeable odor. When pure, it is a colorless liquid, that is highly reactive and polymerizes easily if not inhibited.

The main route to produce acrolein is through the catalyzed air or oxygen oxidation of propylene.

$$CH_3CH=CH_2 + O_2 \rightarrow CH_2=CHCHO + H_2O \quad \Delta H= -340.5 \text{ KJ/mol}$$

Transition metal oxides or their combinations with metal oxides from the lower row 5a elements were found to be effective catalysts for the oxidation of propene to acrolein.[2] Examples of commercially used catalysts are supported CuO (used in the Shell process) and Bi_2O_3/MoO_3 (used in the Sohio process). In both processes, the reaction is carried out at temperature and pressure ranges of 300–360°C and 1–2 atmospheres. In the Sohio process, a mixture of propylene, air, and steam is introduced to the reactor. The hot effluent is quenched to cool the product mixture and to remove the gases. Acrylic acid, a by-product from the oxidation reaction, is separated in a stripping tower where the acrolein-acetaldehyde mixture enters as an overhead stream. Acrolein is then separated from acetaldehyde in a solvent extraction tower. Finally, acrolein is distilled and the solvent recycled.

MECHANISM OF PROPENE OXIDATION

Much work has been invested to reveal the mechanism by which propylene is catalytically oxidized to acrolein over the heterogeneous catalyst surface. Isotope labeling experiments by Sachtler and DeBoer revealed the presence of an allylic intermediate in the oxidation of propylene to acrolein over bismuth molybdate.[3] In these experiments, propylene was tagged once at C_1, another time at C_2 and the third time at C_3.

The formed acrolein was photochemically degraded to ethylene and carbon monoxide. It has been found that radioactivity was exclusively associated with ethylene when propylene tagged with ^{14}C at C_2 was used. Also, carbon monoxide was found to be free from radioactivity:

$$CH_2=^{14}CHCH_3 + O_2 \longrightarrow CH_2=^{14}CHCHO + H_2O$$

$$CH_2=^{14}CHCHO \xrightarrow{hv} CH_2=^{14}CH_2 + CO$$

When propylene tagged with ^{14}C at either C_1 or C_3 was oxidized to acrolein and then degraded, both $CH_2=CH_2$ and CO were radioactive, and the ratio of radioactivity was 1.

A proposed mechanism for the oxidation of propylene to acrolein is by a first step abstraction of an allylic hydrogen from an adsorbed propylene by an oxygen anion from the catalytic lattice to form an allylic intermediate:

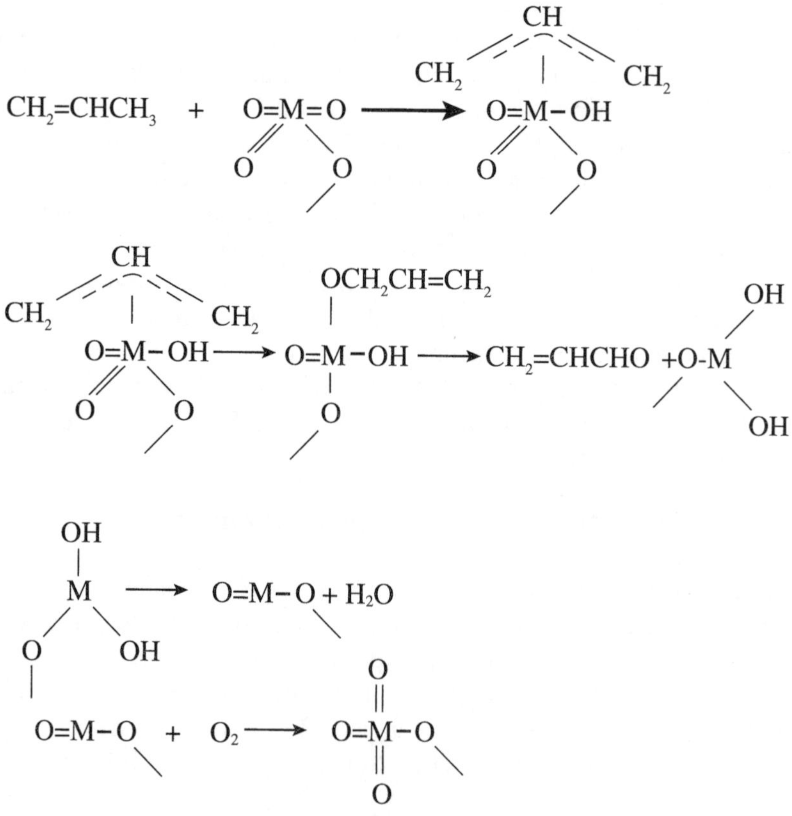

The next step is the insertion of a lattice oxygen into the allylic species. This creates oxide-deficient sites on the catalyst surface accompanied by a reduction of the metal. The reduced catalyst is then reoxidized by adsorbing molecular oxygen, which migrates to fill the oxide-deficient sites. Thus, the catalyst serves as a redox system.[4]

Uses of Acrolein

The main use of acrolein is to produce acrylic acid and its esters. Acrolein is also an intermediate in the synthesis of pharmaceuticals and herbicides. It may also be used to produce glycerol by reaction with iso-propanol (discussed later in this chapter). 2-Hexanedial, which could be a precursor for adipic acid and hexamethylene-diamine, may be prepared from acrolein Tail to tail dimenization of acrolein using ruthenium catalyst produces trans-2-hexanedial. The trimer, trans-6-hydroxy-5-formyl-2,7-octadienal is coproduced.[5] Acrolein, may also be a precursor for 1,3-propanediol. Hydrolysis of acrolein produces 3-hydroxypropionaldehyde which could be hydrogenated to 1,3-propanediol.[6]

$$CH_2=CH\text{-}CHO + H_2O \rightarrow HO\text{-}CH_2\text{-}CH_2\text{-}CHO \xrightarrow{H_2} HOCH_2\text{-}CH_2OH$$

The diol could also be produced from ethylene oxide (Chaper 7).

$$\overset{\overset{\displaystyle O}{\displaystyle \|}}{\textbf{ACRYLIC ACID (CH}_2\textbf{=CHCOH)}}$$

There are several ways to produce acrylic acid. Currently, the main process is the direct oxidation of acrolein over a combination molybdenum-vanadium oxide catalyst system. In many acrolein processes, acrylic acid is made the main product by adding a second reactor that oxidizes acrolein to the acid. The reactor temperature is approximately 250°C:

$$CH_2=CH\overset{\overset{\displaystyle O}{\displaystyle \|}}{C}H + \tfrac{1}{2}\,O_2 \longrightarrow CH_2=CH\overset{\overset{\displaystyle O}{\displaystyle \|}}{C}OH$$

Acrylic acid is usually esterified to acrylic esters by adding an esterification reactor. The reaction occurs in the liquid phase over an ion exchange resin catalyst.

An alternative route to acrylic esters is via a β-propiolactone intermediate. The lactone is obtained by the reaction of formaldehyde and ketene, a dehydration product of acetic acid:

$$CH_2=C=O + H-\overset{\overset{\displaystyle O}{\|}}{C}-H \longrightarrow \begin{matrix} CH_2-C=O \\ | \qquad | \\ CH_2-O \end{matrix}$$

Ketene   β− Propiolactone

The acid-catalyzed ring opening of the four-membered ring lactone in the presence of an alcohol produces acrylic esters:

$$\begin{matrix} CH_2-C=O \\ | \qquad | \\ CH_2-O \end{matrix} + ROH \xrightarrow{H^+} CH_2=CH-\overset{\overset{\displaystyle O}{\|}}{C}OR + H_2O$$

The production of acrylic acid from the oxidative carbonylation of ethylene is described in Chapter 7.

Acrylic acid and its esters are used to produce acrylic resins. Depending on the polymerization method, the resins could be used in the adhesive, paint, or plastic industry.

AMMOXIDATION OF PROPYLENE
(Acrylonitrile [$CH_2=CHCN$])

Ammoxidation refers to a reaction in which a methyl group with allyl hydrogens is converted to a nitrile group using ammonia and oxygen in the presence of a mixed oxides-based catalyst. A successful application of this reaction produces acrylonitrile from propylene:

$$CH_2=CHCH_3 + NH_3 + 1^1/_2O_2 \rightarrow CH_2=CHCN + 3H_2O$$
$$\Delta H = -518 \text{ KJ/mol}$$

As with other oxidation reactions, ammoxidation of propylene is highly exothermic, so an efficient heat removal system is essential.

Acetonitrile and hydrogen cyanide are by-products that may be recovered for sale. Acetonitrile (CH_3CN) is a high polarity aprotic solvent used in DNA synthesizers, high performance liquid chromatography (HPLC), and electrochemistry. It is an important solvent for extracting butadiene from C_4 streams.[7] Table 8-1 shows the specifications of acrylonitrile, HCN, and acetonitrile.[8]

Both fixed and fluid-bed reactors are used to produce acrylonitrile, but most modern processes use fluid-bed systems. The Montedison-UOP process (Figure 8-2) uses a highly active catalyst that gives 95.6% propylene conversion and a selectivity above 80% for acrylonitrile.[8,9] The catalysts used in ammoxidation are similar to those used in propylene oxidation to acrolein. Oxidation of propylene occurs readily at

Table 8-1
Typical analysis of acrylonitrile, HCN and acetonitrile[8]

Acrylonitrile	
Purity (dry basis), wt %	99.9
Hydrogen cyanide, wt-ppm	5
Acetonitrile, wt-ppm	100
Acetaldehyde, wt-ppm	20
Acrolein, wt-ppm	10
Acetone, wt-ppm	40
Peroxides (as H_2O_2), wt-ppm	0.2
Water, wt %	0.2–0.5
Hydrogen Cyanide (HCN)	
Hydrogen cyanide, wt %	99.7
Acrylonitrile, wt %	0.1
Acetonitrile (if recovered as purified product)	
Acetonitrile, wt %	99.0+
Water, wt %	0.1
Acrylonitrile, wt-ppm	500
Acetone, wt-ppm	Absent
HCN, wt-ppm	Absent

322°C over Bi-Mo catalysts. However, in the presence of ammonia, the conversion of propylene to acrylonitrile does not occur until about 402°C. This may be due to the adsorption of ammonia on catalytic sites that block propylene chemisportion. As with propylene oxidation, the first step in the ammoxidation reaction is the abstraction of an alpha hydrogen from propylene and formation of an allylic intermediate. Although the subsequent steps are not well established, it is believed that adsorbed ammonia dissociates on the catalyst surface by reacting with the lattice oxygen, producing water. The adsorbed NH species then reacts with a neighboring allylic intermediate to yield acrylonitrile.

Uses of Acrylonitrile

Acrylonitrile is mainly used to produce acrylic fibers, resins, and elastomers. Copolymers of acrylonitrile with butadiene and styrene are the ABS resins and those with styrene are the styrene-acrylonitrile resins SAN that are important plastics. The 1998 U.S. production of acrylonitrile was approximately 3.1 billion pounds.[10] Most of the production was used for ABS resins and acrylic and modacrylic fibers. Acrylonitrile is also a precursor for acrylic acid (by hydrolysis) and for adiponitrile (by an electrodimerization).

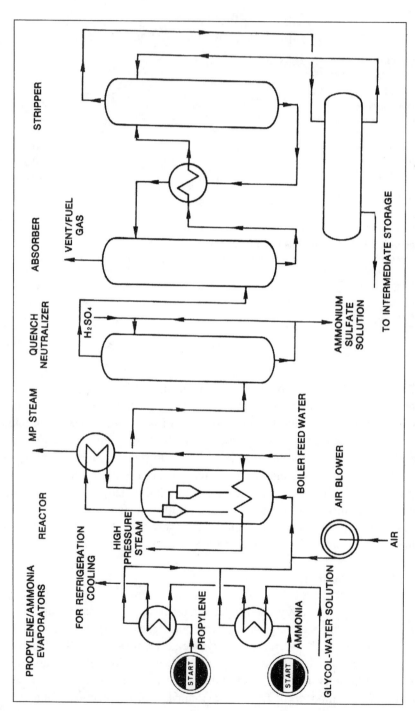

Figure 8-2. A flow diagram of the Montedison-UOP acrylonitrile process.[8]

Adiponitrile (NC(CH₂)₄CN)

Adiponitrile is an important intermediate for producing nylon 66. There are other routes for its production, which are discussed in Chapter 9. The way to produce adiponitrile via propylene is the electrodimerization of acrylonitrile.[11] The following is a representation of the electrochemistry involved:

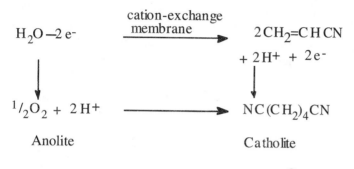

$$H_2O - 2e^-$$ $$\xrightarrow{\text{cation-exchange membrane}}$$ $$2CH_2=CHCN$$
$$+ 2H^+ + 2e^-$$

$$\tfrac{1}{2}O_2 + 2H^+ \longrightarrow NC(CH_2)_4CN$$

Anolite   Catholite

PROPYLENE OXIDE (CH₃CH—CH₂ with epoxide O)

Propylene oxide is similar in its structure to ethylene oxide, but due to the presence of an additional methyl group, it has different physical and chemical properties. It is a liquid that boils at 33.9°C, and it is only slightly soluble in water. (Ethylene oxide, a gas, is very soluble in water).

The main method to obtain propylene oxide is chlorohydrination followed by epoxidation. This older method still holds a dominant role in propylene oxide production. Chlorohydrination is the reaction between an olefin and hypochlorous acid. When propylene is the reactant, propylene chlorohydrin is produced. The reaction occurs at approximately 35°C and normal pressure without any catalyst:

$$CH_3CH=CH_2 + HOCl \rightarrow CH_3CHOHCH_2Cl$$
Propylene chlorohydrin

Approximately 87–90% yield could be achieved. The main by-product is propylene dichloride (6–9%). The next step is the dehydrochlorination of the chlorohydrin with a 5% Ca(OH)₂ solution:

$$2CH_3CHOHCH_2Cl + Ca(OH)_2 \rightarrow 2CH_3CH\overset{O}{-}CH_2 + CaCl_2 + 2H_2O$$

Propylene oxide is purified by steam stripping and then distillation. Byproduct propylene dichloride may be purified for use as a solvent or as a feed to the perchloroethylene process. The main disadvantage of the chlorohydrination process is the waste disposal of $CaCl_2$. Figure 8-3 is a flow diagram of a typical chlorohydrin process.[12]

The second important process for propylene oxide is epoxidation with peroxides. Many hydroperoxides have been used as oxygen carriers for this reaction. Examples are t-butylhydroperoxide, ethylbenzene hydroperoxide, and peracetic acid. An important advantage of the process is that the coproducts from epoxidation have appreciable economic values.

Epoxidation of propylene with ethylbenzene hydroperoxide is carried out at approximately 130°C and 35 atmospheres in presence of molybdenum catalyst. A conversion of 98% on the hydroperoxide has been reported:[13]

$$
\underset{\underset{H}{|}}{\overset{\overset{CH_3}{|}}{C_6H_5COOH}} + CH_3CH{=}CH_2 \longrightarrow CH_3{-}\overset{O}{\overset{/\ \backslash}{CH}}{-}CH_2 + \overset{OH}{\overset{|}{C_6H_5CHCH_3}}
$$

The coproduct α-phenylethyl alcohol could be dehydrated to styrene.

Ethylbenzene hydroperoxide is produced by the uncatalyzed reaction of ethylbenzene with oxygen:

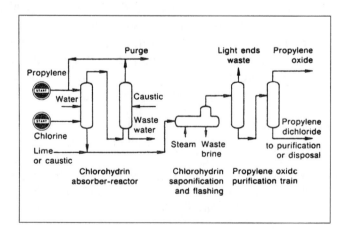

Figure 8-3. A flow diagram of a typical chlorohydrin process for producing propylene oxide.[12]

$$C_6H_5CH_2CH_3 + O_2 \rightarrow C_6H_5CH(CH_3)OOH$$

Table 8-2 shows those peroxides normally used for epoxidation of propylene and the coproducts with economic value.[12]

Epoxidation with hydrogen peroxide has also been tried. The epoxidation reaction is catalyzed with compounds of As, Mo, and B, which are claimed to produce propylene oxide in high yield:

$$CH_3CH{=}CH_2 + H_2O_2 \longrightarrow CH_3CH\overset{\displaystyle O}{\overbrace{\diagdown}}CH_2 + H_2O$$

Deriatives and Uses of Propylene Oxide

Similar to ethylene oxide, the hydration of propylene oxide produces propylene glycol. Propylene oxide also reacts with alcohols, producing polypropylene glycol ethers, which are used to produce polyurethane foams and detergents. Isomerization of propylene oxide produces allyl alcohol, a precursor for glycerol. The 1994 U.S. production of propylene oxide, the 35th highest-volume chemical, was approximately 3.7 billion pounds. Table 8-3 shows the 1992 U.S. propylene oxide capacity of the three firms producing it and the processes used.[14]

The following describes some of the important chemicals based on propylene oxide.

Propylene Glycol (CH$_3$CH(OH)CH$_2$OH)

Propylene glycol (1,2-propanediol) is produced by the hydration of propylene oxide in a manner similar to that used for ethylene oxide:

Table 8-2
Peroxides actually or potentially used to epoxidize propylene[12]

Peroxide feedstock	Epoxidation coproduct	Coproduct derivative
Acetaldehyde	Acetic acid	—
Isobutane	tert-Butyl alcohol	Isobutylene
Ethylbenzene	α-Phenylethyl alcohol	Styrene
Isopentane	Isopentanol	Isopentene and isoprene
Isopropanol	Acetone	Isopropanol

Table 8-3
1992 U.S. propylene oxide capacity[14]

	Location	Annual capacity (millions of lb)	Basic process
Arco Chemical	Bayport, Tex.	1213	Peroxidation (isobutane)
	Channelview, Tex.	1100*	Peroxidation (ethylbenzene)
Dow Chemical	Freeport, Tex.	1100	Chlorohydrin
	Plaquemine, La.	450	Chlorohydrin
Texaco Chemical	Port Neches, Tex.	400**	Peroxidation (isobutane)

*Of this capacity, 500 million lb is slated to come on stream with a new unit in third-quarter 1992.
**Slated to start up in first-quarter 1994.

$$CH_3CH\overset{O}{\overbrace{}}CH_2 + H_2O \longrightarrow CH_3CHOHCH_2OH$$

Propylene glycol

Depending on the propylene oxide/water ratio, di-, tri- and polypropylene glycols can be made the main products.

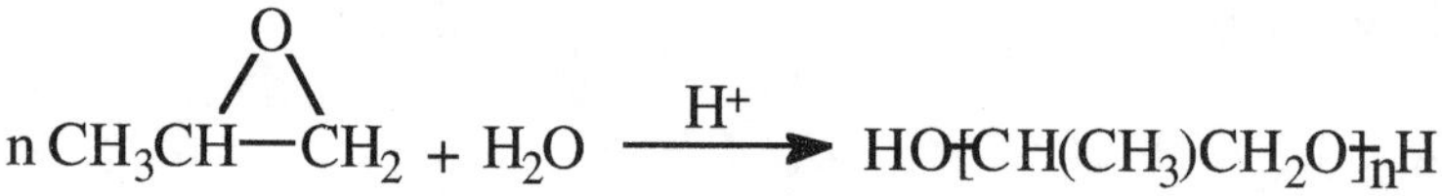

$$n\,CH_3CH\overset{O}{\overbrace{}}CH_2 + H_2O \xrightarrow{H^+} HO[CH(CH_3)CH_2O]_nH$$

Polypropylene glycol

Propylene Carbonate $(\overline{O(CH_3)CHCH_2OC}=O)$

The reaction between propylene oxide and carbon dioxide produces propylene carbonate. The reaction conditions are approximately 200°C and 80 atmospheres. A yield of 95% is anticipated:

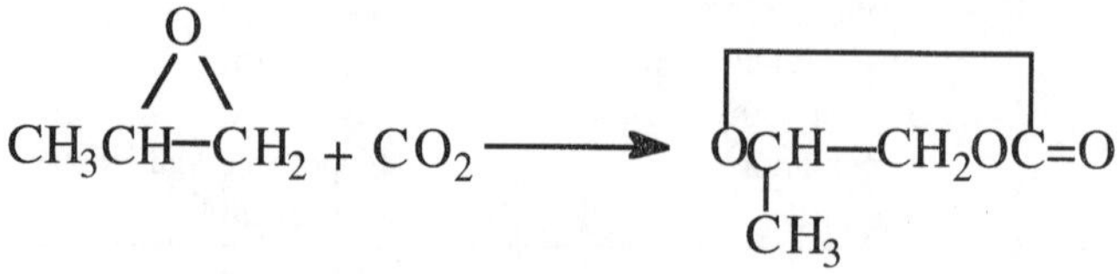

$$CH_3CH\overset{O}{\overbrace{}}CH_2 + CO_2 \longrightarrow \underset{\underset{CH_3}{|}}{\overline{OCH\!-\!CH_2OC}}=O$$

Propylene carbonate is a liquid used as a specialty solvent and a plasticizer.

Allyl Alcohol ($CH_2=CHCH_2OH$)

Allyl alcohol is produced by the catalytic isomerization of propylene oxide at approximately 280°C. The reaction is catalyzed with lithium phosphate. A selectivity around 98% could be obtained at a propylene oxide conversion around 25%:

$$CH_3CH\overset{\displaystyle O}{\overset{\displaystyle /\backslash}{-}}CH_2 \longrightarrow CH_2=CHCH_2OH$$

Allyl alcohol is used in the plasticizer industry, as a chemical intermediate, and in the production of glycerol.

Glycerol via Allyl Alcohol. Glycerol (1,2,3-propanetriol) is a trihydric alcohol of great utility due to the presence of three hydroxyl groups. It is a colorless, somewhat viscous liquid with a sweet odor. Glycerin is the name usually used by pharmacists for glycerol. There are different routes for obtaining glycerol. It is a by-product from the manufacture of soap from fats and oils (a non-petroleum source). Glycerol is also produced from allyl alcohol by epoxidation using hydrogen peroxide or peracids (similar to epoxidation of propylene). The reaction of allyl alcohol with H_2O_2 produces glycidol as an intermediate, which is further hydrolyzed to glycerol:

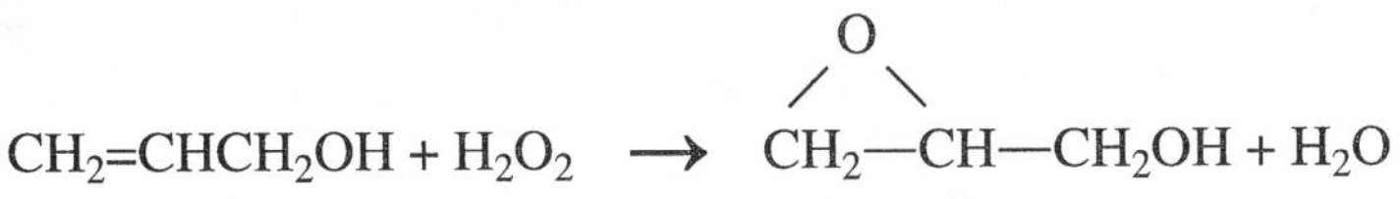

$$CH_2=CHCH_2OH + H_2O_2 \rightarrow CH_2\overset{\displaystyle O}{\overset{\displaystyle /\ \backslash}{-}}CH-CH_2OH + H_2O$$
$$\text{Glycidol}$$

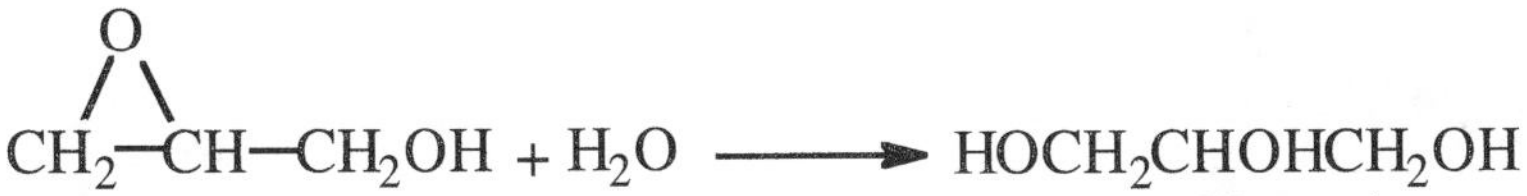

$$CH_2\overset{\displaystyle O}{\overset{\displaystyle /\backslash}{-}}CH-CH_2OH + H_2O \longrightarrow HOCH_2CHOHCH_2OH$$
$$\text{Glycerol}$$

Other routes for obtaining glycerol are also based on propylene. It can be produced from allyl chloride or from acrolein and isopropanol (see following sections).

OXYACYLATION OF PROPYLENE

ALLYL ACETATE ($CH_2=CHCH_2O\overset{\overset{\displaystyle O}{\displaystyle \|}}{C}CH_3$)

Like vinyl acetate from ethylene, allyl acetate is produced by the vapor-phase oxyacylation of propylene. The catalyzed reaction occurs at approximately 180°C and 4 atmospheres over a Pd/KOAc catalyst:

$$CH_3CH=CH_2 + CH_3COOH + \tfrac{1}{2}O_2 \longrightarrow CH_2=CHCH_2O\overset{\overset{\displaystyle O}{\displaystyle \|}}{C}CH_3 + H_2O$$

Allyl acetate is a precursor for 1,4-butanediol via a hydrocarbonylation route, which produces 4-acetoxybutanal. The reaction proceeds with a $Co(CO)_8$ catalyst in benzene solution at approximately 125°C and 3,000 pounds per square inch. The typical mole H_2/CO ratio is 2:1. The reaction is exothermic, and the reactor temperature may reach 180°C during the course of the reaction. Selectivity to 4-acetoxybutanal is approximately 65% at 100% allyl acetate conversion.[15]

CHLORINATION OF PROPYLENE
(Allyl Chloride [$CH_2=CHCH_2Cl$])

Allyl chloride is a colorless liquid, insoluble in water but soluble in many organic solvents. It has a strong pungent odor and an irritating effect on the skin. As a chemical, allyl chloride is used to make allyl alcohol, glycerol, and epichlorohydrin.

The production of allyl chloride could be effected by direct chlorination of propylene at high temperatures (approximately 500°C and one atmosphere). The reaction substitutes an allylic hydrogen with a chlorine atom. Hydrogen chloride is a by-product from this reaction:

$$CH_2=CHCH_3 + Cl_2 \rightarrow CH_2=CHCH_2Cl + HCl$$

The major by-products are cis- and trans- 1,3-dichloropropene, which are used as soil fumigants.

The most important use of allyl chloride is to produce glycerol via an epichlorohydrin intermediate. The epichlorohydrin is hydrolyzed to glycerol:

$$CH_2=CHCH_2Cl + Cl_2 + H_2O \longrightarrow ClCH_2CHOHCH_2Cl + HCl$$

$$2ClCH_2CHOHCH_2Cl + Ca(OH)_2 \longrightarrow 2CH_2\!\!-\!\!CHCH_2Cl + CaCl_2 + 2H_2O$$
$$\underset{O}{\diagdown\diagup}$$

Epichlorohydrin

$$CH_2\!\!-\!\!CHCH_2Cl + 2H_2O \longrightarrow HOCH_2CHOHCH_2OH + HCl$$
$$\underset{O}{\diagdown\diagup}$$

Glycerol

Glycerol, a trihydric alcohol, is used to produce polyurethane foams and alkyd resins. It is also used in the manufacture of plasticizers.

HYDRATION OF PROPYLENE
(Isopropanol [$CH_3CHOHCH_3$])

Isopropanol (2-propanol) is an important alcohol of great synthetic utility. It is the second-largest volume alcohol after methanol (1998 U.S. production was approximately 1.5 billion pounds) and it was the 49th ranked chemical. Isopropanol under the name "isopropyl alcohol" was the first industrial chemical synthesized from a petroleum-derived olefin (1920).

The production of isopropanol from propylene occurs by either a direct hydration reaction (the newer method) or by the older sulfation reaction followed by hydrolysis.

In the direct hydration method, the reaction could be effected either in a liquid or in a vapor-phase process. The slightly exothermic reaction evolves 51.5 KJ/mol.

$$CH_3CH=CH_2 + H_2O \rightarrow CH_3CHOHCH_3$$

In the liquid-phase process, high pressures in the range of 80–100 atmospheres are used. A sulfonated polystyrene cation exchange resin is the catalyst commonly used at about 150°C. An isopropanol yield of 93.5% can be realized at 75% propylene conversion. The only important by-product is diisopropyl ether (about 5%). Figure 8-4 is a flow diagram of the propylene hydration process.[16]

Gas phase hydration, on the other hand, is carried out at temperatures above 200°C and approximately 25 atmospheres. The ICI process employs WO_3 on a silica carrier as catalyst.

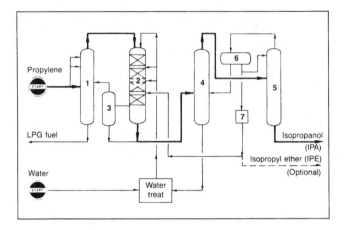

Figure 8-4. A flow diagram for the hydration of propylene to isopropanol:[16] (1) propylene recovery column, (2) reactor, (3) residual gas separation column, (4) aqueous - isopropanol azeotropic distillation column, (5) drying column, (6) isopropyl ether separator, (7) isopropyl ether extraction.

Older processes still use the sulfation route. The process is similar to that used for ethylene in the presence of H_2SO_4, but the selectivity is a little lower than the modern vapor-phase processes. The reaction conditions are milder than those used for ethylene. This manifests the greater ease with which an isopropyl carbocation (a secondary carbonium ion) is formed than a primary ethyl carbonium ion:

$$CH_3CH=CH_2 + H^+ \rightarrow [CH_3\overset{+}{C}HCH_3]$$

$$CH_2=CH_2 + H^+ \rightarrow [CH_3\overset{+}{C}H_2]$$

Table 8-4 compares sulfuric acid concentrations and the temperatures used for the sulfation of different light olefins.[17]

PROPERTIES AND USES OF ISOPROPANOL

Isopropanol is a colorless liquid having a pleasant odor; it is soluble in water. It is more soluble in hydrocarbon liquids than methanol or ethanol. For this reason, small amounts of isopropanol may be mixed with methanol-gasoline blends used as motor fuels to reduce phase-separation problems.[18]

Table 8-4
Acid concentration and temperatures used
for the sulfation of various olefins[17]

Olefins	Formula	Acid conc. range, %	Temperature range °C
Ethylene	$CH_2{=}CH_2$	90–98	60–80
Propylene	$CH_3{-}CH{=}CH_2$	75–85	25–40
Butylenes	$CH_3{-}CH_2{-}CH{=}CH_2$	75–85	15–30
	$CH_3{-}CH{=}CH{-}CH_3$	75–85	15–30
Isobutylene	$CH_3{-}\overset{\displaystyle CH_3}{\underset{\displaystyle \vert}{C}}{=}CH_2$	50–65	0–25

About 50% of isopropanol use is to produce acetone. Other important synthetic uses are to produce esters of many acids, such as acetic (isopropyl acetate, solvent for cellulose nitrate), myristic, and oleic acids (used in lipsticks and lubricants). Isopropylpalmitate is used as an emulsifier for cosmetic materials. Isopropyl alcohol is a solvent for alkaloids, essential oils, and cellulose derivatives.

Acetone Production

Acetone (2-propanone), is produced from isopropanol by a dehydrogenation, oxidation, or a combined oxidation dehydrogenation route.

The dehydrogenation reaction is carried out using either copper or zinc oxide catalyst at approximately 450–550°C. A 95% yield is obtained:

$$CH_3CHOHCH_3 \longrightarrow CH_3\overset{O}{\overset{\|}{C}}CH_3 + H_2$$

The direct oxidation of propylene with oxygen is a noncatalytic reaction occurring at approximately 90–140°C and 15–20 atmospheres. In this reaction hydrogen peroxide is coproduced with acetone. At 15% isopropanol conversion, the approximate yield of acetone is 93% and that for H_2O_2 is 87%:

$$2CH_3CHOHCH_3 + O_2 \longrightarrow 2CH_3\overset{O}{\overset{\|}{C}}CH_3 + H_2O_2$$

The oxidation process uses air as the oxidant over a silver or copper catalyst. The conditions are similar to those used for the dehydrogenation reaction.

Acetone can also be coproduced with allyl alcohol in the reaction of acrolein with isopropanol. The reaction is catalyzed with an MgO and ZnO catalyst combination at approximately 400°C and one atmosphere. It appears that the hydrogen produced from the dehydrogenation of isopropanol and adsorbed on the catalyst surface selectively hydrogenates the carbonyl group of acrolein:

$$CH_3CHOHCH_3 + CH_2=CHCH \longrightarrow CH_3\overset{\overset{\displaystyle O}{\|}}{C}CH_3 + CH_2=CHCH_2OH$$

A direct route for acetone from propylene was developed using a homogeneous catalyst similar to Wacker system ($PdCl_2/CuCl_2$). The reaction conditions are similar to those used for ethylene oxidation to acetaldehyde.[19]

Today, most acetone is obtained via a cumene hydroperoxide process where it is coproduced with phenol. This reaction is noted in Chapter 10.

Propertles and Uses of Acetone

Acetone is a volatile liquid with a distinct sweet odor. It is miscible with water, alcohols, and many hydrocarbons. For this reason, it is a highly desirable solvent for paints, lacquers, and cellulose acetate. Acetone was the 41st highest volume chemical. The 1994 U.S. production was approximately 2.8 billion pounds.

As a symmetrical ketone, acetone is a reactive compound with many synthetic uses. Among the important chemicals based on acetone are methylisobutyl ketone, methyl methacrylate, ketene, and diacetone alcohol.

Mesityl Oxide. This is an alpha-beta unsaturated ketone of high reactivity. It is used primarily as a solvent. It is also used for producing methylisobutyl ketone.

Mesityl oxide is produced by the dehydration of acetone. Hydrogenation of mesityl oxide produces methylisobutyl ketone, a solvent for paints and varnishes:

$$2CH_3\overset{\overset{\displaystyle O}{\|}}{C}CH_3 \longrightarrow (CH_3)_2C=CH\overset{\overset{\displaystyle O}{\|}}{C}CH_3 + H_2O$$

Mesityl Oxide

$$(CH_3)_2C=CHCCH_3 + H_2 \longrightarrow (CH_3)_2CHCH_2CCH_3$$

Methyl Methacrylate ($CH_2=C(CH_3)-COOCH_3$). This is produced by the hydrocyanation of acetone using HCN. The resulting cyanohydrin is then reacted with sulfuric acid and methanol, producing methyl methacrylate:

$$CH_3CCH_3 + HCN \longrightarrow \underset{\underset{OH}{|}}{\overset{\overset{CH_3}{|}}{CH_3C}}-CN$$

$$\underset{\underset{OH}{|}}{\overset{\overset{CH_3}{|}}{CH_3C}}-CN + H_2SO_4 \longrightarrow CH_2 = \overset{\overset{CH_3}{|}}{C}-\overset{O}{\overset{||}{C}}-\overset{+}{N}H_3H\overset{-}{S}O_4$$

$$CH_2=\overset{\overset{CH_3}{|}}{C}-\overset{O}{\overset{||}{C}}-\overset{+}{N}H_3H\overset{-}{S}O_4 + CH_3OH \rightarrow CH_2=C(CH_3)-COOCH_3 + NH_4HSO_4$$

One disadvantage of this process is the waste NH_4HSO_4 stream. Methacrylic acid (MAA) is also produced by the air oxidation of isobutylene or the ammoxidation of isobutylene to methacrylonitrile followed by hydrolysis. These reactions are noted in Chapter 9.

Methacrylic acid and its esters are useful vinyl monomers for producing polymethacrylate resins, which are thermosetting polymers. The extruded polymers are characterized by the transparency required for producing glass-like plastics commercially known as Plexiglas:

Bis-Phenol A $\left[HO-\underset{}{\bigcirc}-\overset{\overset{CH_3}{|}}{\underset{\underset{CH_3}{|}}{C}}-\underset{}{\bigcirc}-OH \right]$

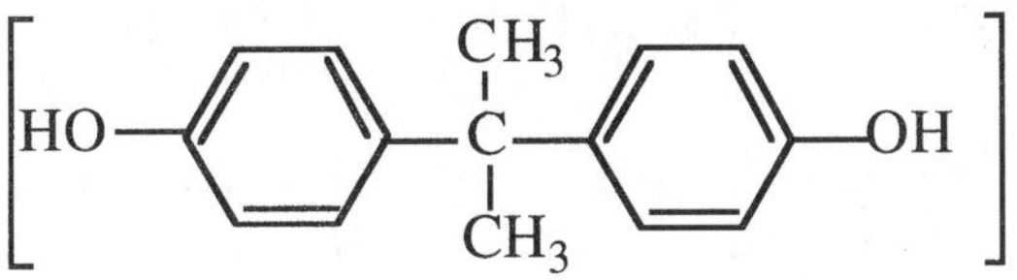

Bisphenol A is a solid material in the form of white flakes, insoluble in water but soluble in alcohols. As a phenolic compound, it reacts with strong alkaline solutions. Bisphenol A is an important monomer for producing epoxy resins, polycarbonates, and polysulfones. It is produced by the condensation reaction of acetone and phenol in the presence of HCI. (See Chapter 10, p. 273)

ADDITION OF ORGANIC ACIDS TO PROPENE

$$\overset{\overset{\displaystyle O}{\|}}{}$$
ISOPROPYL ACETATE ($CH_3COCH(CH_3)_2$)

Isopropyl acetate is produced by the catalytic vapor-phase addition of acetic acid to propylene. A high yield of the ester can be realized (about 99%):

$$CH_3CH{=}CH_2 + CH_3COOH \longrightarrow CH_3\overset{\overset{\displaystyle O}{\|}}{C}OCH(CH_3)_2$$

Isopropyl acetate is used as a solvent for coatings and printing inks. It is generally interchangeable with methylethyl ketone and ethyl acetate.

$$\overset{\overset{\displaystyle O}{\|}}{}$$
ISOPROPYL ACRYLATE ($CH_2 = CHCOCH(CH_3)_2$)

Isopropyl acrylate is produced by an acid catalyzed addition reaction of acrylic acid to propylene. The reaction occurs in the liquid phase at about 100°C:

$$CH_3CH{=}CH_2 + CH_2{=}CH\overset{\overset{\displaystyle O}{\|}}{C}OH \rightarrow CH_2{=}CH\overset{\overset{\displaystyle O}{\|}}{C}OCH(CH_3)_2$$

Due to unsaturation of the ester, it can be polymerized and used as a plasticizer.

HYDROFORMYLATION OF PROPYLENE: THE OXO REACTION (Butyraldehydes)

The catalytic hydroformylation of olefins is discussed in Chapter 5. The reaction of propylene with CO and H_2 produces n-butyraldehyde as the main product. Isobutyraldehyde is a by-product:[20]

$$2\,CH_3CH{=}CH_2 + 2CO + 2H_2 \rightarrow CH_3CH_2CH_2\overset{\overset{\displaystyle O}{\|}}{C}H + CH_3\overset{\overset{\displaystyle CH_3}{|}}{C}H{-}\overset{\overset{\displaystyle O}{\|}}{C}H$$
$$\text{n-Butyraldehyde} \quad \text{Isobutyraldehyde}$$

Figure 8-5 shows the homogeneous Hoechst and Rhone Poulenc process using rhodium catalyst.[21]

Butyraldehydes are usually hydrogenated to the corresponding alcohols. They are also intermediates for other chemicals. The following reviews some of the important chemicals based on butyraldehydes.

n-BUTANOL(CH₃CH₂CH₂CH₂OH)

n-Butanol is produced by the catalytic hydrogenation of n-butyraldehyde. The reaction is carried out at relatively high pressures. The yield is high:

$$CH_3CH_2CH_2CHO + H_2 \rightarrow CH_3CH_2CH_2CH_2OH$$

n-Butanol is primarily used as a solvent or as an esterifying agent. The ester with acrylic acid, for example, is used in the paint, adhesive, and plastic industries.

An alternative route for n-butanol is through the aldol condensation of acetaldehyde (Chapter 7).

2-ETHYLHEXANOL(CH₃(CH₂)₃CH(C₂H₅)CH₂OH)

2-Ethylhexanol is a colorless liquid soluble in many organic solvents. It is one of the chemicals used for producing PVC plasticizers (by react-

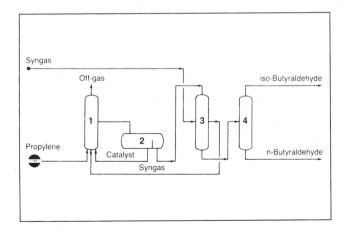

Figure 8-5. The Hoechst AG and Rhone Poulenc process for producing butyraldehydes from propene:[21] (1) reactor, (2) catalyst separation, (3) stripper (using fresh syngas to strip unreacted propylene to recycle), (4) distillation.

ing with phthalic acid; the product is di-2-ethylhexyl phthalate). The 1998 U.S. production of 2-ethylhexanol reached 800 million pounds.

2-Ethylhexanol is produced by the aldol condensation of butyraldehyde. The reaction occurs in presence of aqueous caustic soda and produces 2-ethyl-3-hydroxyhexanal. The aldehyde is then dehydrated and hydrogenated to 2-ethylhexanol:

$$2\,CH_3CH_2CH_2\overset{\overset{\displaystyle O}{\|}}{C}H \longrightarrow CH_3(CH_2)_2\overset{\overset{\displaystyle C_2H_5}{|}}{C}HOHCHCHO$$

$$CH_3(CH_2)_2\overset{\overset{\displaystyle C_2H_5}{|}}{C}HOHCHCHO \rightarrow CH_3(CH_2)_2CH{=}\overset{\overset{\displaystyle C_2H_5}{|}}{C}CHO + H_2O$$

$$CH_3(CH_2)_2CH{=}\overset{\overset{\displaystyle C_2H_5}{|}}{C}CHO + 2H_2 \longrightarrow CH_3(CH_2)_3\overset{\overset{\displaystyle C_2H_5}{|}}{C}HCH_2OH$$

Figure 8-6 shows the Hoechst process.[22]

DISPROPORTIONATION OF PROPYLENE (Metathesis)

Olefins could be catalytically converted into shorter and longer-chain olefins through a catalytic disproportionation reaction. For example, propylene could be disproportionated over different catalysts, yielding ethylene and butylenes. Approximate reaction conditions are 400°C and 8 atmospheres:

$$2CH_3CH{=}CH_2 \rightarrow CH_2{=}CH_2 + CH_3CH{=}CHCH_3$$

Table 8-5 indicates the wide variety of catalysts that can effect this type of disproportionation reaction, and Figure 8-7 is a flow diagram for the Phillips Co. triolefin process for the metathesis of propylene to produce 2-butene and ethylene.[23] Anderson and Brown have discussed in depth this type of reaction and its general utilization.[24] The utility with respect to propylene is to convert excess propylene to olefins of greater economic value. More discussion regarding olefin metathesis is noted in Chapter 9.

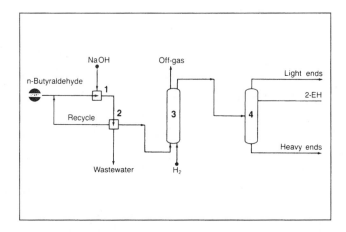

Figure 8-6. The Hoechst AG process for producing 2-ethylhexanol from n-butyraldehyde:[22] (1) Aldol condensation reactor, (2) separation (organic phase from liquid phase), (3) hydrogenation reactor, (4) distillation column.

Table 8-5
Representative disproportionation catalysts

Transition metal compound Heterogeneous	Support
M (CO)$_6$*	Al$_2$O$_3$
MoO$_3$	Al$_2$O$_3$
CoO.MoO$_3$	Al$_2$O$_3$
Re$_2$O$_7$	Al$_2$O$_3$
WO$_3$	SiO$_2$
Homogeneous	**Cocatalyst**
WCl$_6$ (EtOH)	EtALCl$_2$
MX$_2$ (NO)$_2$L$_2$*	R$_3$Al$_2$Cl$_3$
R$_4$N [M (CO)$_5$X]*	RAlX$_2$
ReCl$_5$/O$_2$	RAlCl$_2$

*M = Mo or W; X = halengen (Cl, Br, l); L = Lewis base (e.g., triphenyl-phosphine, pyridien, etc.); R = Allyl groups (butyl)

ALKYLATION USING PROPYLENE

Propylene could be used as an alkylating agent for aromatics. An important reaction with great commercial use is the alkylation of benzene to cumene for phenol and acetone production. The reaction is discussed in Chapter 10.

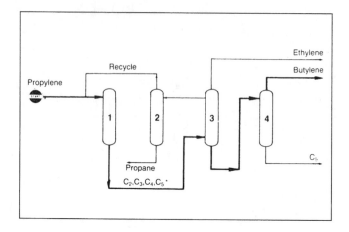

Figure 8-7. The Phillips Petroleum Co. process for producing 2-butene and ethylene from propylene:[23] (1) metathesis reactor, (2) fractionator (to separate propylene recycle from propane), (3, 4) fractionator for separating ethylene, butylenes, and C_5^+.

REFERENCES

1. *Chemical and Engineering News,* March 23, 1998, p. 22.
2. Gates, B. C., Katzer, J. R., and Schuit, G. C., "Chemistry of Catalytic Processes," McGraw-Hill Book Company, 1979, p. 349.
3. Sachtler, W. M., and DeBoer, N. H., Proceeding 3rd Int. Cong. Catal. Amsterdam 1965.
4. Matar, S., Mirbach, M., and Tayim, H., *Catalysis in Petrochemical Processes,* Kluwer Academic Publishers, Dordrecht, The Netherlands, 1989, pp. 93–94.
5. *Chemical and Engineering News,* Oct. 31, 1994, p. 15.
6. *CHEMTECH,* April, 1999, p. 19
7. Borman, S., *Chemical and Engineering News,* Vol. 68, No. 12, 1990, p. 15.
8. Pujada, P. R., Vora, B. V., and Krueding, A. P., "Newest Acrylonitrile Process," *Hydrocarbon Processing,* Vol. 56, No. 5, 1977, pp. 169–172.
9. *Oil and Gas Journal,* June 6, 1977, pp. 171–172.
10. *Chemical and Engineering News,* June 28, 1999, p. 35.
11. Davis, J. C., *Chemical Engineering,* Vol. 82, No. 14, 1975, pp. 44–48.
12. Stobaugh, R. B. et al., *Hydrocarbon Processing,* Vol. 52, No. 1, 1973, pp. 99–108.

13. Landau, R. et al., Proceedings of the 7th World Petroleum Congress, Vol. 5, *Petrochemicals,* 1967, pp. 67–72.
14. Ainsworth S. J., *Chemical and Engineering News,* Vol. 70, No. 9, 1992, pp. 9–11.
15. Brownstein, A. M. and List, H., *Hydrocarbon Processing,* Vol. 56, No. 9, 1977, pp. 159–162.
16. "Petrochemical Handbook," *Hydrocarbon Processing,* Vol. 70, No. 3, 1991, p. 185.
17. Hatch. L. F., *The Chemistry of Petrochemical Reactions,* Gulf Publishing Co., Houston, 1955, p. 76.
18. Matar, S., "Synfuels; Hydrocarbons of the Future," PennWell Publishing Co., Tulsa, OK, 1982, p. 20.
19. "Petrochemical Handbook," *Hydrocarbon Processing,* Vol . 58, No. 11, 1979, p. 122.
20. Cornils, B., *Hydroformylation, Oxo Synthesis, Roelen Reaction: New Synthesis with Carbon Monoxide,* Springer Verlag, Berlin, New York, 1980, pp. 1–224.
21. "Petrochemical Handbook," *Hydrocarbon Processing,* Vol. 70, No. 3, 1991, p. 149.
22. "Petrochemical Handbook," *Hydrocarbon Processing,* Vol. 70, No. 3, 1991, p. 158.
23. "Petrochemical Handbook," *Hydrocarbon Processing,* Vol . 70, No. 3, 1991, p. 144.
24. Anderson, K. L. and Brown, T. D., *Hydrocarbon Processing,* Vol. 55, No. 8, 1976, pp. 119–122.

CHAPTER NINE

C$_4$ Oleffins and Diolefins-Based Chemicals

INTRODUCTION

The C$_4$ olefins produce fewer chemicals than either ethylene or propylene. However, C$_4$ olefins and diolefins are precursors for some significant big-volume chemicals and polymers such as methyl-ter-butyl ether, adiponitrile, 1,4-butanediol, and polybutadiene.

Butadiene is not only the most important monomer for synthetic rubber production, but also a chemical intermediate with a high potential for producing useful compounds such as sulfolane by reaction with SO$_2$, 1,4-hutanediol by acetoxylation-hydrogenation, and chloroprene by chlorination-dehydrochlorination.

CHEMICALS FROM n-BUTENES

The three isomers constituting n-butenes are 1-butene, cis-2-butene, and trans-2-butene. This gas mixture is usually obtained from the olefinic C$_4$ fraction of catalytic cracking and steam cracking processes after separation of isobutene (Chapter 2). The mixture of isomers may be used directly for reactions that are common for the three isomers and produce the same intermediates and hence the same products. Alternatively, the mixture may be separated into two streams, one constituted of 1-butene and the other of cis- and trans-2-butene mixture. Each stream produces specific chemicals. Approximately 70% of 1-butene is used as a comonomer with ethylene to produce linear low-density polyethylene (LLDPE). Another use of 1-butene is for the synthesis of butylene oxide. The rest is used with the 2-butenes to produce other chemicals. n-Butene could also be isomerized to isobutene.[1]

This section reviews important reactions leading to various chemicals from n-butenes.

OXIDATION OF BUTENES

The mixture of n-butenes (1- and 2-butenes) could be oxidized to different products depending on the reaction conditions and the catalyst. The three commercially important oxidation products are acetic acid, maleic anhydride, and methyl ethyl ketone.

Due to the presence of a terminal double bond in 1-butene, oxidation of this isomer via a chlorohydrination route is similar to that used for propylene.

Acetic Acid (CH$_3$COH) with O double-bonded

Currently, the major route for obtaining acetic acid (ethanoic acid) is the carbonylation of methanol (Chapter 5). It may also be produced by the catalyzed oxidation of n-butane (Chapter 6).

The production of acetic acid from n-butene mixture is a vapor-phase catalytic process. The oxidation reaction occurs at approximately 270°C over a titanium vanadate catalyst. A 70% acetic acid yield has been reported.[2] The major by-products are carbon oxides (25%) and maleic anhydride (3%):

$$CH_3CH=CHCH_3 + 2\,O_2 \rightarrow 2\,CH_3\overset{O}{\overset{\|}{C}}OH$$

Acetic acid may also be produced by reacting a mixture of n-butenes with acetic acid over an ion exchange resin. The formed sec-butyl acetate is then oxidized to yield three moles of acetic acid:

$$CH_3CH=CHCH_3 + CH_3CH_2CH=CH_2 + 2CH_3COOH \rightarrow 2CH_3\overset{O}{\overset{\|}{C}}O\overset{CH_3}{\overset{|}{C}}HCH_2CH_3$$

sec-Butyl acetate

$$CH_3\overset{O}{\overset{\|}{C}}O\overset{CH_3}{\overset{|}{C}}HCH_2CH_3 + 2O_2 \longrightarrow 3\,CH_3\overset{O}{\overset{\|}{C}}OH$$

The reaction conditions are approximately 100–120°C and 15–25 atmospheres. The oxidation step is noncatalytic and occurs at approximately 200°C and 60 atmospheres. An acetic acid yield of 58% could be obtained.[3] By-products are formic acid (6%), higher boiling compounds (3%), and carbon oxides (28%). Figure 9-1 shows the Bayer AG two-step process for producing acetic acid from n-butenes.[3]

Acetic acid is a versatile reagent. It is an important esterifying agent for the manufacture of cellulose acetate (for acetate fibers and lacquers), vinyl acetate monomer, and ethyl and butyl acetates. Acetic acid is used to produce pharmaceuticals, insecticides, and dyes. It is also a precursor for chloroacetic acid and acetic anhydride. The 1994 U.S. production of acetic acid was approximately 4 billion pounds.

Acetic Anhydride $(CH_3\overset{\overset{O}{\|}}{C}-O-\overset{\overset{O}{\|}}{C}CH_3)$

Acetic anhydride (acetyl oxide) is a liquid with a strong offensive odor. It is an irritating and corrosive chemical that must be handled with care.

The production of acetic anhydride from acetic acid occurs via the intermediate formation of ketene where one mole of acetic acid loses one mole of water:

$$CH_3\overset{\overset{O}{\|}}{C}OH \longrightarrow CH_2=C=O + H_2O$$
$$\text{Ketene}$$

Ketene further reacts with one mole acetic acid, yielding acetic anhydride:

$$CH_2=C=O + CH_3\overset{\overset{O}{\|}}{C}OH \longrightarrow CH_3\overset{\overset{O}{\|}}{C}-O-\overset{\overset{O}{\|}}{C}CH_3$$

Acetic anhydride is mainly used to make acetic esters and acetyl salicylic acid (aspirin).

Methyl Ethyl Ketone $(CH_3\overset{\overset{O}{\|}}{C}CH_2CH_3)$

Methyl ethyl ketone MEK (2-butanone) is a colorless liquid similar to acetone, but its boiling point is higher (79.5°C). The production of MEK from n-butenes is a liquid-phase oxidation process similar to that used to

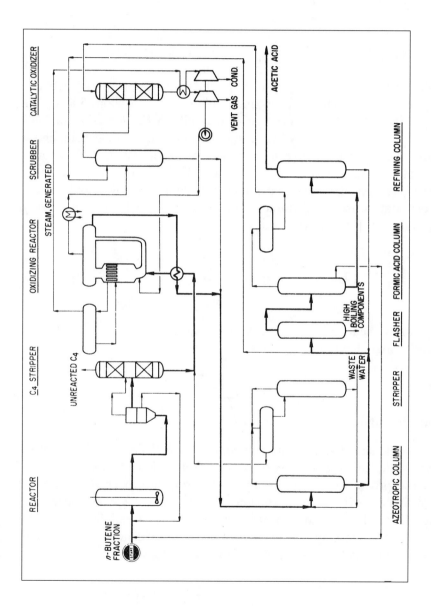

Figure 9-1. The Bayer AG two-step process for producing acetic acid from n-butenes.[3]

produce acetaldehyde from ethylene using a Wacker-type catalyst $(PdCl_2/CuCl_2)$. The reaction conditions are similar to those for ethylene. The yield of MEK is approximately 88%:

$$CH_2=CH-CH_2CH_3 + 1/2\,O_2 \longrightarrow CH_3\overset{\overset{\displaystyle O}{\|}}{C}CH_2CH_3$$

Methyl ethyl ketone may also be produced by the catalyzed dehydrogenation of sec-butanol over zinc oxide or brass at about 500°C. The yield from this process is approximately 95%. MEK is used mainly as a solvent in vinyl and acrylic coatings, in nitrocellulose lacquers, and in adhesives. It is a selective solvent in dewaxing lubricating oils where it dissolves the oil and leaves out the wax. MEK is also used to synthesize various compounds such as methyl ethyl ketone peroxide, a polymerization catalyst used to form acrylic and polyester polymers and methyl pentynol by reacting with acetylene:

$$CH_3\overset{\overset{\displaystyle O}{\|}}{C}CH_2CH_3 + HC\equiv CH \longrightarrow CH_3CH_2\overset{\overset{\displaystyle CH_3}{|}}{\underset{\underset{\displaystyle OH}{|}}{C}}-C\equiv CH$$

Methyl pentynol is a solvent for polyamides, a corrosion inhibitor, and an ingredient in the synthesis of hypnotics.

Maleic Anhydride

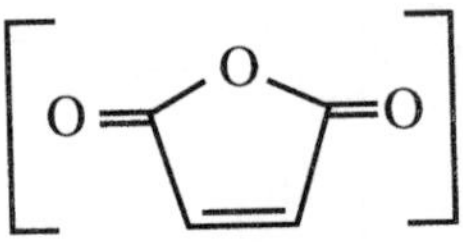

Maleic anhydride, a solid compound that melts at 53°C, is soluble in water, alcohol, and acetone, but insoluble in hydrocarbon solvents.

The production of maleic anhydride from n-butenes is a catalyzed reaction occurring at approximately 400–440°C and 2–4 atmospheres. A special catalyst, constituted of an oxide mixture of molybdenum, vanadium, and phosphorous, may be used. Approximately 45% yield of maleic anhydride could be obtained from this route:

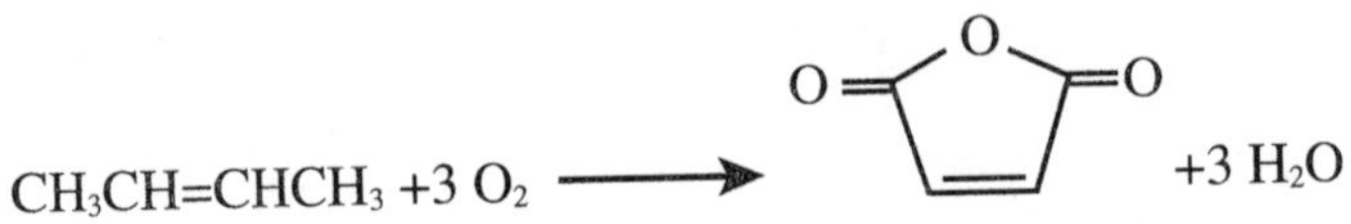

$$CH_3CH=CHCH_3 + 3\,O_2 \longrightarrow \text{(maleic anhydride)} + 3\,H_2O$$

Other routes to maleic anhydride are the oxidation of n-butane, a major source for this compound (Chapter 6), and the oxidation of benzene (Chapter 10).

Maleic anhydride is important as a chemical because it polymerizes with other monomers while retaining the double bond, as in unsaturated polyester resins. These resins, which represent the largest end use of maleic anhydride, are employed primarily in fiber-reinforced plastics for the construction, marine, and transportation industries. Maleic anhydride can also modify drying oils such as linseed and sunflower.

As an intermediate, maleic anhydride is used to produce malathion, an important insecticide, and maleic hydrazide, a plant growth regulator:

$$(CH_3O)_2-\overset{\overset{\text{S}}{\|}}{P}S-\underset{\underset{\text{O}}{\|}}{\overset{\overset{\text{O}}{\|}}{C}H}-\overset{\overset{\text{O}}{\|}}{C}OC_2H_5$$

Malathion   Maleic hydrazide

Maleic anhydride is also a precursor for 1,4-butanediol through an esterification route followed by hydrogenation.[4] In this process, excess ethyl alcohol esterifies maleic anhydride to monoethyl maleate. In a second step, the monoester catalytically esterifies to the diester. Excess ethanol and water are then removed by distillation. The ethanol-water mixture is distilled to recover ethanol, which is recycled:

$$O=\overset{O}{\diagdown}=O \quad +C_2H_5OH \longrightarrow \overset{\overset{\text{O}}{\|}}{HOC}CH=CH\overset{\overset{\text{O}}{\|}}{C}OC_2H_5$$

$$\overset{\overset{\text{O}}{\|}}{HOC}CH=CH\overset{\overset{\text{O}}{\|}}{C}OC_2H_5 + C_2H_5OH \longrightarrow C_2H_5\overset{\overset{\text{O}}{\|}}{OC}CH=CH\overset{\overset{\text{O}}{\|}}{C}OC_2H_5+H_2O$$

Diethylmaleate

Hydrogenation of diethylmaleate in the vapor phase over a nonprecious metal catalyst produces diethyl succinate. Successive hydrogenation produces γ-butyrolactone, butanediol, and tetrahydrofuran.

$$C_2H_5OCCH=CHCOC_2H_5 + 3\,H_2 \longrightarrow \quad + 2\,C_2H_5OH$$

γ-Butyrolactone

$$2 \quad + 4\,H_2 \longrightarrow HOCH_2CH_2CH_2CH_2OH + \quad + H_2O$$

1,4-Buytanediol

Selectivity to the coproducts is high, but the ratios of the coproducts may be controlled with appropriate reactor operating conditions. Figure 9-2 is a block diagram for the butane diol process.[4] 1,4-Butanediol from butadiene is discussed later in this chapter.

Butylene Oxide (CH$_3$CH$_2$CH—CH$_2$)

Butylene oxide, like propylene oxide, is produced by the chlorohydrination of 1-butene with HOCl followed by epoxidation. The reaction conditions are similar to those used for propylene:

$$CH_3CH_2CH=CH_2 + HOCl \rightarrow CH_3CH_2CHOHCH_2Cl$$

Butylene chlorohydrin

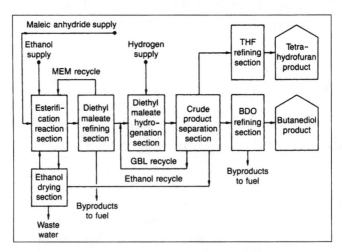

Figure 9-2. A block diagram for producing 1,4-butanediol from maleic anhydride.[4]

$$2CH_3CH_2CHOHCH_2Cl + Ca(OH)_2 \longrightarrow 2CH_3CH_2CH\overset{O}{\overset{\diagup \diagdown}{-}}CH_2 + CaCl_2 + 2H_2O$$

Butylene oxide may be hydrolyzed to butylene glycol, which is used to make plasticizers. 1,2-Butylene oxide is a stabilizer for chlorinated solvents and also an intermediate in organic synthesis such as in surfactants and pharmaceuticals.

Hydration of n-Butenes (sec-Butanol [$CH_3CHOHCH_2CH_3$])

sec-Butanol (2-butanol, sec-butyl alcohol), a liquid, has a strong characteristic odor. Its normal boiling point is 99.5°C, which is near water's. The alcohol is soluble in water but less so than isopropyl and ethyl alcohols.

sec-Butanol is produced by a reaction of sulfuric acid with a mixture of n-butenes followed by hydrolysis. Both 1-butene and cis- and trans-2-butenes yield the same carbocation intermediate, which further reacts with the HSO_4^{1-} or SO_4^{2-} ions, producing a sulfate mixture:

$$\begin{matrix} CH_2=CH-CH_2CH_3 \\ or \\ CH_3CH=CH-CH_3 \end{matrix} + H^+ \longrightarrow [CH_3\overset{+}{C}H-CH_2CH_3]$$

The sulfation reaction occurs in the liquid phase at approximately 35°C. An 85 wt% alcohol yield could be realized. The reaction is similar to the sulfation of ethylene or propylene and results in a mixture of sec-butyl hydrogen sulfate and di-sec-butyl sulfate. The mixture is further hydrolyzed to sec-butanol and sulfuric acid:

$$3\ CH_3CH_2CH=CH_2 + 2\ H_2SO_4 \longrightarrow \overset{CH_3}{\overset{|}{CH_3CH_2CHOSO_3H}} + \overset{CH_3}{\overset{|}{(CH_3CH_2CH)_2OSO_3}}$$

$$+3\ H_2O \longrightarrow 3\ CH_3CH_2CHOHCH_3 + 2H_2SO_4$$

The only important by-product is di-sec-butyl ether, which may be recovered.

The major use of sec-butanol is to produce MEK by dehydrogenation, as mentioned earlier. 2-Butanol is also used as a solvent, a paint remover, and an intermediate in organic synthesis.

Isomerization of n-Butenes

n-Butene could be isomerized to isobutene using Shell FER catalyst which is active and selective. n-Butene mixture from steam cracker or

FCC after removal of C_5 olefins via selective hydrogenation step passes to the isomerization unit. It has been proposed that after the formation of a butyl carbocation, a cyclopropyl carbocation is formed which gives a primary carbenium ion that produces isobutene[1]:

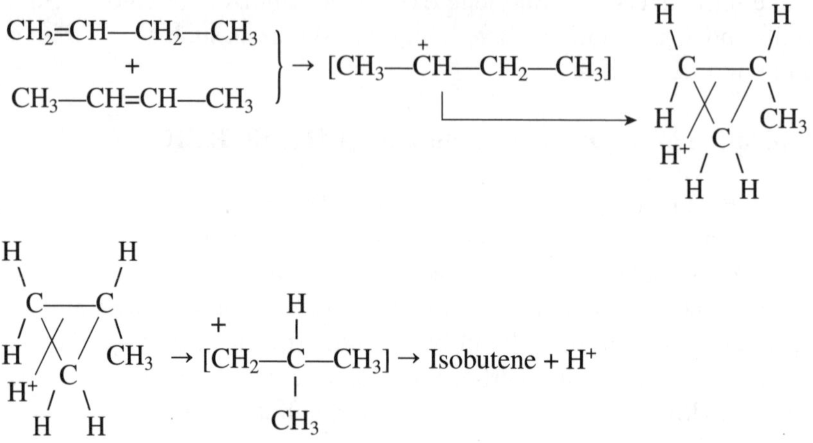

$$CH_2=CH-CH_2-CH_3$$
$$+$$
$$CH_3-CH=CH-CH_3$$
$$\rightarrow [CH_3-\overset{+}{C}H-CH_2-CH_3]$$

METATHESIS OF OLEFINS

Metathesis is a catalyzed reaction that converts two olefin molecules into two different olefins. It is an important reaction for which many mechanistic approaches have been proposed by scientists working in the fields of homogenous catalysis and polymerization.[5, 6] One approach is the formation of a fluxional five-membered metallocycle. The intermediate can give back the starting material or the metathetic products via a concerted mechanism:

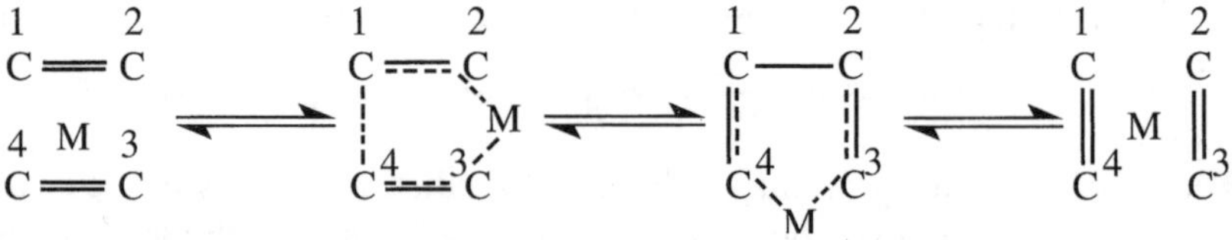

Another approach is a stepwise mechanism that involves the initial formation of a metal carbene followed by the formation of a four-membered metallocycle species:[7]

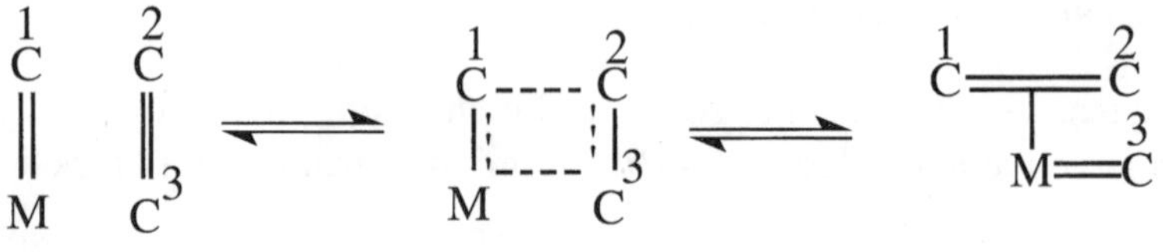

Olefin metatheses are equilibrium reactions among the two-reactant and two-product olefin molecules. If chemists design the reaction so that one product is ethylene, for example, they can shift the equilibrium by removing it from the reaction medium.[8] Because of the statistical nature of the metathesis reaction, the equilibrium is essentially a function of the ratio of the reactants and the temperature. For an equimolar mixture of ethylene and 2-butene at 350°C, the maximum conversion to propylene is 63%. Higher conversions require recycling unreacted butenes after fractionation.[9] This reaction was first used to produce 2-butene and ethylene from propylene (Chapter 8). The reverse reaction is used to prepare polymer-grade propylene form 2-butene and ethylene:[10]

$$CH_3CH=CHCH_3 + CH_2=CH_2 \leftrightarrows 2CH_3CH=CH_2$$

The metathetic reaction occurs in the gas phase at relatively high temperatures (150°–350°C) with molybdenum or tungsten supported catalysts or at low temperature (≈50°C) with rhenium-based catalyst in either liquid or gas-phase. The liquid-phase process gives a better conversion. Equilibrium conversion in the range of 55–65% could be realized, depending on the reaction temperature.[8]

In this process, which has been jointly developed by Institute Francais du Petrole and Chinese Petroleum Corp., the C$_4$ feed is mainly composed of 2-butene (1-butene does not favor this reaction but reacts differently with olefins, producing metathetic by-products). The reaction between 1-butene and 2-butene, for example, produces 2-pentene and propylene. The amount of 2-pentene depends on the ratio of 1-butene in the feedstock. 3-Hexene is also a by-product from the reaction of two butene molecules (ethylene is also formed during this reaction). The properties of the feed to metathesis are shown in Table 9-1.[11] Table 9-2 illustrates the results from the metatheses reaction at two conversions. The main by-product was 2-pentene. Olefins in the range of C$_6$–C$_8$ and higher were present, but to a much lower extent than C$_5$.

Figure 9-3 shows a simplified flow diagram for the olefin metathesis.[11]

Table 9-1
Properties of feed to the metathesis process[11]

Composition	Wt%
n-Butane	2.8
Butene- 1	7.2
Butene-2	90.0

Table 9-2
Results of metathesis of 2-butene at two conversion levels[11]

Item	Case 1	Case 2
Ethylene feed, kg/h	8.1	8.1
Total C_4 feed, kg/h	14.3	13.4
C_4 recycle, kg/h	4.4	9.6
Butene-2 conversion		
% per pass	62.3	59.6
% overall	87.8	94.6
Propylene product		
% selectivity	93.8	96.6
% yield from butene-2	82.4	91.3

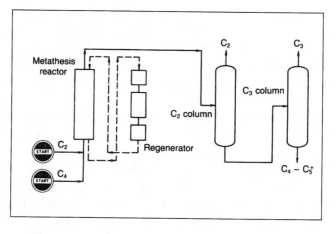

Figure 9-3. A flow diagram showing the metathesis process for producing polymer grade propylene from ethylene and 2-butene.[11]

OLIGOMERIZATION OF BUTENES

2-Butenes (after separation of 1-butene) can be oligomerized in the liquid phase on a heterogeneous catalyst system to yield mainly C_8 and C_{12} olefins.[12] The reaction is exothermic, and requires a multitubular carbon steel reactor. The exothermic heat is absorbed by water circulating around the reactor shell. Either a single- or a two-stage system is used. The process can be made to produce either more linear or more branched oligomers. Linear oligomers are used to produce nonyl alcohols for plasticizers, alkyl phenols for surfactants, and tridecyl alcohols for detergent intermediates. Branched oligomers are valuable gasoline components. Figure 9-4 shows the Octol oligomerization process.[13] A typical analysis of A-type oligomers (branched) is shown in Table 9-3.[12]

CHEMICALS FROM ISOBUTYLENE

Isobutylene ($CH_2=C(CH_3)_2$) is a reactive C_4 olefin. Until recently, almost all isobutylene was obtained as a by-product with other C_4 hydrocarbons from different cracking processes. It was mainly used to produce alkylates for the gasoline pool. A small portion was used to produce chemicals such as isoprene and diisobutylene. However, increasing demand for oxygenates from isobutylene has called for other sources.

n-Butane is currently used as a precursor for isobutylene. The first step is to isomerize n-butane to isobutane, then dehydrogenate it to isobutylene. This serves the dual purpose of using excess n-butane (that must be removed from gasolines due to new rules governing gasoline vapor pressure) and producing the desired isobutylene. Currently, the major use of iosbutylene is to produce methyl-ter-butyl ether.

The following section reviews the chemistry of isobutylene and its important chemicals.

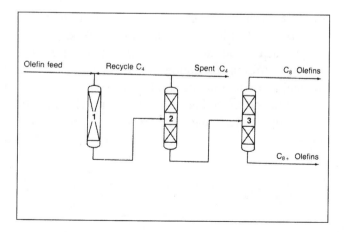

Figure 9-4. The Octol Oligomerization process for producing C_8's and C_{12}'s and C_{16}'s olefins from n-butenes:[13] (1) multitubular reactor, (2) debutanizer column, (3) fractionation tower.

Table 9-3
Typical analysis of branched oligomers (Type A)[12]

Densily (20°C), kg/l	0.755	
Flash point °C	–4	
Ignition temperature, °C	240	
Pour point °C	below –40	
Hydrocarbon no. distribution		**% by mass**
C_6		0.7
C_7		1.0
C_8		66.2
C_9		2.0
C_{10}		3.0
C_{11}		1.2
C_{12}		16.6
C_{13} to C_{15}		0.5
C_{16}		7.8
C_{16+}		1.0
	RON	**MON**
Gasoline hase stock		
(unleaded, low in olefins)	97.0	85.7
+5% oligomers	97.0	85.3
+ 10% oligomers	96.8	85.0

OXIDATION OF ISOBUTYLENE
(Methacrolein and Methacrylic Acid)

Much like the oxidation of propylene, which produces acrolein and acrylic acid, the direct oxidation of isobutylene produces methacrolein and methacrylic acid. The catalyzed oxidation reaction occurs in two steps due to the different oxidation characteristics of isobutylene (an olefin) and methacrolein (an unsaturated aldehyde). In the first step, isobutylene is oxidized to methacrolein over a molybdenum oxide-based catalyst in a temperature range of 350–400°C. Pressures are a little above atmospheric:

$$\underset{}{CH_2{=}\overset{\overset{\textstyle CH_3}{|}}{C}{-}CH_3} + O_2 \longrightarrow \underset{}{CH_2{=}\overset{\overset{\textstyle CH_3}{|}}{C}{-}CHO} + H_2O$$

Methacrolein

In the second step, methacrolein is oxidized to methacrylic acid at a relatively lower temperature range of 250–350°C. A molybdenum-supported compound with specific promoters catalyzes the oxidation.

$$CH_2=\underset{\underset{CH_3}{|}}{C}-CHO + \tfrac{1}{2}O_2 \longrightarrow CH_2=\underset{\underset{CH_3}{|}}{C}-COOH$$

Methacrylic acid

Methacrylic acid is esterified with methanol to produce methyl methacrylate monomer.

Methacrylic acid and methacrylates are also produced by the hydrocyanation of acetone followed by hydrolysis and esterification (Chapter 8).

Ammoxidation of isobutylene to produce methacrylonitrile is a similar reaction to ammoxidation of propylene to acrylonitrile. However, the yield is low.

EPOXIDATION OF ISOBUTYLENE
(Isobutylene Oxide Production)

Isobutylene oxide is produced in a way similar to propylene oxide and butylene oxide by a chlorohydrination route followed by reaction with Ca(OH)$_2$. Direct catalytic liquid-phase oxidation using stoichiometric amounts of thallium acetate catalyst in aqueous acetic acid solution has been reported. An isobutylene oxide yield of 82% could be obtained.[14]

Direct non-catalytic liquid-phase oxidation of isobutylene to isobutylene oxide gave low yield (28.7%) plus a variety of oxidation products such as acetone, ter-butyl alcohol, and isobutylene glycol:

$$(CH_3)_2C=CH_2 + \tfrac{1}{2}O_2 \longrightarrow CH_3-\underset{\underset{CH_3}{|}}{\overset{\overset{O}{\diagup\diagdown}}{C}}-CH_2$$

Isobutylene oxide

Hydrolysis of isobutylene oxide in the presence of an acid produces isobutylene glycol:

$$CH_3-\underset{\underset{CH_3}{|}}{\overset{\overset{O}{\diagup\diagdown}}{C}}-CH_2 + H_2O \longrightarrow CH_3\underset{\underset{CH_3}{|}}{C}(OH)CH_2OH$$

Isobutylene glycol

Isobutylene glycol may also be produced by a direct catalyzed liquid phase oxidation of isobutylene with oxygen in presence of water. The catalyst is similar to the Wacker-catalyst system used for the oxidation of ethylene to acetaldehyde. Instead of $PdCl_2/CuCl_2$ used with ethylene, a $TlCl_3/CuCl_2$ catalyst is employed[15]:

$$CH_3-\overset{\overset{\displaystyle CH_3}{|}}{C}=CH_2 + \tfrac{1}{2}O_2 + H_2O \longrightarrow CH_3\overset{\overset{\displaystyle CH_3}{|}}{C}(OH)CH_2OH$$

Liquid-phase oxidation of isobutylene glycol produces othydroxyisobutyric acid. The reaction conditions are 70–80°C at pH 2–7 in presence of a catalyst (5% pt/C)[16]:

$$CH_3\overset{\overset{\displaystyle CH_3}{|}}{C}(OH)CH_2OH + O_2 \longrightarrow CH_3\overset{\overset{\displaystyle CH_3}{|}}{C}(OH)COOH + H_2O$$

$$\alpha\text{-Hydroxyisobutyric acid}$$

Dehydration of the acid produces 95% yield of methacrylic acid:

$$CH_3-\overset{\overset{\displaystyle CH_3}{|}}{C}(OH)COOH \longrightarrow CH_2=\overset{\overset{\displaystyle CH_3}{|}}{C}-COOH + H_2O$$

ADDITION OF ALCOHOLS TO ISOBUTYLENE
(Methyl- and Ethyl-Ter-Butyl Ether)

The reaction between isobutylene and methyl and ethyl alcohols is an addition reaction catalyzed by a heterogeneous sulfonated polystyrene resin. When methanol is used a 98% yield of methyl-ter-butyl ether MTBE is obtained:

$$CH_3OH + CH_3-\overset{\overset{\displaystyle CH_3}{|}}{C}=CH_2 \longrightarrow (CH_3)_3C-OCH_3$$

$$\text{MTBE}$$

The reaction conditions have been noted in Chapter 5.

Ethyl-ter-butyl ether (ETBE) is also produced by the reaction of ethanol and isobutylene under similar conditions with a heterogeneous acidic ion-exchange resin catalyst (similar to that with MTBE):

$$CH_3CH_2OH + CH_3-\overset{\overset{\displaystyle CH_3}{|}}{C}=CH_2 \longrightarrow (CH_3)_3C-OCH_2CH_3$$

$$\text{ETBE}$$

MTBE and ETBE constitute a group of oxygenates that are currently in high demand for gasoline octane-number boosters. Both MTBE and ETBE have a similar research octane number of 118, but the latter ether has a motor octane number of 102 versus 100 for MTBE.[17] However, the oxygen content of MTBE is 18.2% compared to 15.7% for ETBE. The lower oxygen content of ETBE is related to the air/fuel ratio, which may not require a change in the automobile carburetors. A comparison between the two ethers regarding phase separation, antiknock behavior, and fuel economy has been reviewed by Iborra et al.[18]

HYDRATION OF ISOBUTYLENE
(Ter-Butyl Alcohol [(CH$_3$)$_3$COH])

The acid-catalyzed hydration of isobutylene produces ter-butyl alcohol. The reaction occurs in the liquid phase in the presence of 50–65% H$_2$SO$_4$ at mild temperatures (10–30°C). The yield is approximately 95%:

$$CH_3-\overset{\overset{\displaystyle CH_3}{|}}{C}=CH_2 + H_2O \xrightarrow{\ H^+\ } CH_3-\overset{\overset{\displaystyle CH_3}{|}}{\underset{\underset{\displaystyle CH_3}{|}}{C}}-OH$$

ter-Butyl alcohol (TBA) is used as a chemical intermediate because a tertiary butyl carbocation forms easily. It is also used as a solvent in pharmaceutical formulations, a paint remover, and a high-octane gasoline additive. The alcohol is a major by-product from the synthesis of propylene oxide using tertiary butyl hydroperoxide. Surplus ter-butyl alcohol could be used to synthesize highly pure isobutylene for MTBE production by a dehydration step. The reaction conditions, the catalyst used in a pilot-scale unit, and the yield are reviewed by Abraham and Prescott.[19] It was concluded that MTBE conversion increases from 8 wt% to 88 wt% as the temperature increases from 400°F to 600°F at about 40 LHSV (liquid hourly space velocity). At a lower space velocity (≈20 LHSV), conversion increased from 12 wt% to 99 wt% for the same temperature range. Figure 9-5 shows the effect of temperature and LHSV on the conversion[19]:

$$CH_3-\overset{\overset{\displaystyle CH_3}{|}}{\underset{\underset{\displaystyle CH_3}{|}}{C}}-OH \longrightarrow CH_3-\overset{\overset{\displaystyle CH_3}{|}}{C}=CH_2 + H_2O$$

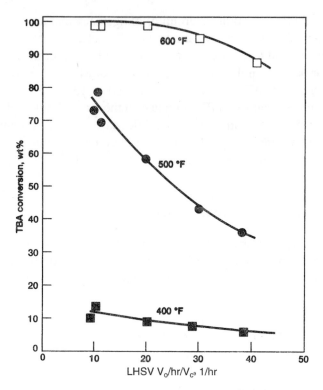

Figure 9-5. Effect of temperature and liquid hourly space velocity on conversion.[19]

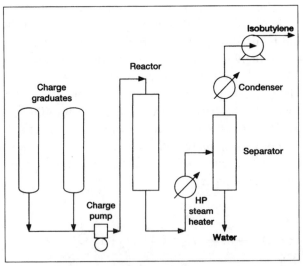

Figure 9-6. A simplified flow diagram of a tertiary butyl alcohol pilot plant.[19]

Figure 9-6 is a simplified flow diagram of a TBA dehydration pilot unit.[19]

CARBONYLATION OF ISOBUTYLENE (Neopentanoic Acid)

$$[(CH_3)_3C-\overset{\overset{\displaystyle O}{\|}}{C}OH)]$$

The addition of carbon monoxide to isobutylene under high pressures and in the presence of an acid produces a carbon monoxide-olefin complex, an acyl carbocation. Hydrolysis of the complex at lower pressures yields neopentanoic acid:

$$CH_3-\overset{\overset{\displaystyle CH_3}{|}}{C}=CH_2 \;+\; H^+ + CO \longrightarrow \left[CH_3-\overset{\overset{\displaystyle CH_3}{|}}{\underset{\underset{\displaystyle CH_3}{|}}{C}}-CO\right]^+$$

$$\left[CH_3-\overset{\overset{\displaystyle CH_3}{|}}{\underset{\underset{\displaystyle CH_3}{|}}{C}}-CO\right]^+ \;+\; H_2O \longrightarrow (CH_3)_3C-\overset{\overset{\displaystyle O}{\|}}{C}OH + H^+$$

Neopentanoic acid (trimethylacetic acid) is an intermediate and an esterifying agent used when a stable neo structure is needed.

DIMERIZATION OF ISOBUTYLENE

Isobutylene could be dimerized in the presence of an acid catalyst to diisobutylene. The product is a mixture of diisobutylene isomers, which are used as alkylating agents in the plasticizer industry and as a lube oil additive (dimerization of olefins is noted in Chapter 3).

CHEMICALS FROM BUTADIENE

Butadiene is a diolefinic hydrocarbon with high potential in the chemical industry. In 1955, it was noticed that "the assured future of butadiene (CH_2=CH-CH=CH_2) lies with synthetic rubber . . . the potential of butadiene is in its chemical versatility . . . its low cost, ready availability, and great activity tempt researchers."[20]

Butadiene is a colorless gas, insoluble in water but soluble in alcohol. It can be liquefied easily under pressure. This reactive compound polymerizes readily in the presence of free radical initiators.

Butadiene is mainly obtained as a byproduct from the steam cracking of hydrocarbons and from catalytic cracking. These two sources account for over 90% of butadiene demand. The remainder comes from dehydrogenation of n-butane or n-butene streams (Chapter 3). The 1998 U.S. production of butadiene was approximately 4 billion pounds, and it was the 36th highest-volume chemical. Worldwide butadiene capacity was nearly 20 billion pounds.

Butadiene is easily polymerized and copolymerized with other monomers. It reacts by addition to other reagents such as chlorine, hydrocyanic acid, and sulfur dioxide, producing chemicals of great commercial value.

ADIPONITRILE ($NC(CH_2)_4CN$)

Adiponitrile, a colorless liquid, is slightly soluble in water but soluble in alcohol. The main use of adiponitrile is to make nylon 6/6.

The production of adiponitrile from butadiene starts by a free radical chlorination, which produces a mixture of 1,4-dichloro-2-butene and 3,4-dichloro-1-butene:

$$2CH_2=CH—CH=CH_2 + 2Cl_2 \rightarrow ClCH_2CH=CHCH_2Cl$$
$$+CH_2=CHCHClCH_2Cl$$

The vapor-phase chlorination reaction occurs at approximately 200–300°C. The dichlorobutene mixture is then treated with NaCN or HCN in presence of copper cyanide. The product 1,4-dicyano-2-butene is obtained in high yield because allylic rearrangement to the more thermodynamically stable isomer occurs during the cyanation reaction:

$$ClCH_2CH=CHCH_2Cl + CH_2=CHCHClCH_2Cl + 4NaCN$$
$$\rightarrow 2NCCH_2CH=CHCH_2CN + 4NaCl$$

The dicyano compound is then hydrogenated over a platinum catalyst to adiponitrile.

$$NCCH_2CH=CHCH_2CN + H_2 \rightarrow NC—(CH_2)_4—CN$$
$$\text{Adiponitrile}$$

Adiponitrile may also be produced by the electrodimerization of acrylonitrile (Chapter 8) or by the reaction of ammonia with adipic acid followed by two-step dehydration reactions:

$$\text{HOOC(CH}_2)_4\text{COOH} + 2\text{NH}_3 \rightarrow \overset{\overset{\displaystyle O}{\displaystyle \|}}{\text{H}_4\text{NOC}}\text{(CH}_2)_4\overset{\overset{\displaystyle O}{\displaystyle \|}}{\text{CONH}_4}$$

$$\xrightarrow{\text{H}_3\text{PO}_4} \overset{\overset{\displaystyle O}{\displaystyle \|}}{\text{H}_2\text{NC}}\text{(CH}_2)_4\overset{\overset{\displaystyle O}{\displaystyle \|}}{\text{CNH}_2} + 2\text{H}_2\text{O}$$

$$\xrightarrow{\text{H}_3\text{PO}_4} \text{NC}-\text{(CH}_2)_4-\text{CN} + 2\text{H}_2\text{O}$$

HEXAMETHYLENEDIAMINE (H$_2$N-(CH$_2$)$_6$-NH$_2$)

Hexamethylenediamine (1,6-hexanediamine) is a colorless solid, soluble in both water and alcohol. It is the second monomer used to produce nylon 6/6 with adipic acid or its esters.

The main route for the production of hexamethylene diamine is the liquid-phase catalyzed hydrogenation of adiponitrile:

$$\text{NC}-\text{(CH}_2)_4-\text{CN} + 4\text{H}_2 \rightarrow \text{H}_2\text{N}-\text{(CH}_2)_6-\text{NH}_2$$

The reaction conditions are approximately 200°C and 30 atmospheres over a cobalt-based catalyst.

ADIPIC ACID (HOOC(CH$_2$)$_4$COOH)

Adipic acid may be produced by a liquid-phase catalytic carbonylation of butadiene.[21] A catalyst of RhCl$_2$ and CH$_3$I is used at approximately 220°C and 75 atmospheres. Adipic acid yield is about 49%. Both α-gultaric acid (25%) and valeric acid (26%) are coproduced:

$$\text{CH}_2=\text{CH}-\text{CH}=\text{CH}_2 + 2\text{CO} + 2\text{H}_2\text{O} \rightarrow \text{HOOC(CH}_2)_4\text{COOH}$$

BASF is operating a semicommercial plant for the production of adipic acid via this route.[22] A new route to adipic acid occurs via a sequential carbonylation, isomerization, hydroformylation reactions.[23] The following illustrates these steps:

$$\text{CH}_2=\text{CH}-\text{CH}=\text{CH}_2 + \text{CO} + \text{CH}_3\text{OH} \rightarrow \text{CH}_3\text{CH}=\text{CH}-\text{CH}_2\overset{\overset{\displaystyle O}{\displaystyle \|}}{\text{COCH}_3}$$

$$\text{CH}_3\text{CH}=\text{CHCH}_2\overset{\overset{\displaystyle O}{\displaystyle \|}}{\text{COCH}_3} + 2\text{CO} + 3\text{H}_2 \rightarrow \text{CH}_3\overset{\overset{\displaystyle O}{\displaystyle \|}}{\text{C}}\text{(CH}_2)_4\overset{\overset{\displaystyle O}{\displaystyle \|}}{\text{COCH}_3} + \text{H}_2\text{O}$$

$$\underset{\text{O}}{\overset{\text{O}}{\parallel}}\quad\underset{\text{O}}{\overset{\text{O}}{\parallel}} \qquad \underset{\text{O}}{\overset{\text{O}}{\parallel}}\quad\underset{\text{O}}{\overset{\text{O}}{\parallel}}$$

$$CH_3-C-(CH_2)_4-C-OCH_3 + O_2 \rightarrow HOC(CH_2)_4-COCH_3 \overset{\text{Hydr.}}{\rightarrow} \text{Adipic}$$
$$\text{acid}$$

The main process for obtaining adipic acid is the catalyzed oxidation of cyclohexane (Chapter 10).

BUTANEDIOL (HO-(CH$_2$)$_4$-OH)

The production of 1,4-butanediol (1,4-BDO) from propylene via the carbonylation of allyl acetate is noted in Chapter 8. 1,4-Butanediol from maleic anhydride is discussed later in this chapter. An alternative route for the diol is through the acetoxylation of butadiene with acetic acid followed by hydrogenation and hydrolysis.

The first step is the liquid phase addition of acetic acid to butadiene. The acetoxylation reaction occurs at approximately 80°C and 27 atmospheres over a Pd-Te catalyst system. The reaction favors the 1,4-addition product (1,4-diacetoxy-2-butene). Hydrogenation of diacetoxybutene at 80°C and 60 atmospheres over a Ni/Zn catalyst yields 1,4-diacetoxybutane. The latter compound is hydrolyzed to 1,4-butanediol and acetic acid:

$$CH_2=CH-CH=CH_2 + 2CH_3\overset{\overset{\text{O}}{\parallel}}{C}-OH \rightarrow CH_3\overset{\overset{\text{O}}{\parallel}}{C}OCH_2CH=CHCH_2O\overset{\overset{\text{O}}{\parallel}}{C}CH_3$$

Main product

$$CH_3\overset{\overset{\text{O}}{\parallel}}{C}OCH_2CH=CHCH_2O\overset{\overset{\text{O}}{\parallel}}{C}CH_3 + H_2 \rightarrow CH_3\overset{\overset{\text{O}}{\parallel}}{C}O(CH_2)_4O\overset{\overset{\text{O}}{\parallel}}{C}CH_3$$

$$CH_3\overset{\overset{\text{O}}{\parallel}}{C}O(CH_2)_4O\overset{\overset{\text{O}}{\parallel}}{C}CH_3 + 2H_2O \rightarrow HO(CH_2)_4OH + 2CH_3\overset{\overset{\text{O}}{\parallel}}{C}OH$$

Acetic acid is then recovered and recycled. Butanediol is mainly used for the production of thermoplastic polyesters.

CHLOROPRENE (CH$_2$=C̈—CH=CH$_2$)
$$\overset{\text{Cl}}{\underset{|}{}}$$
CHLOROPRENE (CH$_2$=C—CH=CH$_2$)

Chloroprene (2-chloro 1,3-butadiene), a conjugated non-hydrocarbon diolefin, is a liquid that boils at 59.2°C and while only slightly soluble in water it is soluble in alcohol. The main use of chloroprene is to polymerize it to neoprene rubber.

Butadiene produces chloroprene through a high temperature chlorination to a mixture of dichlorobutenes, which is isomerized to 3,4-dichloro-1-butene. This compound is then dehydrochlorinated to chloroprene:

$$CH_2{=}CHCHClCH_2Cl \rightarrow CH_2{=}\overset{\overset{\displaystyle Cl}{|}}{C}{-}CH{=}CH_2 + HCl$$

Sulfolane

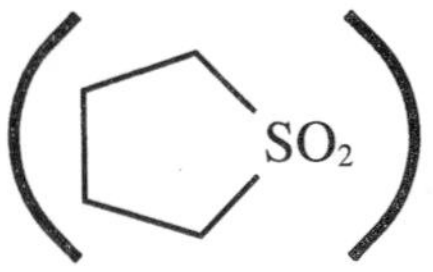

Sulfolane (tetramethylene sulfone) is produced by the reaction of butadiene and sulfur dioxide followed by hydrogenation:

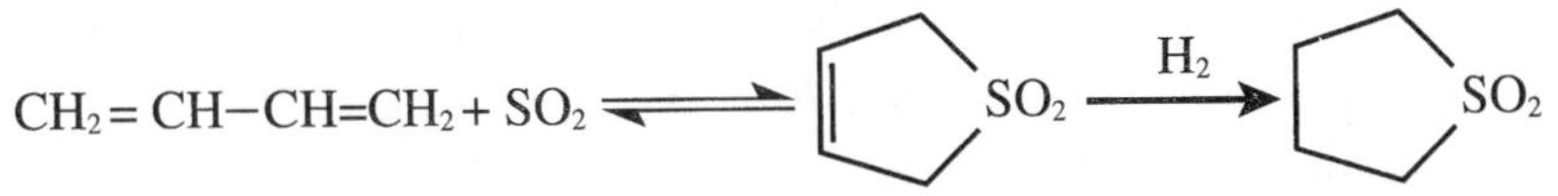

$$CH_2{=}CH{-}CH{=}CH_2 + SO_2 \rightleftharpoons$$

Optimum temperature for highest sulfolene yield is approximately 75°C. At approximately 125°C, sulfolene decomposes to butadiene and SO_2. This simple method could be used to separate butadiene from a mixture of C_4 olefins because the olefins do not react with SO_2.

Sulfolane is a water-soluble biodegradable and highly polar compound valued for its solvent properties. Approximately 20 million pounds of sulfolane are consumed annually in applications that include delignification of wood, polymerization and fiber spinning, and electroplating bathes.[25] It is a solvent for selectively extracting aromatics from reformates and coke oven products.

CYCLIC OLIGOMERS OF BUTADIENE

Butadiene could be oligomerized to cyclic dienes and trienes using certain transition metal complexes. Commercially, a mixture of $TiCl_4$ and $Al_2Cl_3(C_2H_5)_3$ is used that gives predominantly cis, trans, trans-1,5,9-cyclododecatriene along with approximately 5% of the dimer 1,5-cyclooctadiene[24]:

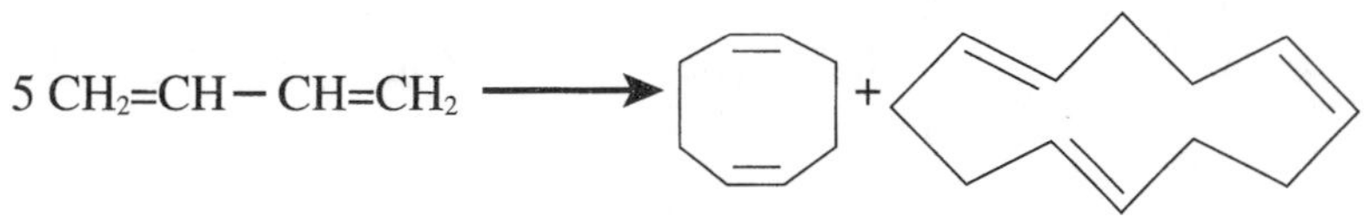

$$5\ CH_2{=}CH{-}CH{=}CH_2 \longrightarrow$$

1,5,9-Cyclododecatriene is a precursor for dodecane-dioic acid through a hydrogenation step followed by oxidation. The diacid is a monomer for the production of nylon 6/12.

Cyclododecane from cyclododecatriene may also be converted to the C_{12} lactam, which is polymerized to nylon 12.

REFERENCES

1. *Chemical and Engineering News,* Oct. 25, 1993, p. 30.
2. Brockhaus, R., German Patent, 1279, 011, 1968.
3. "Petrochemical Handbook," *Hydrocarbon Processing,* Vol. 58, No. 11, 1979, p. 120.
4. Harris, N. and Tuck, M. W., "Butanediol via Maleic Anhydride," *Hydrocarbon Processing,* Vol. 69, No. 5, 1990, pp. 79–82.
5. Grubbs, R. H. et al., *J. Am. Chem. Soc.,* Vol. 98, 1976, p. 3478.
6. Katz, T. J. *Adv. Organomet. Chem.,* Vol. 16, 1977, p. 283.
7. Herisson, J. L. and Chaurin, Y., *Makromol. Chem.,* 141, 1970, p. 161; Tsonis, C. P., *Journal of Applied Polymer Science,* Vol. 26, 1981, pp. 3525–3536.
8. Stinson, S., "New Rhenium Catalyst for Olefin Chemistry," *Chemical and Engineering News,* Vol. 70, No. 6, 1992, p. 29.
9. Cosyns, J. et al., *Hydrocarbon Processing,* Vol. 77, No. 3, 1998, p. 61.
10. Patton, P. A. and McCarthy, T. J., "Running the Impossible Reaction, Metathesis of Cyclohexene," *CHEMTECH,* July 1987, pp. 442–446.
11. Amigues, P. et al., "Propylene From Ethylene and Butene-2," *Hydrocarbon Processing,* Vol. 69, No. 10, 1990, pp. 79–80.
12. Nierlich, F. "Oligomerize for Better Gasoline," *Hydrocarbon Processing,* Vol. 71, No. 2, 1992, pp. 45–46.
13. "Petrochemical Handbook," *Hydrocarbon Processing,* Vol. 70, No. 3, 1991, p. 166.
14. Hucknall, D. J., "Selective Oxidation of Hydrocarbons," Academic Press Inc., New York 1974, pp. 55–69.
15. British Patent, 1, 182, 273 to Tejin.
16. West German Offen, 2, 354, 331 to Atlantic Richfield.
17. Unzelman, G. H., "U.S. Clean Air Act Expands Role for Oxygenates," *Oil and Gas Journal,* April 15, 1991.
18. Iborra, M., Izquierdo J. F., Tejero, J. and Cunil, F., *CHEMTECH,* Vol. 18, No. 2, 1988, pp. 120–122.

19. Abraham, O. C. and Prescott, G. F., "Make Isobutene from TBA," *Hydrocarbon Processing,* Vol. 71, No. 2, 1992, p. 51.
20. Hatch, L. F., *The Chemistry of Petrochemical Reactions,* Houston, Gulf Publishing Co., 1955, p. 149.
21. Belgian Patent 770, 615 to BASF, 1971.
22. *CHEMTECH,* April 1999, p. 19.
23. Heaton, C. A., ed. "An Introduction to Industrial Chemistry," 2nd ed. Blacki and Son Ltd., London, 1991, p. 395.
24. Parshall, G. W. and Nuget, W. A., "Functional Chemicals via Homogeneous Catalysis," *CHEMTECH,* May 1988, pp. 314–320.
25. *Chemical and Engineering News,* Sept 5, 1994, p. 26.

CHAPTER TEN

Chemicals Based on Benzene, Toluene, and Xylenes

INTRODUCTION

The primary sources of benzene, toluene, and xylenes (BTX) are refinery streams, especially from catalytic reforming and cracking, and pyrolysis gasoline from steam cracking and from coal liquids. BTX and ethyl benzene are extracted from these streams using selective solvents such as sulfolene or ethylene glycol. The extracted components are separated through lengthy fractional distillation, crystallization, and isomerization processes (Chapter 2).

The reactivity of C_6, C_7, C_8 aromatics is mainly associated with the benzene ring. Aromatic compounds in general are liable for electrophilic substitution. Most of the chemicals produced directly from benzene are obtained from its reactions with electrophilic reagents. Benzene could be alkylated, nitrated, or chlorinated to important chemicals that are precursors for many commercial products.

Toluene and xylenes (methylbenzenes) are substituted benzenes. Although the presence of methyl substituents activates the benzene ring for electrophilic attack, the chemistry of methyl benzenes for producing commercial products is more related to reactions with the methyl than with the phenyl group. As an electron-withdrawing substituent (of methane), the phenyl group influences the methyl hydrogens and makes them more available for chemical attack. The methyl group could be easily oxidized or chlorinated as a result of the presence of the phenyl substituent.

REACTIONS AND CHEMICALS OF BENZENE

Benzene (C_6H_6) is the most important aromatic hydrocarbon. It is the precursor for many chemicals that may be used as end products or intermediates. Almost all compounds derived directly from benzene are converted to other chemicals and polymers. For example, hydrogenation of benzene produces cyclohexane. Oxidation of cyclohexane produces cyclohexanone, which is used to make caprolactam for nylon manufacture. Due to the resonance stabilization of the benzene ring, it is not easily polymerized. However, products derived from benzene such as styrene, phenol, and maleic anhydride can polymerize to important commercial products due to the presence of reactive functional groups. Benzene could be alkylated by different alkylating agents, hydrogenated to cyclohexane, nitrated, or chlorinated.

The current world benzene capacity is approximately 35 million tons. The 1994 U.S. production of benzene was about 14.7 million pounds.[1]

The chemistry for producing the various chemicals from benzene is discussed in this section. Figure 10-1 shows the important chemicals derived from benzene.

ALKYLATION OF BENZENE

Benzene can be alkylated in the presence of a Lewis or a Bronsted acid catalyst. Olefins such as ethylene, propylene, and C_{12}–C_{14} alpha olefins are used to produce benzene alkylates, which have great commercial value. Alkyl halides such as monochloroparaffins in the C_{12}–C_{14} range also serve this purpose.

The first step in alkylation is the generation of a carbocation (carbonium ion). When an olefin is the alkylating agent, a carbocation intermediate forms.

$$RCH{=}CH_2 \xrightarrow{\ H^+\ } [R\overset{+}{C}HCH_3]$$

Carboncations also form from an alkyl halide when a Lewis acid catalyst is used. Aluminum chloride is the commonly used Friedel-Crafts alkylation catalyst. Friedel-Crafts alkylation reactions have been reviewed by Roberts and Khalaf:[2]

$$RCI + AlCl_3 \rightarrow [R^+ {\_\ \_\ \_\ \_} AlCl_4^-]$$

The next step is an attack by the carbocation on the benzene ring, followed by the elimination of a proton and the formation of a benzene alkylate:

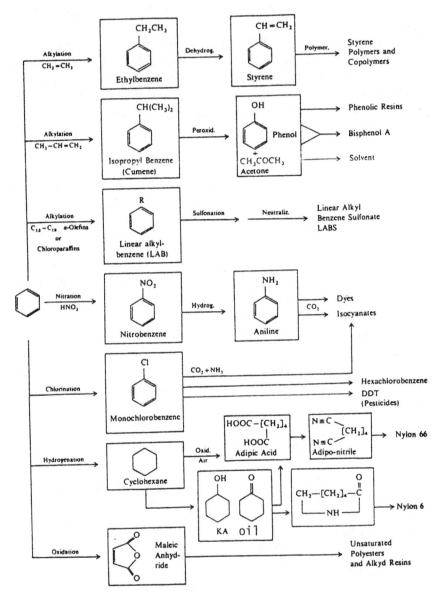

Figure 10-1. Important chemicals based on benzene.

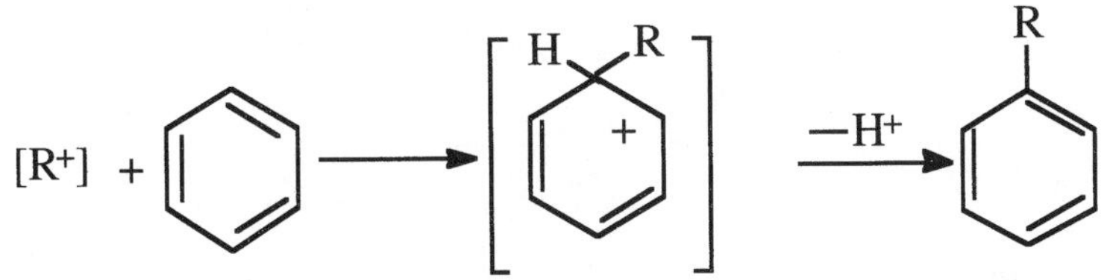

$$[R^+] \; + \; \bighexagon \longrightarrow \left[\bighexagon\overset{H \; R}{\underset{+}{}} \right] \xrightarrow{\;-H^+\;} \bighexagon\!\!R$$

Ethylbenzene $\left(\bighexagon\!\!CH_2CH_3 \right)$

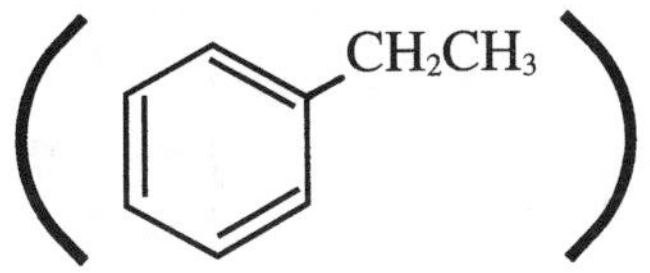

Ethylbenzene (EB) is a colorless aromatic liquid with a boiling point of 136.2°C, very close to that of *p*-xylene. This complicates separating it from the C_8 aromatic equilibrium mixture obtained from catalytic reforming processes. (See Chapter 2 for separation of C_8 aromatics). Ethylbenzene obtained from this source, however, is small compared to the synthetic route.

The main process for producing EB is the catalyzed alkylation of benzene with ethylene:

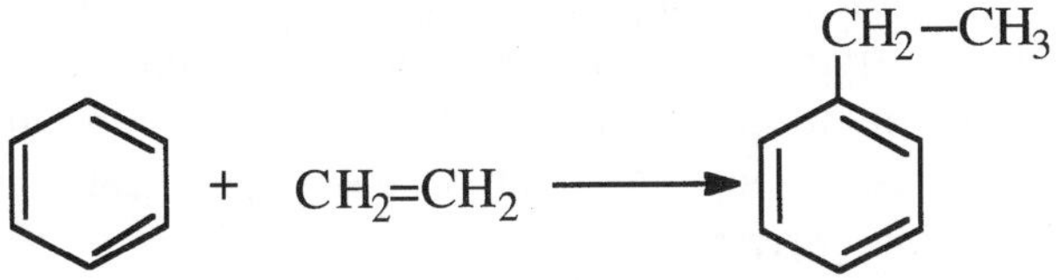

$$\bighexagon \; + \; CH_2{=}CH_2 \longrightarrow \bighexagon\!\!\overset{CH_2-CH_3}{}$$

Many different catalysts are available for this reaction. $AlCl_3$-HCl is commonly used. Ethyl chloride may be substituted for HCl in a mole-for-mole basis. Typical reaction conditions for the liquid-phase $AlCl_3$ catalyzed process are 40–100°C and 2–8 atmospheres. Diethylbenzene and higher alkylated benzenes also form. They are recycled and dealkylated to EB.

The vapor-phase Badger process (Figure 10-2), which has been commercialized since 1980, can accept dilute ethylene streams such as those produced from FCC off gas.[3] A zeolite type heterogeneous catalyst is used in a fixed bed process. The reaction conditions are 420°C and 200–300 psi. Over 98% yield is obtained at 90% conversion.[4,5] Polyethylbenzene (polyalkylated) and unreacted benzene are recycled and join the fresh feed to the reactor. The reactor effluent is fed to the benzene fractionation system to recover unreacted benzene. The bottoms

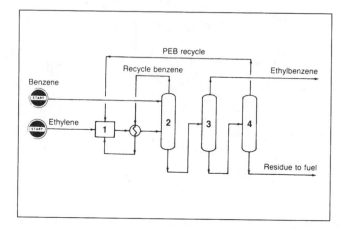

Figure 10-2. The Badger process for producing ethylbenzene:[3] (1) reactor, (2) fractionator (for recovery of unreacted benzene), (3) EB fractionator, (4) poly-ethylbenzene recovery column.

containing ethylbenzene and heavier polyalkylates are fractionated in two columns. The first column separates the ethylbenzene product, and the other separates polyethylbenzene for recycling. An optimization study of EB plants by constraint control was conducted by Hummel et al. They concluded that optimum operation could be maintained through a control system when conditions such as catalyst activity and heat transfer coefficients vary during operation.[6]

Ethylbenzene is mainly used to produce styrene. Over 90% of the 12.7 billion pounds of EB produced in the U.S. during 1998 was dehydrogenated to styrene.

Styrene

Styrene (vinylbenzene) is a liquid (b.p. 145.2°C) that polymerizes easily when initiated by a free radical or when exposed to light. The 1998 U.S. production of styrene was approximately 11 billion pounds.

Dehydrogenation of ethylbenzene to styrene occurs over a wide variety of metal oxide catalysts. Oxides of Fe, Cr, Si, Co, Zn, or their mixtures can be used for the dehydrogenation reaction. Typical reaction conditions for the vapor-phase process are 600–700°C, at or below atmospheric pressure. Approximately 90% styrene yield is obtained at 30–40% conversion:

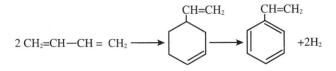

In the Monsanto/Lummus Crest process (Figure 10-3), fresh ethylbenzene with recycled unconverted ethylbenzene are mixed with superheated steam. The steam acts as a heating medium and as a diluent. The endothermic reaction is carried out in multiple radial bed reactors filled with proprietary catalysts. Radial beds minimize pressure drops across the reactor. A simulation and optimization of styrene plant based on the Lummus Monsanto process has been done by Sundaram et al.[7] Yields could be predicted, and with the help of an optimizer, the best operating conditions can be found. Figure 10-4 shows the effect of steam-to-EB ratio, temperature, and pressure on the equilibrium conversion of ethylbenzene.[7]

Alternative routes for producing styrene have been sought. One approach is to dimerize butadiene to 4-vinyl-1-cyclohexene, followed by catalytic dehydrogenation to styrene:[8]

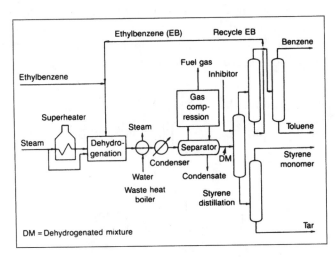

Figure 10-3. Schematic diagram of the Monsanto/Lummus Crest styrene plant.[7]

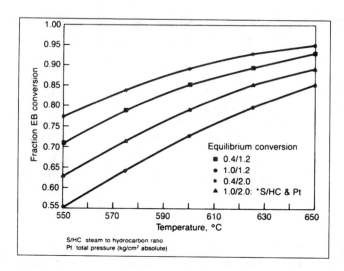

Figure 10-4. Effect of steam/EB, temperature, and pressure on the conversion of ethylbenzene.[7]

The process which was developed by DOW involves cyclodimerization of butadiene over a proprietary copper-loaded zeolite catalyst at moderate temperature and pressure (100°C and 250 psig). To increase the yield, the cyclodimerization step takes place in a liquid phase process over the catalyst. Selectivity for vinylcyclohexene (VCH) was over 99%. In the second step VCH is oxidized with oxygen over a proprietary oxide catalyst in presence of steam. Conversion over 90% and selectivity to styrene of 92% could be achieved.[9]

Another approach is the oxidative coupling of toluene to stilbene followed by disproportionation to styrene and benzene:

High temperatures are needed for this reaction, and the yields are low.

Cumene

$$\left(\begin{array}{c} CH_3\!-\!CHCH_3 \\ \end{array} \right)$$

Cumene (isopropylbenzene), a liquid, is soluble in many organic solvents but not in water. It is present in low concentrations in light refinery streams (such as reformates) and coal liquids. It may be obtained by distilling (cumene's B.P. is 152.7°C) these fractions.

The main process for producing cumene is a synthetic route where benzene is alkylated with propylene to isopropylbenzene.

Either a liquid or a gas-phase process is used for the alkylation reaction. In the liquid-phase process, low temperatures and pressures (approximately 50°C and 5 atmospheres) are used with sulfuric acid as a catalyst.

$$\bigcirc + CH_2\!\!=\!\!CHCH_3 \longrightarrow \bigcirc^{CH(CH_3)_2} \quad \Delta H^o_{298} = -113 \text{ KJ/mol}$$

Small amounts of ethylene can be tolerated since ethylene is quite unreactive under these conditions. Butylenes are relatively unimportant because butylbenzene can be removed as bottoms from the cumene column.

In the vapor-phase process, the reaction temperature and pressure are approximately 250°C and 40 atmospheres. Phosphoric acid on Kieselguhr is a commonly used catalyst. To limit polyalkylation, a mixture of propene-propane feed is used. Propylene can be as low as 40% of the feed mixture. A high benzene/propylene ratio is also used to decrease polyalkylation. A selectivity of about 97% based on benzene can be obtained.

In the UOP process (Figure 10-5), fresh propylene feed is combined with fresh and recycled benzene, then passed through heat exchangers and a steam preheater before being charged to the reactor.[10] The effluent is separated, and excess benzene recycled. Cumene is finally clay treated and fractionated. The bottom product is mainly diisopropyl benzene, which is reacted with benzene in a transalkylation section:

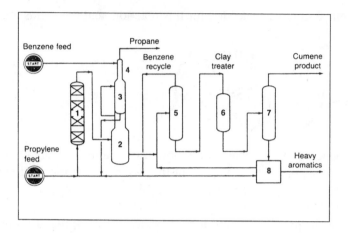

Figure 10-5. A flow diagram of the UOP cumene process:[10] (1) reactor, (2,3) two-stage flash system, (4) depropanizer, (5) benzene column, (6) clay treatment, (7) fractionator, (8) transalkylation section.

To reduce pollution, Dow developed a new catalyst system from the mordenite-zeolite group to replace phosophoric acid or aluminum chloride catalysts. The new catalysts eliminates the disposal of acid wastes and handling corrosive materials.[11]

The 1998 U.S. cumene production was approximately 6.7 billion pounds and was mainly used to produce phenol and acetone. A small amount of cumene is used to make α-methylstyrene by dehydrogenation.

α-Methylstyrene

α-Methylstyrene is used as a monomer for polymer manufacture and as a solvent.

Phenol and Acetone from Cumene

Phenol, C_6H_5OH (hydroxybenzene), is produced from cumene by a two-step process. In the first step, cumene is oxidized with air to cumene hydroperoxide. The reaction conditions are approximately 100–130°C and 2–3 atmospheres in the presence of a metal salt catalyst:

CH$_3$CHCH$_3$ (CH$_3$)$_2$COOH

⬡ + O$_2$ ⟶ ⬡   ΔH = −116 KJ/mol

In the second step, the hydroperoxide is decomposed in the presence of an acid to phenol and acetone. The reaction conditions are approximately 80°C and slightly below atmospheric:

(CH$_3$)$_2$COOH   OH   O
                                                         ‖
⬡ ⟶ H⁺ ⟶ ⬡   +   CH$_3$CCH$_3$

In this process (Figure 10-6), cumene is oxidized in the liquid phase.[12] The oxidation product is concentrated to 80% cumene hydroperoxide by

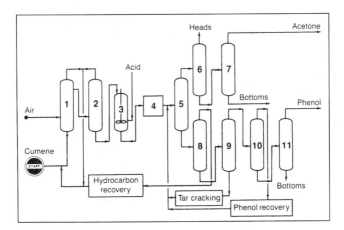

Figure 10-6. The Mitsui Petrochemical Industries process for producing phenol and acetone from cumene:[12] (1) autooxidation reactor, (2) vacuum tower, (3) cleavage reactor, (4) neutralizer, (5–11) purification train.

vacuum distillation. To avoid decomposition of the hydroperoxide, it is transferred immediately to the cleavage reactor in the presence of a small amount of H_2SO_4. The cleavage product is neutralized with alkali before it is finally purified.

After an initial distillation to split the coproducts phenol and acetone, each is purified in separate distillation and treating trains. An acetone finishing column distills product acetone from an acetone/water/oil mixture. The oil, which is mostly unreacted cumene, is sent to cumene recovery. Acidic impurities, such as acetic acid and phenol, are neutralized by caustic injection. Figure 10-7 is a simplified flow diagram of an acetone finishing column, and Table 10-1 shows the feed composition to the acetone finishing column.[13]

Cumene processes are currently the major source for phenol and coproduct acetone. Chapter 8 notes other routes for producing acetone.

Previously, phenol was produced from benzene by sulfonation followed by caustic fusion to sodium phenate. Phenol is released from the sodium salt of phenol by the action of carbon dioxide or sulfur dioxide.

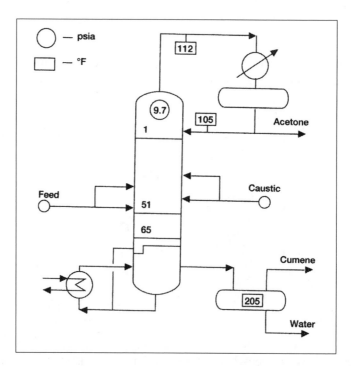

Figure 10-7. A simplified process flow chart of an acetone finishing column.[13]

Table 10-1
Feed composition of acetone finishing column[13]

Component	wt%
Acetone	48%
Water	22%
Cumene	24%
Alpha-methylstyrene and other heavy hydrocarbons	4%
Neutralized organics (sodium acetate, sodium phenate, etc.)	1%
Free caustic	1%

Direct hydroxylation of benzene to phenol could be achieved using zeolite catalysts containing rhodium, platinum, palladium, or irridium. The oxidizing agent is nitrous oxide, which is unavoidable a byproduct from the oxidation of KA oil (see KA oil, this chapter) to adipic acid using nitric acid as the oxidant.[14]

Phenol is also produced from chlorobenzene and from toluene via a benzoic acid intermediate (see "Reactions and Chemicals from Toluene").

Properties and Uses of Phenol

Phenol, a white crystalline mass with a distinctive odor, becomes reddish when subjected to light. It is highly soluble in water, and the solution is weakly acidic.

Phenol was the 33rd highest-volume chemical. The 1994 U.S. production of phenol was approximately 4 billion pounds. The current world capacity is approximately 15 billion pounds. Many chemicals and polymers derive from phenol. Approximately 50% of production goes to phenolic resins. Phenol and acetone produce bis-phenol A, an important monomer for epoxy resins and polycarbonates. It is produced by condensing acetone and phenol in the presence of HCI, or by using a cation exchange resin. Figure 10-8 shows the Chiyoda Corp. bisphenol A process.[15]

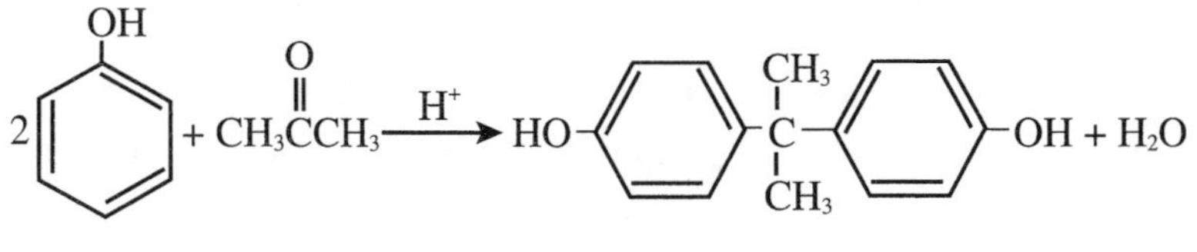

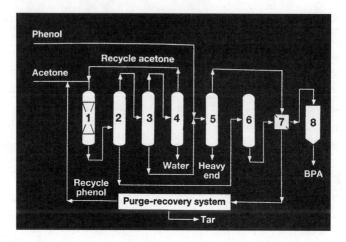

Figure 10-8. The CT-BISA (Chiyoda Corp.) process for producing bis-phenol A from acetone and phenol.[15] (1) reactor, (2–4) distillation columns, (5) phenol distillation column, (6) crystallizer, (7) solid/liquid separator, (8) prilling tower.

Important chemicals derived from phenol are salicylic acid; acetylsalicyclic acid (aspirin); 2,4-dichlorophenoxy acetic acid (2,4-D), and 2,4,5-triphenoxy acetic acid (2,4,5-T), which are selective herbicides; and pentachlorophenol, a wood preservative:

Salicyclic acid   Aspirin   2,4,5-T

Other halophenols are miticides, bactericides, and leather preservatives. Halophenols account for about 5% of phenol uses.

About 12% of phenol demand is used to produce caprolactam, a monomer for nylon 6. The main source for caprolactam, however, is toluene.

Phenol can be alkylated to alkylphenols. These compounds are widely used as nonionic surfactants, antioxidants, and monomers in resin polymer applications:

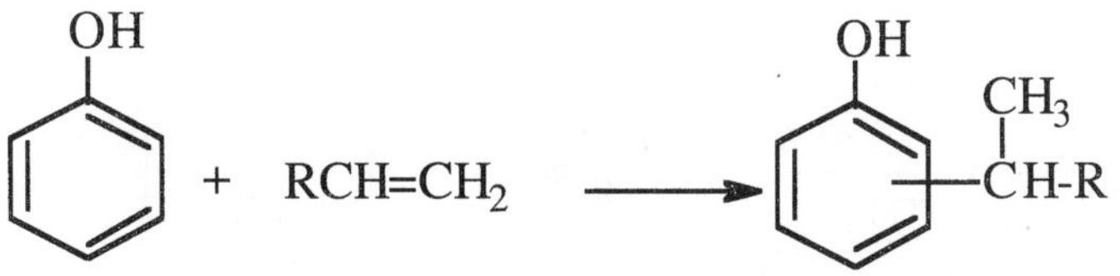

An alkylphenol

Phenol is also a precursor for aniline. The major process for aniline ($C_6H_5NH_2$) is the hydrogenation of nitrobenzene (see "Nitration of Benzene").

Linear Alkylbenzene

Linear alkylbenzene (LAB) is an alkylation product of benzene used to produce biodegradable anionic detergents. The alkylating agents are either linear C_{12}–C_{14} mono-olefins or monochloroalkanes. The linear olefins (alpha olefins) are produced by polymerizing ethylene using Ziegler catalysts (Chapter 7) or by dehydrogenating n-paraffins extracted from kerosines. Monochloroalkanes, on the other hand, are manufactured by chlorinating the corresponding n-paraffins. Dehydrogenation of n-paraffins to monoolefins using a newly developed dehydrogenation catalyst by UOP has been reviewed by Vora et al.[16] The new catalyst is highly active and allows a higher per-pass conversion to monoolefins. Because the dehydrogenation product contains a higher concentration of olefins for a given alkylate production rate, the total hydrocarbon feed to the HF alkylation unit is substantially reduced.[16]

Alkylation of benzene with linear monoolefins is industrially preferred. The Detal process (Figure 10-9) combines the dehydrogenation of n-paraffins and the alkylation of benzene.[17] Monoolefins from the dehydrogenation section are introduced to a fixed-bed alkylation reactor over a heterogeneous solid catalyst. Older processes use HF catalysts in a liquid phase process at a temperature range of 40–70°C. The general alkylation reaction of benzene using alpha olefins could be represented as:

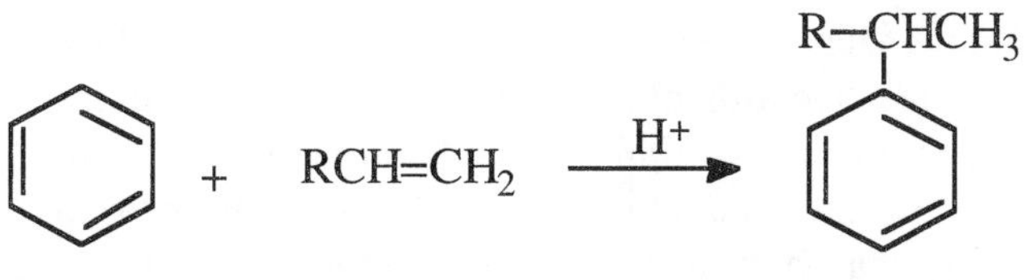

Linear alkylbenzene (LAB)

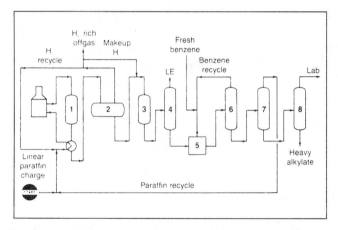

Figure 10-9. The UOP (Detal) process for producing linear alkylbenzene:[17] (1) pacol dehydrogenation reactor, (2) gas-liquid separation, (3) reactor for converting diolefins to monoolefins, (4) stripper, (5) alkylation reactor, (6,7,8) fractionators.

Typical properties of detergent alkylate are shown in Table 10-2.[16] Detergent manufacturers buy linear alkylbenzene, sulfonate it with SO_3, and then neutralize it with NaOH to produce linear alkylbenzene sulfonate (LABS), the active ingredient in detergents:

CHLORINATION OF BENZENE

Chlorination of benzene is an electrophilic substitution reaction in which Cl^+ serves as the electrophile. The reaction occurs in the presence of a Lewis acid catalyst such as $FeCl_3$. The products are a mixture of mono- and dichlorobenzenes. The *ortho-* and the *para*-dichlorobenzenes are more common than meta-dichlorobenzene. The ratio of the mono-chloro to dichloro products essentially depends on the benzene/chlorine ratio and the residence time. The ratio of the dichloro-isomers (*o-* to *p-* to *m*-dichlorobenzenes) mainly depends on the reaction temperature and residence time:

Table 10-2
Typical properties of detergent alkylate[16]

	Linear detergent alkylate
Bromine number	0.02
Saybolt color	+30
Alkylbenzene content, wt%	97.4
Doctor test	NEGATIVE
Unsulfonatable content, wt%	1.0
Water, wt%	0.1
Specific gravity at 60°F	0.8612
Refractive index, n^{20}_D	1.4837
Flash point (ASTM D-93), °F	280
Average molecular weight	240
Distillation (ASTM D-86), °F	
IBP	538
10 vol%	547
30 vol%	550
50 vol%	554
70 vol%	559
90 vol%	569
95 vol%	576
EP	589
Saybolt color of a 5% sodium alkylbenzene sulfonate solution	+26
Normal alkylbenzene, wt%	93
2-Phenyl isomer, wt%	20.0
Paraffin, wt%	0.1
Biodegradability (ASTM D-2667), %	>95.0

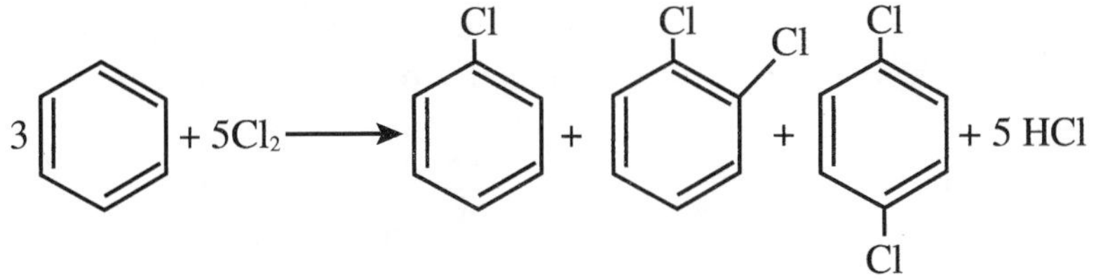

Typical liquid-phase reaction conditions for the chlorination of benzene using $FeCl_3$ catalyst are 80–100°C and atmospheric pressure. When a high benzene/Cl_2 ratio is used, the product mixture is approximately 80% monochlorobenzene, 15% p-dichlorobenzene and 5% o-dichlorobenzene.

Continuous chlorination processes permit the removal of mono-chlorobenzene as it is formed, resulting in lower yields of higher chlorinated benzene.

Monochlorobenzene is also produced in a vapor-phase process at approximately 300°C. The by-product HCl goes into a regenerative oxychlorination reactor. The catalyst is a promoted copper oxide on a silica carrier:

$$4 \, HCl + O_2 \longrightarrow 2 \, Cl_2 + 2 \, H_2O$$

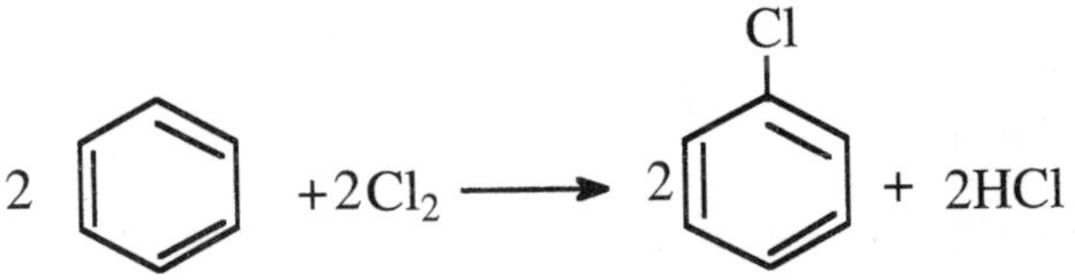

Higher conversions have been reported when temperatures of 234–315°C and pressures of 40–80 psi are used.[18]

Monochlorobenzene is the starting material for many compounds, including phenol and aniline. Others, such as DDT, chloronitrobenzenes, polychlorobenzenes, and biphenyl, do not have as high a demand for monochlorobenzene as aniline and phenol.

NITRATION OF BENZENE (Nitrobenzene [$C_6H_5NO_2$])

Similar to the alkylation and the chlorination of benzene, the nitration reaction is an electrophilic substitution of a benzene hydrogen (a proton) with a nitronium ion (NO_2^+). The liquid-phase reaction occurs in presence of both concentrated nitric and sulfuric acids at approximately 50°C. Concentrated sulfuric acid has two functions: it reacts with nitric acid to form the nitronium ion, and it absorbs the water formed during the reaction, which shifts the equilibrium to the formation of nitrobenzene:

$$HNO_3 + 2H_2SO_4 \longrightarrow 2 \, HSO_4^- + H_3O^+ + NO_2^+$$

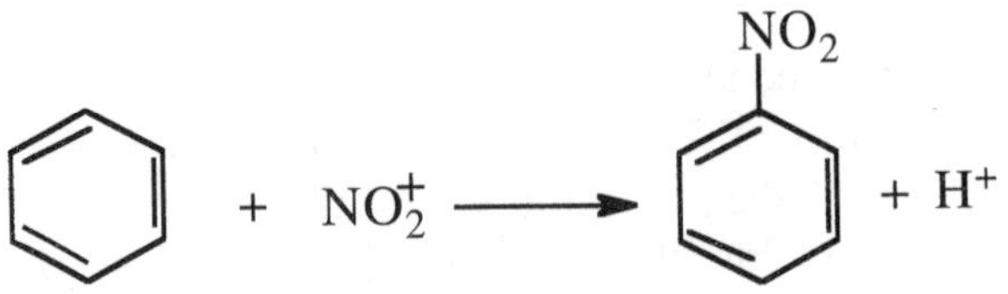

Most of the nitrobenzene ($\approx$97%) produced is used to make aniline. Other uses include synthesis of quinoline, benzidine, and as a solvent for cellulose ethers.

Aniline ($C_6H_5NH_2$)

Aniline (aminobenzene) is an oily liquid that turns brown when exposed to air and light. The compound is an important dye precursor.

The main process for producing aniline is the hydrogenation of nitrobenzene:

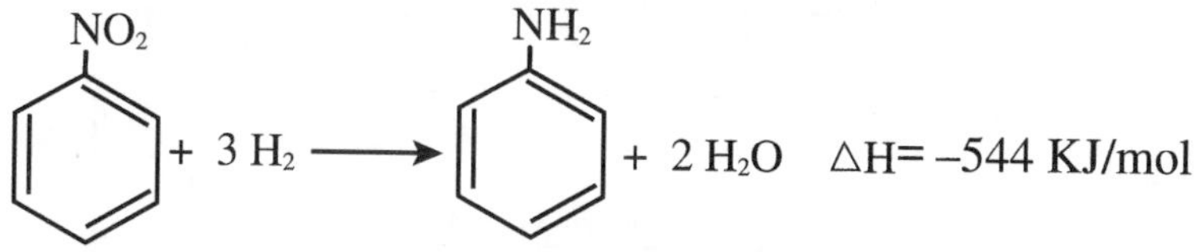

$$+ \; 3 \; H_2 \longrightarrow \quad + \; 2 \; H_2O \quad \Delta H = -544 \; KJ/mol$$

The hydrogenation reaction occurs at approximately 270°C and slightly above atmospheric over a Cu/Silica catalyst. About a 95% yield is obtained.

An alternative way to produce aniline is through ammonolysis of either chlorobenzene or phenol. The reaction of chlorobenzene with aqueous ammonia occurs over a copper salt catalyst at approximately 210°C and 65 atmospheres. The yield of aniline from this route is also about 96%:

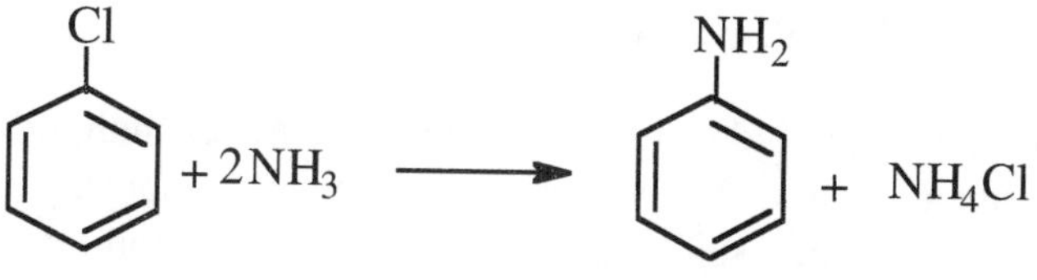

$$+ \; 2NH_3 \longrightarrow \quad + \; NH_4Cl$$

Ammonolysis of phenol occurs in the vapor phase. In the Scientific Design Co. process (Figure 10-10), a mixed feed of ammonia and phenol is heated and passed over a heterogeneous catalyst in a fixed-bed system.[19] The reactor effluent is cooled, the condensed material distilled, and the unreacted ammonia recycled. Aniline produced this way should be very pure:

$$+ \; NH_3 \longrightarrow \quad + \; H_2O$$

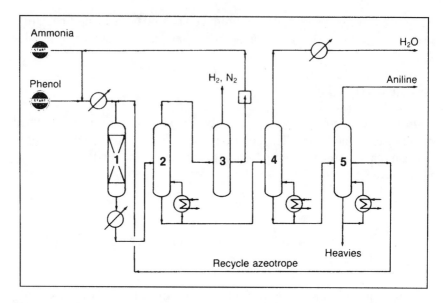

Figure 10-10. The Scientific Co. process for producing aniline from phenol:[19] (1) fixed-bed reactor, (2) liquid-gas separator, (3) ammonia compression and recycling, (4) drier, (5) fractionator.

OXIDATION OF BENZENE

Benzene oxidation is the oldest method to produce maleic anhydride. The reaction occurs at approximately 380°C and atmospheric pressure. A mixture of V_2O_5/MO_3 is the usual catalyst. Benzene conversion reaches 90%, but selectivity to maleic anhydride is only 50–60%; the other 40–50% is completely oxidized to CO_2:[20]

$$\text{C}_6\text{H}_6 + \%\,O_2 \longrightarrow \text{(maleic anhydride)} + 2\,CO_2 + 2H_2O$$

Currently, the major route to maleic anhydride, especially for the newly-erected processes, is the oxidation of butane (Chapter 6). Maleic anhydride also comes from oxidation of n-butenes. Properties and chemicals derived from maleic anhydride are noted in Chapter 9.

HYDROGENATION OF BENZENE

Cyclohexane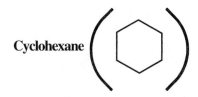

The hydrogenation of benzene produces cyclohexane. Many catalyst systems, such as Ni/alumina and Ni/Pd, are used for the reaction. General reaction conditions are 160–220°C and 25–30 atmospheres. Higher temperatures and pressures may also be used with sulfided catalysts:

$$\text{benzene} + 3H_2 \longrightarrow \text{cyclohexane} \qquad \Delta H = -266 \text{ KJ/mol}$$

Older methods use a liquid phase process (Figure 10-11).[10] New gas-phase processes operate at higher temperatures with noble metal catalysts. Using high temperatures accelerates the reaction (faster rate).[21] The hydrogenation of benzene to cyclohexane is characterized by a highly exothermic reaction and a significant decrease in the product volume

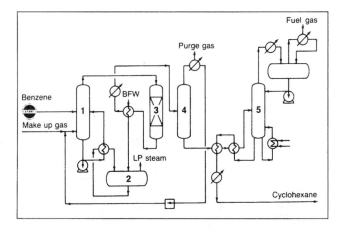

Figure 10-11. The Institut Francais du Petrole process for the hydrogenation of benzene to cyclohexane:[10] (1) liquid-phase reactor, (2) heat exchanger, (3) catalytic pot (acts as a finishing reactor when conversion of the main reactor drops below the required level), (4) high-pressure separator, (5) stabilizer.

(from 4 to 1). Equilibrium conditions are therefore strongly affected by temperature and pressure. Figure 10-12 shows the effect of H_2/benzene mole ratio on the benzene content in the products.[21] It is clear that benzene content in the product decreases with an increase of the reactor inlet pressure.

Another nonsynthetic source for cyclohexane is natural gasoline and petroleum naphtha. However, only a small amount is obtained from this source. The 1994 U.S. production of cyclohexane was approximately 2.1 billion pounds (the 45th highest chemical volume).

Properties and Uses of Cyclohexane

Cyclohexane is a colorless liquid, insoluble in water but soluble in hydrocarbon solvents, alcohol, and acetone. As a cyclic paraffin, it can be easily dehydrogenated to benzene. The dehydrogenation of cyclohexane

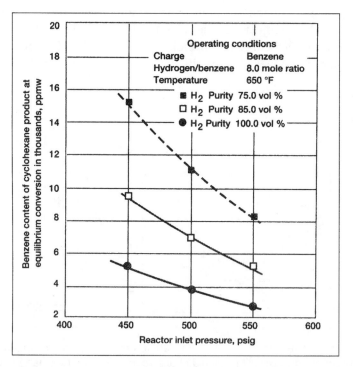

Figure 10-12. Effect of hydrogen purity and pressure on benzene conversion to cyclohexane.[21]

and its derivatives (present in naphthas) to aromatic hydrocarbons is an important reaction in the catalytic reforming process.

Essentially, all cyclohexane is oxidized either to a cyclohexanone-cyclohexanol mixture used for making caprolactam or to adipic acid. These are monomers for making nylon 6 and nylon 6/6.

Oxidation of Cyclohexane (Cyclohexanone-Cyclohexanol and Adipic Acid)

Cyclohexane is oxidized in a liquid-phase process to a mixture of cyclohexanone and cyclohexanol (KA oil). The reaction conditions are 95–120°C at approximately 10 atmospheres in the presence of a cobalt acetate and orthoboric acid catalyst system. About 95% yield can be obtained:

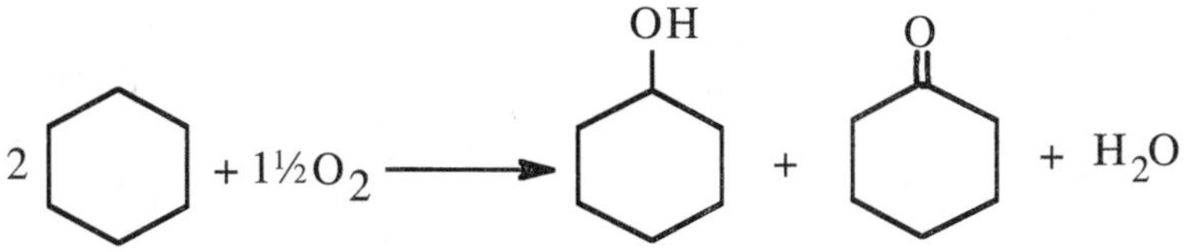

$$2\ \bigcirc + 1\tfrac{1}{2}O_2 \longrightarrow \overset{OH}{\bigcirc} + \overset{O}{\bigcirc} + H_2O$$

KA oil is used to produce caprolactam, the monomer for nylon 6. Caprolactam is also produced from toluene through the intermediate formation of cyclohexane carboxylic acid.

Cyclohexane is also a precursor for adipic acid. Oxidizing cyclohexane in the liquid-phase at lower temperatures and for longer residence times (than for KA oil) with a cobalt acetate catalyst produces adipic acid:

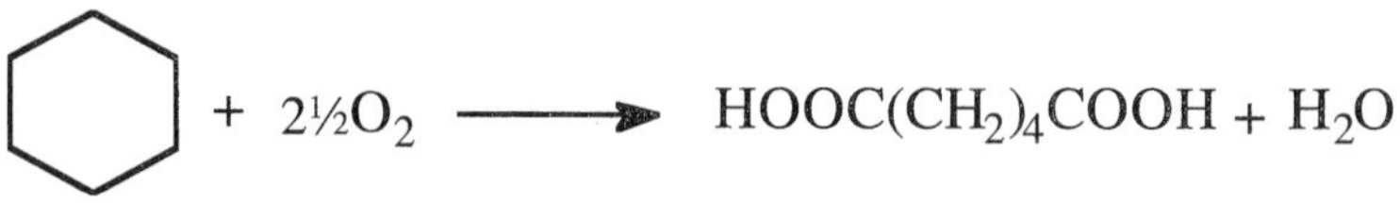

$$\bigcirc + 2\tfrac{1}{2}O_2 \longrightarrow HOOC(CH_2)_4COOH + H_2O$$

Adipic acid may also be produced from butadiene via a carbonylation route (Chapter 9).

Adipic acid and its esters are used to make nylon 6/6. It may also be hydrogenated to 1,6-hexanediol, which is further reacted with ammonia to hexamethylenediamine.

$$HOOC(CH_2)_4COOH + 4H_2 \rightarrow HO(CH_2)_6OH + 2H_2O$$

$$HO(CH_2)_6OH + 2NH_3 \rightarrow H_2N(CH_2)_6NH_2 + 2H_2O$$

Hexamethylenediamine is the second monomer for nylon 6/6.

REACTIONS AND CHEMICALS OF TOLUENE

Toluene (methylbenzene) is similar to benzene as a mononuclear aromatic, but it is more active due to presence of the electron-donating methyl group. However, toluene is much less useful than benzene because it produces more polysubstituted products. Most of the toluene extracted for chemical use is converted to benzene via dealkylation or disproportionation. The rest is used to produce a limited number of petrochemicals. The main reactions related to the chemical use of toluene (other than conversion to benzene) are the oxidation of the methyl substituent and the hydrogenation of the phenyl group. Electrophilic substitution is limited to the nitration of toluene for producing mononitrotoluene and dinitrotoluenes. These compounds are important synthetic intermediates.

The 1994 U.S. toluene production (of all grades) was approximately 6.8 billion pounds. Hydrodealkylating toluene to benzene was the largest end use in United States and West Europe, followed by solvent applications.

DEALKYLATION OF TOLUENE

Toluene is dealkylated to benzene over a hydrogenation-dehydrogenation catalyst such as nickel. The hydrodealkylation is essentially a hydrocracking reaction favored at higher temperatures and pressures. The reaction occurs at approximately 700°C and 40 atmospheres. A high benzene yield of about 96% or more can be achieved:

Hydrodealkylation of toluene and xylenes with hydrogen is noted in Chapter 3.

Dealkylation also can be effected by steam. The reaction occurs at 600–800°C over Y, La, Ce, Pr, Nd, Sm, or Th compounds, $Ni-Cr_2O_3$ catalysts, and $Ni-Al_2O_3$ catalysts at temperatures between 320–630°C.[22] Yields of about 90% are obtained. This process has the advantage of producing, rather than using, hydrogen.

DISPROPORTIONATION OF TOLUENE

The catalytic disproportionation of toluene (Figure 10-13)[23] in the presence of hydrogen produces benzene and a xylene mixture. Disproportionation is an equilibrium reaction with a 58% conversion per pass theoretically possible. The reverse reaction is the transalkylation of xylenes with benzene:

Typical conditions for the disproportionation reaction are 450–530°C and 20 atmospheres. A mixture of $CoO\text{-}MoO_3$ on aluminosilicates/alumina catalysts can be used. Conversions of approximately 40% are normally used to avoid more side reactions and faster catalyst deactivation.[24] The equilibrium constants for this reaction are not significantly changed by shifting from liquid to vapor phase or by large temperature changes.[25]

Currently, zeolites, especially those of ZSM-5 type, are preferred for their higher activities and selectivities. They are also more stable thermally. Modifying ZSM-5 zeolites with phosphorous, boron, or

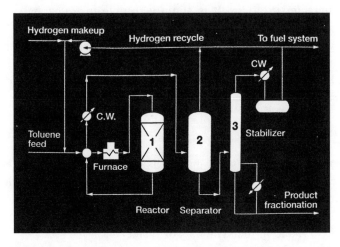

Figure 10-13. The Mobil Oil Corp., IFP process for the disproportionation of toluene to mixed xylenes.[23]

magnesium compounds produces xylene mixtures rich in the p-isomer (70–90%). It has been proposed that the oxides of these elements, present in zeolites, reduce the dimensions of the pore openings and channels and so favor formation and outward diffusion of p-xylene, the isomer with the smallest minimum dimension.[26,27]

OXIDATION OF TOLUENE

Benzoic Acid

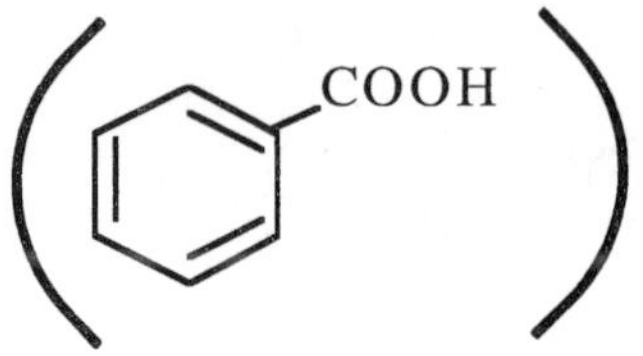

$$\left(\quad \text{COOH} \quad \right)$$

Oxidizing toluene in the liquid phase over a cobalt acetate catalyst produces benzoic acid. The reaction occurs at about 165°C and 10 atmospheres. The yield is over 90%:

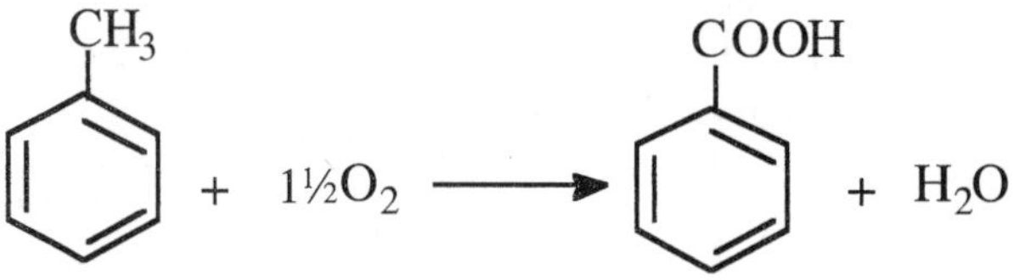

$$\text{CH}_3 \quad + \quad 1\frac{1}{2}\text{O}_2 \longrightarrow \text{COOH} \quad + \quad \text{H}_2\text{O}$$

Benzoic acid (benzene carboxylic acid) is a white crystalline solid with a characteristic odor. It is slightly soluble in water and soluble in most common organic solvents.

Though much benzoic acid gets used as a mordant in calico printing, it also serves to season tobacco, preserve food, make dentifrices, and kill fungus. Furthermore it is a precursor for caprolactam, phenol, and terephthalic acid.

Caprolactam Production

Caprolactam, a white solid that melts at 69°C, can be obtained either in a fused or flaked form. It is soluble in water, ligroin, and chlorinated hydrocarbons. Caprolactam's main use is to produce nylon 6. Other minor uses are as a crosslinking agent for polyurethanes, in the plasticizer industry, and in the synthesis of lysine.

The first step in producing caprolactam from benzoic acid is its hydrogenation to cyclohexane carboxylic acid at approximately 170°C and 16 atmospheres over a palladium catalyst:[28]

COOH

+ 3 H$_2$ ⟶

COOH

The resulting acid is then converted to caprolactam through a reaction with nitrosyl-sulfuric acid:

COOH

+ (NO)HSO$_4$ ⟶

C=O

NH + H$_2$SO$_4$ + CO$_2$

Nitrosyl sulfuric acid   Caprolactam

Figure 10-14 shows an integrated caprolactam production process.[28] Toluene, the feed, is first oxidized to benzoic acid. Benzoic acid is then hydrogenated to cyclohexane carboxylic acid, which reacts with nitrosyl-sulfuric acid yielding caprolactam. Nitrosyl sulfuric acid comes from reacting nitrogen oxides with oleum. Caprolactam comes as an acidic solution that is neutralized with ammonia and gives ammonium sulfate as

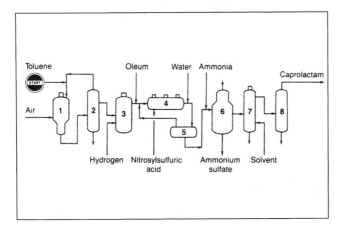

Figure 10-14. The SNIA BPD process for producing caprolactam:[28] (1) toluene oxidation reactor, (2) fractionator, (3) hydrogenation reactor (stirred autoclave), (4) multistage reactor (conversion to caprolactam), (5) water dilution, (6) crystallizer, (7) solvent extraction, (8) fractionator.

a by-product of commercial value. Recovered caprolactam is purified through solvent extraction and fractionation.

Phenol from Benzoic Acid

The action of a copper salt converts benzoic acid to phenol. The copper, reoxidized by air, functions as a real catalyst. The Lummus process operates in the vapor phase at approximately 250°C. Phenol yield of 90% is possible:

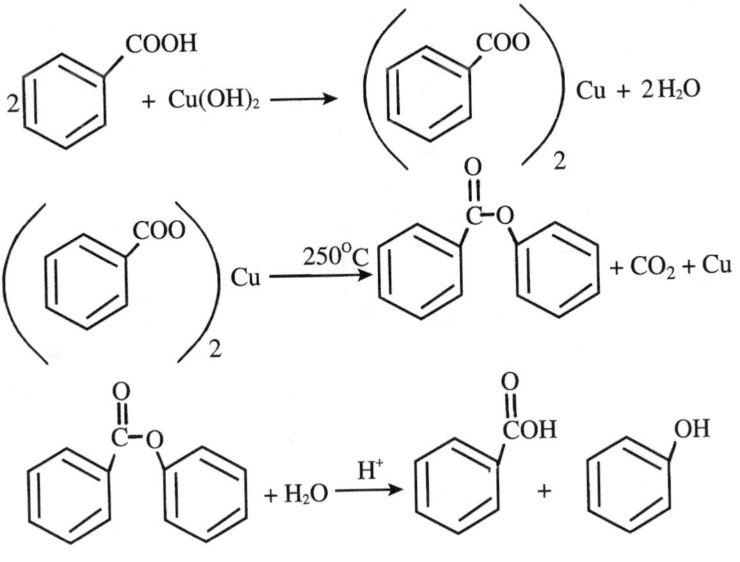

$$Cu + \tfrac{1}{2}O_2 + H_2O \longrightarrow Cu(OH)_2$$

The overall reaction is

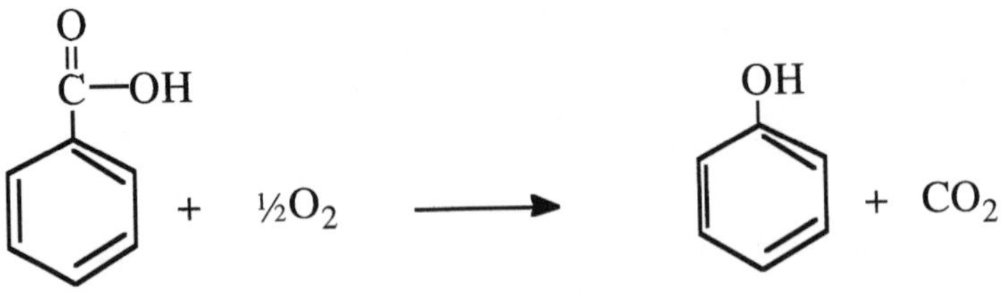

In the Lummus process (Figure 10-15), the reaction occurs in the liquid phase at approximately 220–240°C over Mg^{2+} + Cu^{2+} benzoate.[29] Magnesium benzoate is an initiator, with the Cu^{2+} reduced to Cu^{1+}. The copper (1) ions are reoxidized to copper (II) ions.

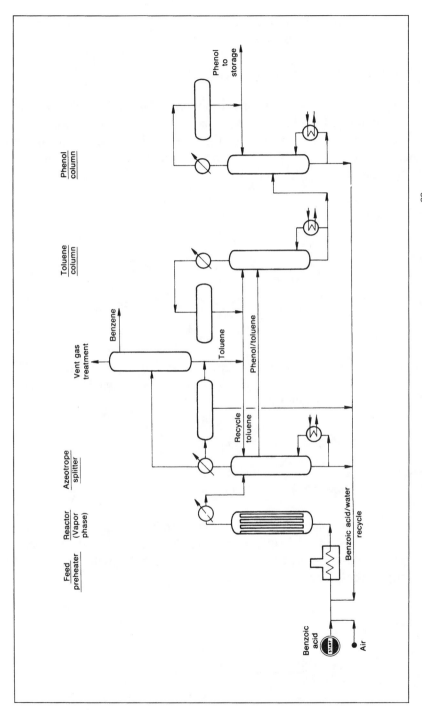

Figure 10-15. The Lummus benzoic-acid-to-phenol process.[29]

Phenol can also be produced from chlorobenzene and from cumene, the major route for this commodity.

Terephthalic Acid from Benzoic Acid

Terephthalic acid is an important monomer for producing polyesters. The main route for obtaining the acid is the catalyzed oxidation of paraxylene. It can also be produced from benzoic acid by a disproportionation reaction of potassium benzoate in the presence of carbon dioxide. Benzene is the coproduct:

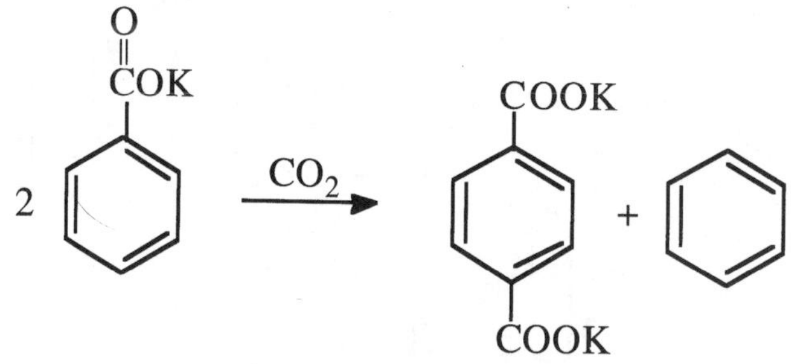

The reaction occurs in a liquid-phase process at approximately 400°C using ZnO or CdO catalysts. Terephthalic acid is obtained from an acid treatment; the potassium salt is recycled.[30,31]

Benzaldehyde

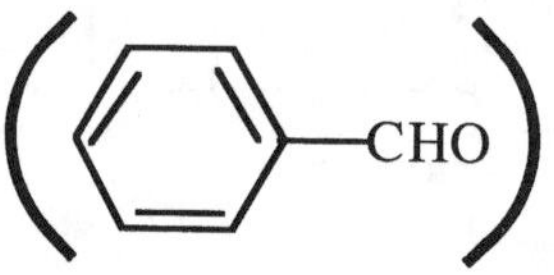

Oxidizing toluene to benzaldehyde is a catalyzed reaction in which a selective catalyst limits further oxidation to benzoic acid. In the first step, benzyl alcohol is formed and then oxidized to benzaldehyde. Further oxidation produces benzoic acid:

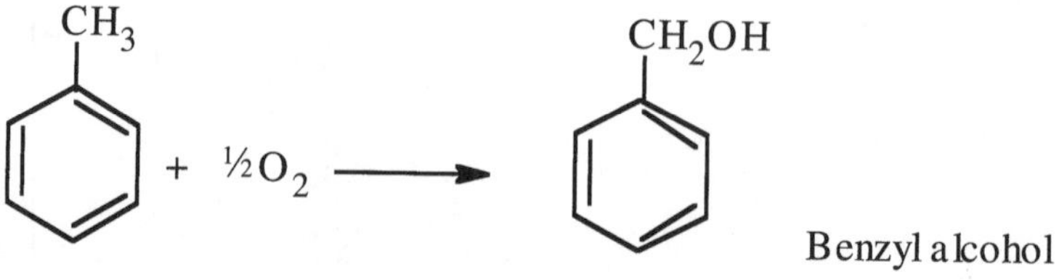

Benzyl alcohol

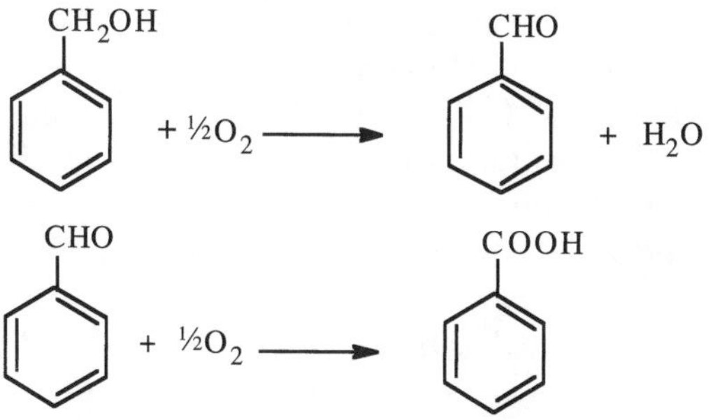

The problem with this reaction is that each successive oxidation occurs more readily than the preceding one (more acidic hydrogens after introducing the oxygen hetero atom, which facilitates the oxidation reaction to occur). In addition to using a selective catalyst, the reaction can be limited to the production of the aldehyde by employing short residence times and a high toluene-to-oxygen ratio. In one process, a mixture of UO_2 (93%) and MnO_2 (7%) is the catalyst. A yield of 30–50% could be obtained at low conversions of 10–20%. The reaction temperature is approximately 500°C. In another process, the reaction goes forward in the presence of methanol over an $FeBr_2$—$CoBr_2$ catalyst mixture at approximately 100–140°C.

Benzaldehyde has limited uses as a chemical intermediate. It is used as a solvent for oils, resins, cellulose esters, and ethers. It is also used in flavoring compounds and in synthetic perfumes.

CHLORINATION OF TOLUENE

The chlorination of toluene by substituting the methyl hydrogens is a free radical reaction. A mixture of three chlorides (benzyl chloride, benzal chloride and benzotrichloride) results.

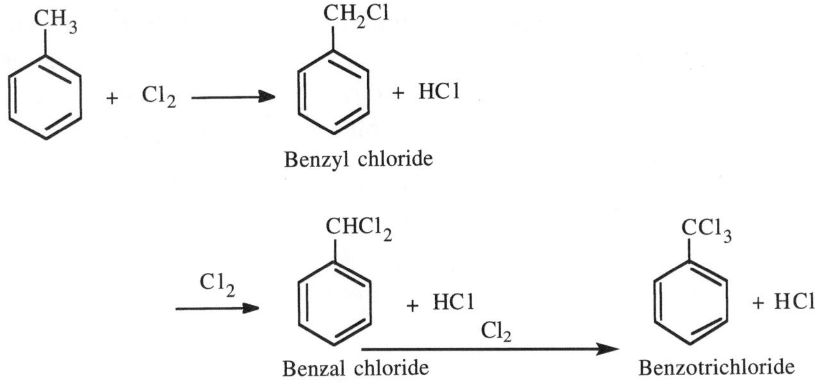

The ratio of the chloride mixture mainly derives from the toluene/chlorine ratio and the contact time. Benzyl chloride is produced by passing dry chlorine into boiling toluene (110°C) until reaching a density of 1.283. At this density, the concentration of benzyl chloride reaches the maximum. Light can initiate the reaction.

Benzyl chloride can produce benzyl alcohol by hydrolysis:

CH$_2$Cl

$+ H_2O \longrightarrow$

CH$_2$OH

$+ HCl$

Benzyl alcohol

Benzyl alcohol is a precursor for butylbenzyl phthalate,

$$\underset{\text{C}_4\text{H}_9\text{OCC}_6\text{H}_4\text{COCH}_2\text{C}_6\text{H}_5,}{\overset{\text{O} \quad\ \ \text{O}}{\overset{\|\quad\ \ \|}{}}}$$

a vinyl chloride plasticizer. Benzyl chloride is also a precursor for phenylacetic acid via the intermediate benzyl cyanide. Phenylacetic acid is used to make phenobarbital (a sedative) and penicillin G.

Benzal chloride is hydrolyzed to benzaldehyde, and benzotrichloride is hydrolyzed to benzoic acid.

Chlorinated toluenes are not large-volume chemicals, but they are precursors for many synthetic chemicals and pharmaceuticals.

NITRATION OF TOLUENE

Nitration of toluene is the only important reaction that involves the aromatic ring rather than the aliphatic methyl group. The nitration reaction occurs with an electrophilic substitution by the nitronium ion. The reaction conditions are milder than those for benzene due to the activation of the ring by the methyl substituent. A mixture of nitrotoluenes results. The two important monosubstituted nitrotoluenes are o- and p-nitrotoluenes:

CH$_3$

NO$_2$

o-Nitrotoluene

CH$_3$

NO$_2$

p-Nitrotoluene

Mononitrotoluenes are usually reduced to corresponding toluidines, which make dyes and rubber chemicals:

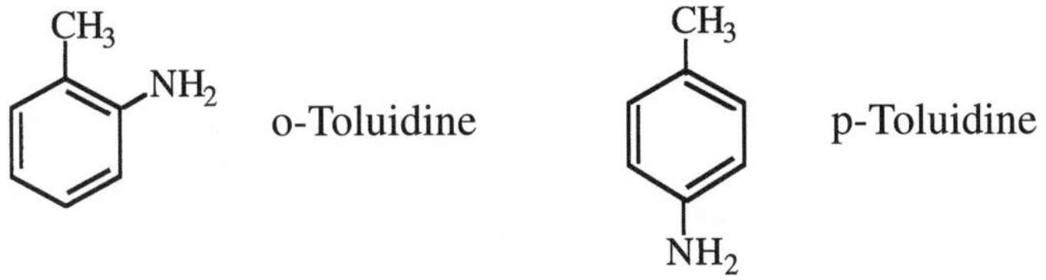

o-Toluidine   p-Toluidine

Dinitrotoluenes are produced by nitration of toluene with a mixture of concentrated nitric and sulfuric acid at approximately 80°C. The main products are 2,4- and 2,6-dinitrotoluenes:

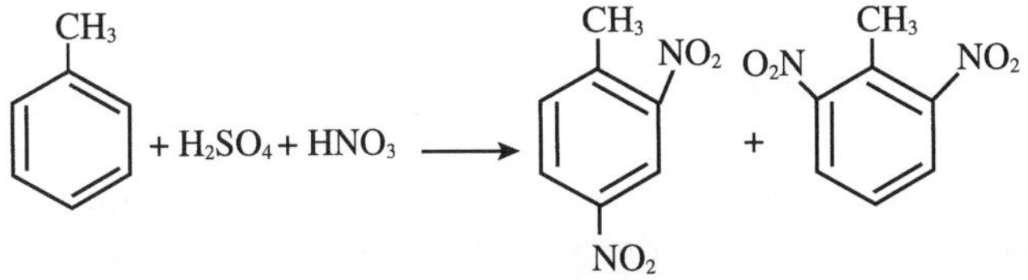

2,4-Dinitrotoluene   2,6-Dinitrotoluene
80%   20%

The dinitrotoluenes are important precursors for toluene diisocyanates (TDI), monomers used to produce polyurethanes.

The TDI mixture is synthesized from dinitrotoluenes by a first-step hydrogenation to the corresponding diamines. The diamines are then treated with phosgene to form TDI. The yield from toluene is approximately 85%:

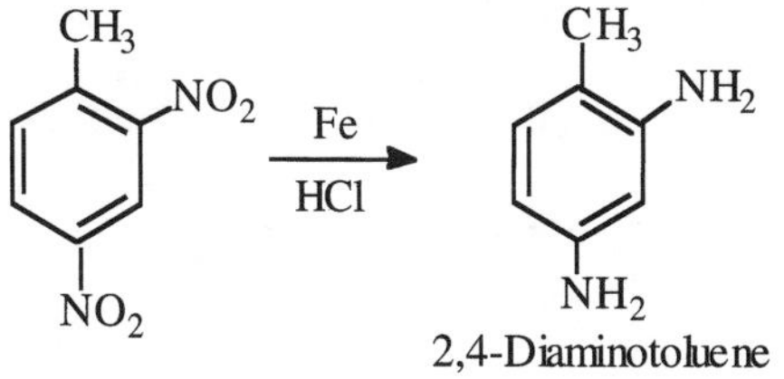

2,4-Diaminotoluene

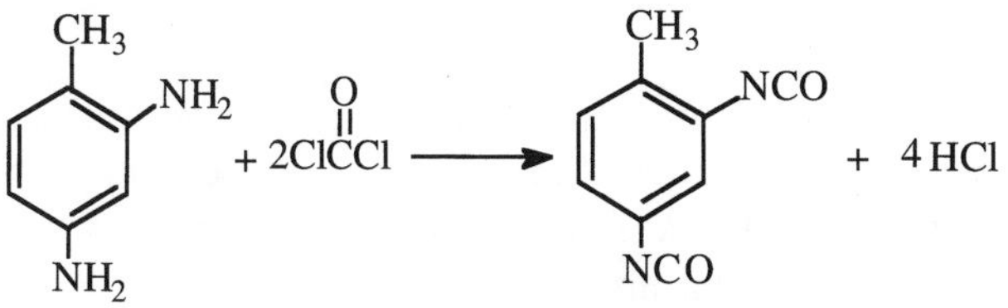

2,4-Toluene diisocyanate

An alternative route for TDI is through a liquid-phase carbonylation of dinitrotoluene in presence of PdCl$_2$ catalyst at approximately 250°C and 200 atmospheres:

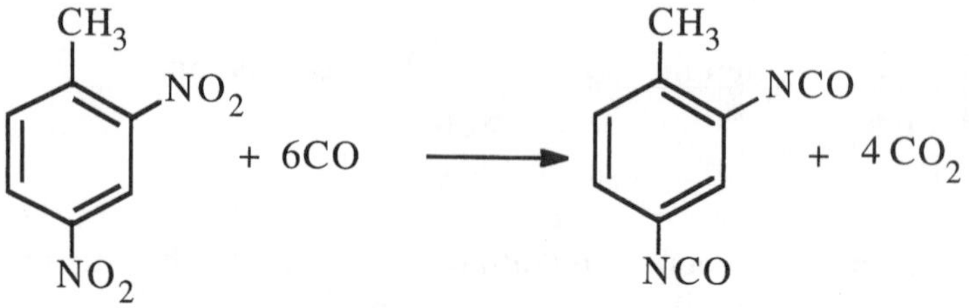

Trinitrotoluene TNT is a well-known explosive obtained by further nitration of the dinitrotoluenes.

CARBONYLATION OF TOLUENE

The carbonylation reaction of toluene with carbon monoxide in the presence of HF/BF$_3$ catalyst produces p-tolualdehyde. A high yield results (96% based on toluene and 98% based on CO). p-Tolualdehyde could be further oxidized to terephthalic acid, an important monomer for polyesters:

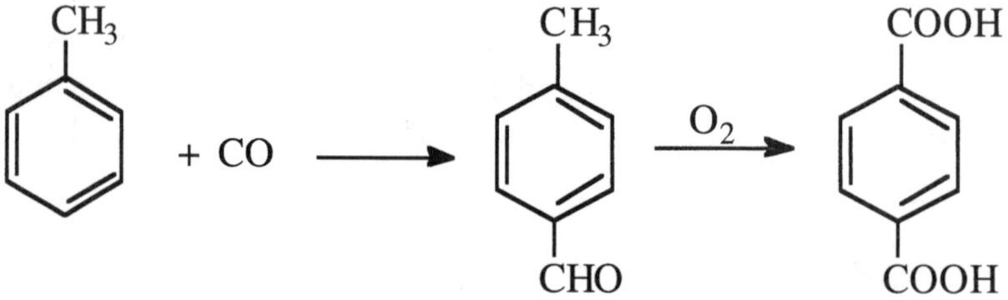

p-Tolualdehyde is also an intermediate in the synthesis of perfumes, dyes and pharmaceuticals.

CHEMICALS FROM XYLENES

Xylenes (dimethylbenzenes) are an aromatic mixture composed of three isomers (o-, m-, and p-xylene). They are normally obtained from catalytic reforming and cracking units with other C$_6$, C$_7$, and C$_8$ aromatics. Separating the aromatic mixture from the reformate is done by extraction-distillation and isomerization processes (Chapter 2).

para-Xylene is the most important of the three isomers for producing terephthalic acid to manufacture polyesters. *m*-Xylene is the least used of the three isomers, but the equilibrium mixture obtained from catalytic reformers has a higher ratio of the meta isomer. Table 10-3 shows the thermodynamic composition of C_8 aromatics at three temperatures.[32] *m*-Xylene is usually isomerized to the more valuable *p*-xylene.

As mentioned earlier, xylene chemistry is primarily related to the methyl substituents, which are amenable to oxidation.

Approximately 65% of the isolated xylenes are used to make chemicals. The rest are either used as solvents or blended with gasolines. The 1998 U.S. production of mixed xylenes for chemical use was approximately 9.5 million pounds. *p*-Xylene alone was about 7.7 million pounds that year.

TEREPHTHALIC ACID (HOOCC$_6$H$_4$COOH)

The catalyzed oxidation of *p*-xylene produces terephthalic acid (TPA). Cobalt acetate promoted with either NaBr or HBr is used as a catalyst in an acetic acid medium. Reaction conditions are approximately 200°C and 15 atmospheres. The yield is about 95%:

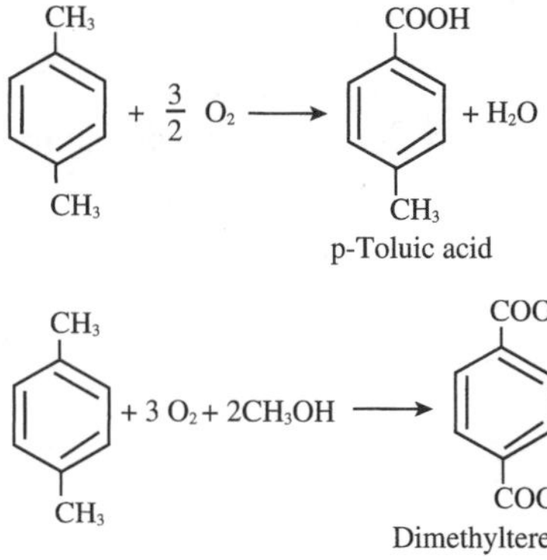

p-Toluic acid

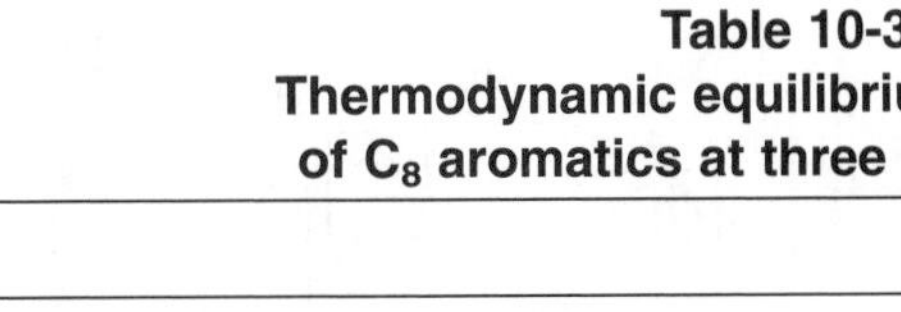

Dimethylterephthalate

Table 10-3
Thermodynamic equilibrium composition
of C$_8$ aromatics at three temperatures[32]

	Composition		
Aromatics wt%	200°C	300°C	500°C
p-Xylene	21.8	21.1	18.9
o-Xylene	20.6	21.6	23.0
m-Xylene	53.5	51.1	47.1
Ethylbenzene	4.1	6.2	11.0

Special precautions must be taken so that the reaction does not stop at the *p*-toluic acid stage. One approach is to esterify toluic acid as it is formed with methanol. This facilitates the oxidation of the second methyl group. The resulting dimethyl terephthalate (DMT) may be hydrolyzed to terephthalic acid.

Another approach is to use an easily oxidized substance such as acetaldehyde or methylethyl ketone, which, under the reaction conditions, forms a hydroperoxide. These will accelerate the oxidation of the second methyl group. The DMT process encompasses four major processing steps: oxidation, esterification, distillation, and crystallization. Figure 10-16 shows a typical *p*-xylene oxidation process to produce terephthalic acid or dimethyl terephthalate.[33] The main use of TPA and DMT is to produce polyesters for synthetic fiber and film.

Phthalic Anhydride

Currently, phthalic anhydride is mainly produced through catalyzed oxidation of *o*-xylene. A variety of metal oxides are used as catalysts. A typical one is $V_2O_5 + TiO_2/Sb_2O_3$. Approximate conditions for the vapor-phase oxidation are 375–435°C and 0.7 atmosphere. The yield of phthalic anhydride is about 85%:

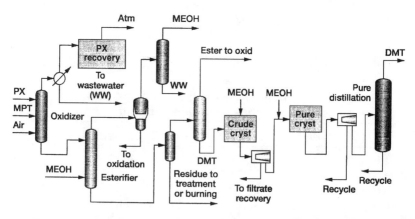

Figure 10-16. A typical *p*-xylene to dimethyl terephthalate process.[33]

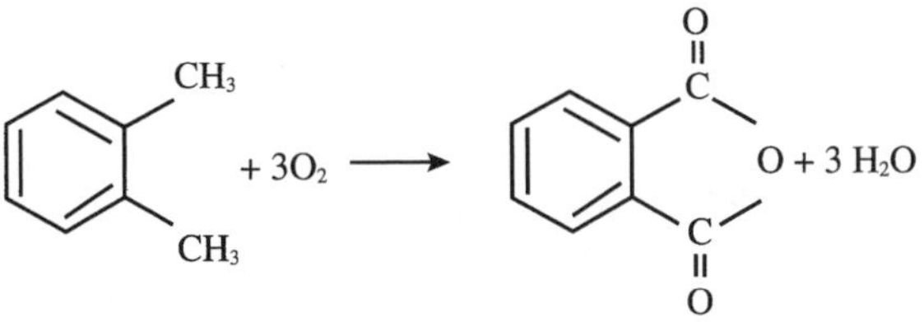

Liquid-phase oxidation of *o*-xylene also works at approximately 150°C. Cobalt or manganese acetate in acetic acid medium serves as a catalyst.

The major by-products of this process are maleic anhydride, benzoic acid, and citraconic anhydride (methylmaleic anhydride). Maleic anhydride could be recovered economically.[34]

Phthalic anhydride's main use is for producing plasticizers by reactions with C_4–C_{10} alcohols. The most important polyvinyl chloride plasticizer is formed by the reaction of 2-ethylhexanol (produced via butyraldehyde, Chapter 8) and phthalic anhydride:

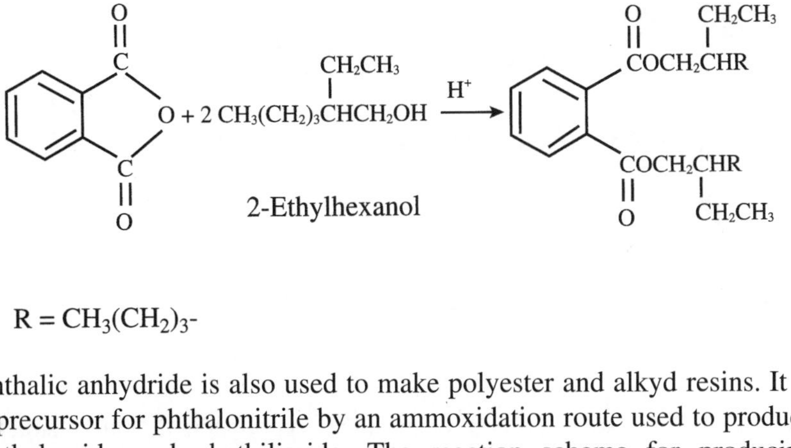

$$R = CH_3(CH_2)_3-$$

Phthalic anhydride is also used to make polyester and alkyd resins. It is a precursor for phthalonitrile by an ammoxidation route used to produce phthalamide and phathilimide. The reaction scheme for producing phthalonitrile, phthalamide, and phathilimide is shown in Figure 10-17.[34]

Isophthalic Acid

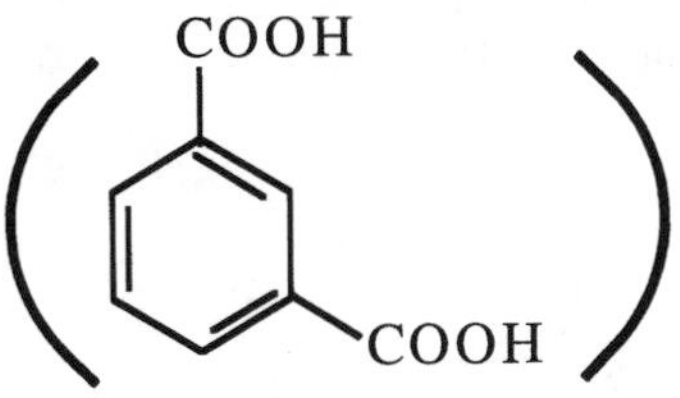

The oxidation of *m*-xylene produces isophthalic acid. The reaction occurs in the liquid-phase in presence of ammonium sulfite:

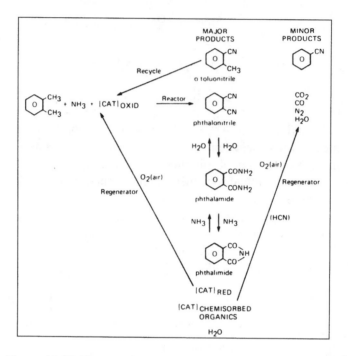

Figure 10-17. The reaction scheme for *o*-xylene to phthalonitrile.[34]

$$\text{(m-xylene)} + 2\,(NH_4)_2SO_3 \longrightarrow \text{(isophthalic acid)} + 2\,H_2S + 4NH_3 + 2\,H_2O$$

Isophthalic acid's main use is for producing polyesters that are characterized by a higher abrasion resistance than those using other phthalic acids.
Polyesters from isophthalic acid are used for pressure molding applications.

Ammoxidation of isophthalic acid produces isophthalonitrile. The
reaction resembles the one used for ammoxidation of phthalic anhydride:

$$\text{(m-xylene)} + 2\,NH_3 + 3\,O_2 \longrightarrow \text{(isophthalonitrile)} + 6\,H_2O$$

Isophthalonitrile

Isophthalonitrile serves as a precursor for agricultural chemicals. It is readily hydrogenated to the corresponding diamine, which can form polyamides or be converted to isocyanates for polyurethanes.

REFERENCES

1. *Chemical and Engineering News,* April 10, 1995, p. 17.
2. Roberts, R. and Khalaf, A., *Friedel-Crafts Alkylation Chemistry,* Marcel Dekker, Inc. New York, 1984, Chapter 2.
3. "Petrochemical Handbook," *Hydrocarbon Processing,* Vol. 70, No. 3, 1991, p. 154.
4. Lewis, P. J. and Dwyer, F. G., *Oil and Gas Journal,* Sept. 26, 1977, pp. 55–58.
5. Dwyer, F. G., Lewis, P. J., and Schneider, F. H., *Chemical Engineering,* Jan. 5, 1976, pp. 90–91.
6. Hummel, H. K., DeWit G. B., and Maarleveld, A., "The Optimization of EB Plant by Constraint Control," *Hydrocarbon Processing,* Vol. 70, No. 3, 1991, pp. 67–71.
7. Sundaram, K. M. et al., "Styrene Plant Simulation and Optimization," *Hydrocarbon Processing,* Vol. 70, No. 1, 1991, pp. 93–97.
8. *CHEMTECH,* Vol. 7, No. 6, 1977, pp. 334–451.
9. *Chemical and Engineering News,* June 20, 1994, p. 31.
10. "Petrochemical Handbook," *Hydrocarbon Processing,* Vol. 70, No. 3, 1991, p. 152.
11. Illman, D. "Environmentally Benign Chemistry Aims for Processes That Don't Pollute," *Chemical and Engineering News,* Sept. 5, 1994, p. 26.
12. "Petrochemical Handbook," *Hydrocarbon Processing,* Vol. 70, No. 3, 1991, p. 168.
13. Fulmer, J. W. and Graf, K. C. "Distill Acetone in Tower Packing," *Hydrocarbon Processing,* Vol. 70. No. 10, 1991, pp. 87–91.
14. Platkin, J. and Fitzgerald, M. "Patent Watch," *CHEMTECH,* June 1999, p. 39. U.S. patent 5874646, Feb. 23, 1999.
15. "Petrochemical Handbook," *Hydrocarbon Processing,* Vol. 78, No. 3, 1999, p. 98.
16. Vora, B. V. et al., "Latest LAB Developments," *Hydrocarbon Processing,* Vol. 63, No.11, 1984, pp. 86–90.
17. "Petrochemical Handbook," *Hydrocarbon Processing,* Vol. 70, No. 3, 1991, p. 130.

18. Frontier Chemical Co., U.S. Patent 3, 148, 222 (1964).
19. "Petrochemical Handbook," *Hydrocarbon Processing,* Vol. 70, No. 3, 1991, p. 136.
20. Matar, S., Mirbach, M., and Tayim, H., *Catalysis in Petrochemical Processes,* Kluwer Academic Publishers, The Netherlands, 1989, pp. 84–108.
21. Abraham, O. C. and Chapman, G. L. "Hydrogenate benzene," *Hydrocarbon Processing,* Vol. 70, No. 10, 1991, pp. 95–97.
22. Ohsumi, Y. and Komatsuzaki, Y., U.S. Patent 3, 903, 186, Sept. 2, 1975 to Mitsubishi Chemical Industries, Ltd. and Asis Oil Co. Ltd. Japan.
23. "Petrochemical Handbook," *Hydrocarbon Processing,* Vol. 78, No. 3, 1999, p. 122.
24. Vora, B. V., Jensen, R. H., and Rockett, K. W., Paper No. 20 (P-I) Second Arab Conference on Petrochemicals, Abu Dhabi, March 15–22, 1976.
25. Hasting, S. H. and Nicholson D. E., *J. Chem. Eng. Data,* Vol. 6, 1961, p. 1.
26. Kaeding, W., Chu, C., Young, L. and Butter, S. "Selective Disproportionation of Toluene to Produce Benzene and p-Xylene," *Journal of Catalysis,* Vol. 69, No. 2, 1981, pp. 392–398.
27. Meisel, S. L. "Catalysis Research Bears Fruit," *CHEMTECH,* January 1988, pp. 32–37.
28. "Petrochemical Handbook," *Hydrocarbon Processing,* Vol. 70, No. 3, 1991, p. 150.
29. Gelbein, A. D. and Nislick, A. S., *Hydrocarbon Processing,* Vol. 57, No. 11, 1978, pp. 125–128.
30. Cines, M. R., U.S. Patent 3, 746, 754, July 17, 1973 to Phillips Petroleum Co.; U.S. Patent 2, 905, 709 and 2, 794, 830.
31. Sittig, M., *Aromatic Hydrocarbons, Manufacture and Technology,* Park Ridge, N.J.: Noyes Data Corp., 1976, pp. 303–306.
32. Masseling, J. H., *CHEMTECH,* Vol. 6, No. 11, 1976, p. 714.
33. Braggiato, C. and Gualy, R., "Improve DMT Production," *Hydrocarbon Processing,* Vol. 77, No. 6, 1998, pp. 61–65.
34. Sze, M. C. and Gelbein, A. P., *Hydrocarbon Processing,* Vol. 55, No. 2, 1976, pp. 103–106.

CHAPTER ELEVEN

Polymerization

INTRODUCTION

Polymerization is a reaction in which chain-like macromolecules are formed by combining small molecules (monomers).

Monomers are the building blocks of these large molecules called polymers. One natural polymer is cellulose (the most abundant organic compound on earth), a molecule made of many simple glucose units (monomers) joined together through a glycoside linkage.[1] Proteins, the material of life, are polypeptides made of α-amino acids attached by an amide

$$\begin{matrix} O \\ \parallel \\ +CNH+ \end{matrix}$$ linkage.

The polymer industry dates back to the 19th century, when natural polymers, such as cotton, were modified by chemical treatment to produce artificial silk (rayon). Work on synthetic polymers did not start until the beginning of the 20th century. In 1909, L. H. Baekeland prepared the first synthetic polymeric material using a condensation reaction between formaldehyde and phenol. Currently, these polymers serve as important thermosetting plastics (phenol formaldehyde resins). Since Baekeland's discovery, many polymers have been synthesized and marketed. Many modern commercial products (plastics, fibers, rubber) derive from polymers. The huge polymer market directly results from extensive work in synthetic organic compounds and catalysts. Ziegler's discovery of a coordination catalyst in the titanium family paved the road for synthesizing many stereoregular polymers with improved properties. This chapter reviews the chemistry involved in the synthesis of polymers.

MONOMERS, POLYMERS, AND COPOLYMERS

A *monomer* is a reactive molecule that has at least one functional group (e.g. -OH, -COOH, -NH$_2$, -C=C-). Monomers may add to themselves as in the case of ethylene or may react with other monomers having different functionalities. A monomer initiated or catalyzed with a specific catalyst polymerizes and forms a macromolecule—a polymer. For example, ethylene polymerized in presence of a coordination catalyst produces a linear homopolymer (linear polyethylene):

n CH$_2$=CH$_2$ → ‍$+$CH$_2$–CH$_2$$+_n$ (Linear polyethylene)

A *copolymer,* on the other hand, results from two different monomers by addition polymerization. For example, a thermoplastic polymer with better properties than an ethylene homopolymer comes from copolymerizing ethylene and propylene:

$$n\ CH_2{=}CH_2 + n\ CH_3CH{=}CH_2 \longrightarrow\ +CH_2CH_2\ CH_2CH\overset{\underset{|}{CH_3}}{}+_n$$

Block copolymers are formed by reacting two different *prepolymers,* which are obtained by polymerizing the molecules of each monomer separately. A block copolymer made of styrene and butadiene is an important synthetic rubber:

Alternating copolymers have the monomers of one type alternating in a regular manner with the monomers of the other, regardless of the composition of the reactants. For example, an alternate copolymer of vinyl acetate and vinyl chloride could be represented as:

$$\overset{\overset{OAC}{|}}{-CH_2C}\ \overset{\overset{Cl}{|}}{HCH_2C}\ \overset{\overset{OAC}{|}}{HCH_2C}\ \overset{\overset{Cl}{|}}{HCH_2C}\ H \overline{}$$

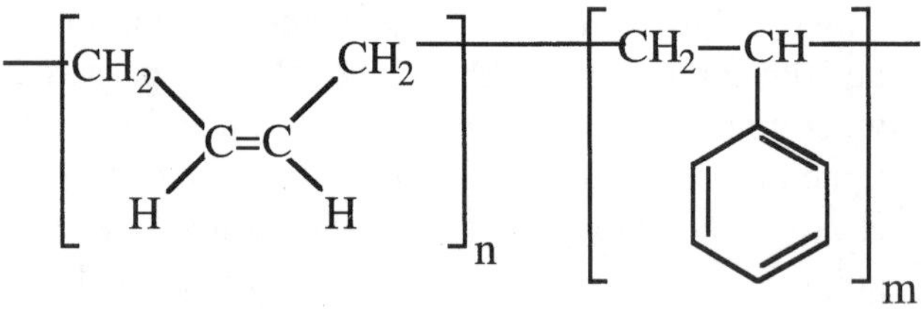

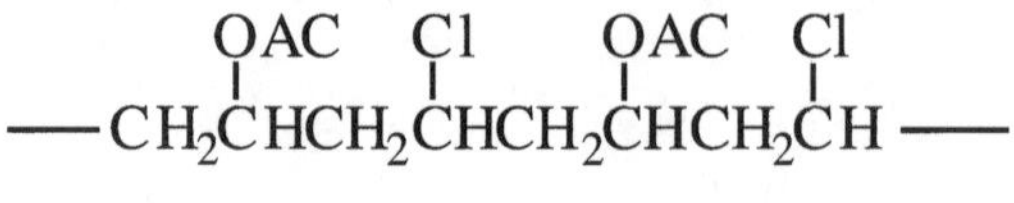

Random copolymers have the different monomer molecules distributed randomly along the polymer chain.

A polymer molecule may have just a linear chain or one or more branches protruding from the polymer backbone. Branching results mainly from chain transfer reactions (see "Chain Transfer Reactions" later in this chapter) and affects the polymer's physical and mechanical properties. Branched polyethylene usually has a few long branches and many more short branches

$$\text{---} CH_2 - CH_2 - CH - CH_2 \text{---}$$
$$\qquad\qquad | $$
$$\qquad\qquad CH_2$$
$$\qquad\qquad |$$
$$\qquad\qquad CH_2$$
$$\qquad\qquad \}$$

Intentional branching may improve the properties of the product polymer through grafting. A *graft copolymer* can be obtained by creating active sites on the polymer backbone. The addition of a different monomer then reacts at the active site and forms a branch. For example, polyethylene irradiated with gamma rays and then exposed to a reactive monomer, such as acrylonitrile, produces a polyethylene-polymer with acrylonitrile branches:[2,3]

$$\text{---} CH_2 - CH - CH_2 \text{---}$$
$$\qquad\qquad | $$
$$\qquad\qquad CH_2$$
$$\qquad\qquad |$$
$$\qquad\qquad CHCN$$
$$\qquad\qquad \}$$

Crosslinked polymers have two or more polymer chains linked together at one or more points other than their ends. The network formed improves the mechanical and physical properties of the polymer.

Crosslinking may occur during the polymerization reaction when multifunctional groups are present (as in phenol-formaldehyde resins) or through outside linking agents (as in the vulcanization of rubber with sulfur).

POLYMERIZATION REACTIONS

Two general reactions form synthetic polymers: chain addition and condensation.

Addition polymerization requires a chain reaction in which one monomer molecule adds to a second, then a third and so on to form a macromolecule. Addition polymerization monomers are mainly low molecular-weight olefinic compounds (e.g., ethylene or styrene) or conjugated diolefins (e.g., butadiene or isoprene).

Condensation polymerization can occur by reacting either two similar or two different monomers to form a long polymer. This reaction usually releases a small molecule like water, as in the case of the esterification of a diol and a diacid. In condensation polymerization where ring opening occurs, no small molecule is released (see "Condensation Polymerization" later in this chapter).

ADDITION POLYMERIZATION

Addition polymerization is employed primarily with substituted or unsubstituted olefins and conjugated diolefins. Addition polymerization initiators are free radicals, anions, cations, and coordination compounds. In addition polymerization, a chain grows simply by adding monomer molecules to a propagating chain. The first step is to add a free radical, a cationic or an anionic initiator (I^Z) to the monomer. For example, in ethylene polymerization (with a special catalyst), the chain grows by attaching the ethylene units one after another until the polymer terminates. This type of addition produces a linear polymer:

$$I^Z + CH_2{=}CH_2 \rightarrow I{+}CH_2{-}CH_2{+}^z$$

Branching occurs especially when free radical initiators are used due to chain transfer reactions (see following section, "Free Radical Polymerizations"). For a substituted olefin (such as vinyl chloride), the addition primarily produces the most stable intermediate (I). Intermediate (II) does not form to any appreciable extent:

$$I^Z + CH_2{=}CHR \longrightarrow \begin{cases} [I-CH_2CH]^Z \quad (\text{I}) \\ \qquad\quad | \\ \qquad\quad R \\ [I-CH-CH_2]^Z \ (\text{II}) \\ \quad\ \ | \\ \quad\ \ R \end{cases}$$

I^Z = a free radical $I^•$, cation I^+, or an anion I^-
R = alkyl, phenyl, Cl, etc.

Propagation then occurs by successive monomer molecules additions to the intermediates. Three addition modes are possible: (a) Head to tail; (b) Head to head, and (c) tail to tail.

The head-to-tail addition mode produces the most stable intermediate. For example, styrene polymerization mainly produces the head-to-tail intermediate:

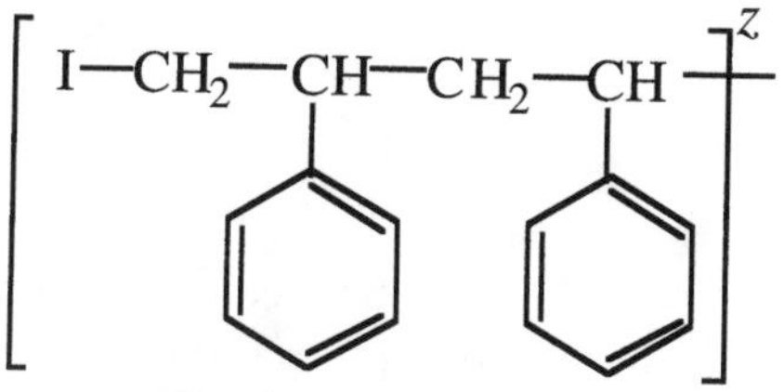

$$\left[\text{I}-CH_2-CH-CH_2-CH \right]_z$$

Head to-tail mode

Head-to-head or tail-to-tail modes of addition are less likely because the intermediates are generally unstable:

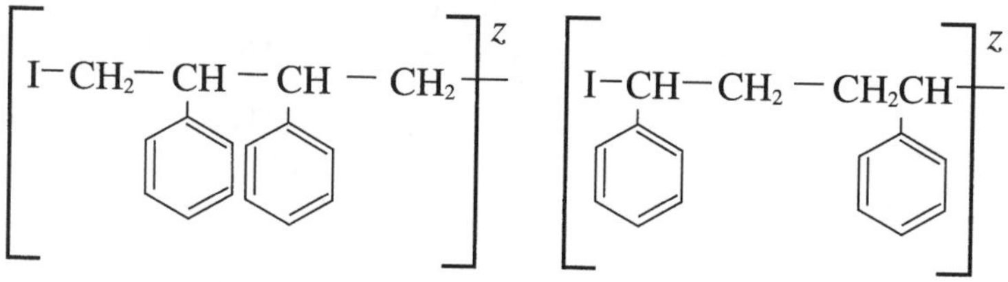

$$\left[\text{I}-CH_2-CH-CH-CH_2 \right]_z \qquad \left[\text{I}-CH-CH_2-CH_2CH \right]_z$$

Head-to-head mode   Tail-to-tail mode

Chain growth continues until the propagating polymer chain terminates.

Free Radical Polymerization

Free radical initiators can polymerize olefinic compounds. These chemical compounds have a weak covalent bond that breaks easily into two free radicals when subjected to heat. Peroxides, hydroperoxides and azo compounds are commonly used. For example, heating peroxybenzoic acid forms two free radicals, which can initiate the polymerization reaction:

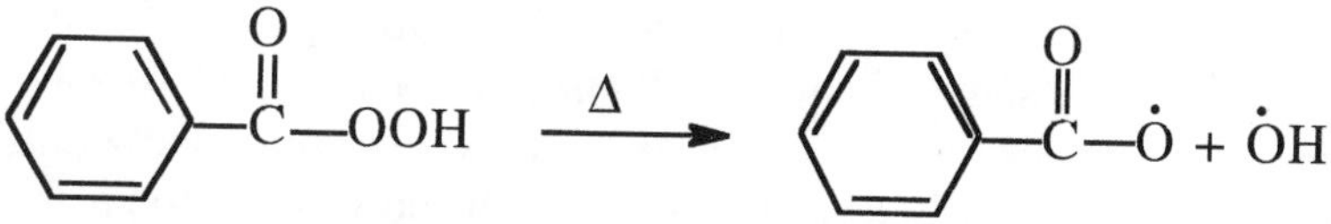

$$\text{C}_6H_5-\overset{\overset{\displaystyle O}{\|}}{C}-OOH \xrightarrow{\Delta} \text{C}_6H_5-\overset{\overset{\displaystyle O}{\|}}{C}-\dot{O} + \dot{O}H$$

Free radicals are highly reactive, short lived, and therefore not selective. Chain transfer reactions often occur and result in a highly branched product polymer. For example, the polymerization of ethylene using an organic peroxide initiator produces highly branched polyethylene. The branches result from the abstraction of a hydrogen atom by a propagating polymer intermediate, which creates a new active center. The new center can add more ethylene molecules, forming a long branch:

$$-CH_2\overset{\bullet}{C}H_2 + -CH_2CH_2CH_2CH_2- \longrightarrow -CH_2CH_3 + -CH_2\overset{\bullet}{C}HCH_2CH_2-$$

$$-CH_2CH(CH_2CH_2)_{n-1}CH_2\overset{\bullet}{C}H_2 \quad\quad\quad \underset{\longleftarrow}{nCH_2=CH_2|}$$
$$\quad\quad\quad |$$
$$\quad\quad\quad CH_2$$
$$\quad\quad\quad |$$
$$\quad\quad\quad CH_2$$
$$\quad\quad\quad |$$

Intermolecular chain transfer reactions may occur between two propagating polymer chains and result in the termination of one of the chains. Alternatively, these reactions take place by an intramolecular reaction by the coiling of a long chain. Intramolecular chain transfer normally results in short branches:[4]

$$\overset{\bullet}{C}H_2 \longrightarrow CH_2 \longrightarrow -\overset{\bullet}{C}H \overset{CH_2=CH_2}{\longrightarrow} -\overset{\bullet}{C}HCH_2CH_2$$
$$\quad\quad\quad | \quad\quad\quad\quad\quad | \quad\quad\quad\quad\quad |$$
$$-CH_2 \quad CH_2 \quad\quad C_4H_9 \quad\quad\quad C_4H_9$$
$$\quad\quad \diagdown \quad \diagup$$
$$\quad\quad\quad CH_2$$

Free radical polymers may terminate when two propagating chains combine. In this case, the tail-to-tail addition mode is most likely.

Polymer propagation stops with the addition of a chain transfer agent. For example, carbon tetrachloride can serve as a chain transfer agent:

$$-CH_2-\overset{\bullet}{C}H_2 + CCl_4 \rightarrow -CH_2-CH_2Cl + \overset{\bullet}{C}Cl_3$$

The $\overset{\bullet}{C}Cl_3$ free radical formed can initiate a new polymerization reaction.

Cationic Polymerization

Strong protonic acids can affect the polymerization of olefins (Chapter 3). Lewis acids, such as $AlCl_3$ or BF_3, can also initiate polymerization. In this case, a trace amount of a proton donor (cocatalyst), such as water or methanol, is normally required. For example, water combined with BF_3 forms a complex that provides the protons for the polymerization reaction.

An important difference between free radical and ionic polymerization is that a counter ion only appears in the latter case. For example, the intermediate formed from the initiation of propene with BF_3-H_2O could be represented as

$$H^+[BF_3OH]^- + CH_2=CH-CH_3 \longrightarrow (CH_3)_2\overset{+}{C}H[BF_3OH]^-$$

The next step is the insertion of the monomer molecules between the ion pair.

$$(CH_3)_2\overset{+}{C}H[BF_3OH]^- + n\ CH_2=CH-CH_3$$

$$\downarrow$$

$$(CH_3)_2CH-(CH_2-\underset{\underset{\displaystyle H}{|}}{C}H)_{n-1}-CH_2-\overset{CH_3}{\underset{|}{C}}H[BF_3OH]^-$$

$$\text{CH}_3 \qquad\qquad \text{CH}_3$$

In ionic polymerizations, reaction rates are faster in solvents with high dielectric constants, which promote the separation of the ion pair.

Cationic polymerizations work better when the monomers possess an electron-donating group that stabilizes the intermediate carbocation. For example, isobutylene produces a stable carbocation, and usually copolymerizes with a small amount of isoprene using cationic initiators. The product polymer is a synthetic rubber widely used for tire inner tubes:

$$\left[-CH_2-\underset{\underset{\displaystyle CH_3}{|}}{\overset{\overset{\displaystyle CH_3}{|}}{C}}-CH_2-\overset{\overset{\displaystyle CH_3}{|}}{C}=CH-CH_2-\right]_n$$

Cationic initiators can also polymerize aldehydes. For example, BF_3 helps produce commercial polymers of formaldehyde. The resulting polymer, a polyacetal, is an important thermoplastic (Chapter 12):

$$-\!\!\left[CH_2-O\right]\!\!-$$

In general, the activation energies for both cationic and anionic polymerization are small. For this reason, low-temperature conditions are normally used to reduce side reactions.[5] Low temperatures also minimize chain transfer reactions. These reactions produce low-molecular weight polymers by disproportionation of the propagating polymer:

$$-\overset{+}{C}H_2C(CH_3)_2X^- + CH_2=\overset{\overset{\displaystyle CH_3}{|}}{C}-CH_3 \rightarrow -CH_2-\overset{\overset{\displaystyle CH_3}{|}}{C}=CH_2 + (CH_3)_3\overset{+}{C}X^-$$

X^- represents the counter ion.

Cationic polymerization can terminate by adding a hydroxy compound such as water:

$$\overset{+}{-CH_2C(CH_3)_2X^-} + H_2O \longrightarrow -CH_2\overset{\overset{\displaystyle OH}{|}}{C}(CH_3)_2 + HX$$

Anionic Polymerization

Anionic polymerization is better for vinyl monomers with electron withdrawing groups that stabilize the intermediates. Typical monomers best polymerized by anionic initiators include acrylonitrile, styrene, and butadiene. As with cationic polymerization, a counter ion is present with the propagating chain. The propagation and the termination steps are similar to cationic polymerization.

Many initiators, such as alkyl and aryllithium and sodium and lithium suspensions in liquid ammonia, effect the polymerization. For example, acrylonitrile combined with n-butyllithium forms a carbanion intermediate:

$$CH_2{=}CHCN + n\ C_4H_9^-Li^+ \longrightarrow C_4H_9-CH_2\overset{\overset{\displaystyle CN}{|}}{\underset{\underset{\displaystyle H}{|}}{C}}{}^-Li^+$$

Chain growth occurs through a nucleophilic attack of the carbanion on the monomer. As in cationic polymerizations, lower temperatures favor anionic polymerizations by minimizing branching due to chain transfer reactions.

Solvent polarity is also important in directing the reaction bath and the composition and orientation of the products. For example, the polymerization of butadiene with lithium in tetrahydrofuran (a polar solvent) gives a high 1,2 addition polymer.[6] Polymerization of either butadiene or isoprene using lithium compounds in nonpolar solvent such as n-pentane produces a high cis-1,4 addition product. However, a higher cis-1,4-poly-isoprene isomer was obtained than when butadiene was used. This occurs because butadiene exists mainly in a transoid conformation at room temperature[7] (a higher cisoid conformation is anticipated for isoprene):

$$\underset{\begin{array}{c}\text{A complex of}\\\text{isoprene with}\\\text{RLi (cisoid)}\end{array}}{CH_2{=}\overset{\overset{\displaystyle CH_3}{|}}{C}-CH{=}CH_2} + \overset{d-\ d+}{R\ Li} \longrightarrow R\ Li \qquad \underset{\text{cis- 1,4-Polyisoprene}}{}$$

Coordination Polymerization

Polymerizations catalyzed with coordination compounds are becoming more important for obtaining polymers with special properties (linear and stereospecific). The first linear polyethylene polymer was prepared from a mixture of triethylaluminum and titanium tetrachloride (Ziegler catalyst) in the early 1950s. Later, Natta synthesized a stereoregular polypropylene with a Ziegler-type catalyst. These catalyst combinations are now called Zieglar-Natta catalysts.

In coordination polymerization, the bonds are appreciably covalent but with a certain percentage of ionic character. Bonding occurs between a transition metal central ion and the ligand (perhaps an olefin, a diolefin or carbon monoxide) to form a coordination complex. The complex reacts further with the ligand to be polymerized by an insertion mechanism. Different theories about the formation of coordination complexes have been reviewed by Huheey.[8] In recent years, much interest has been centered on using late transition metals such as iron and cobalt for polymerization. Due to their lower electrophilicity, they have greater tolerence for polar functionality. It was found that the catalyst activity and the polymer branches could be modified by altering the bulk of the ligand that surrounds the central metal. Such a protection reduces chain-transfer reactions and results in a high molecular-weight polymer. An example of these catalysts are pyridine bis-imine ligands complexed with iron and cobalt salts.[9]

Ziegler-Natta catalysts currently produce linear polyethylene (non-branched), stereoregular polypropylene, cis-polybutadiene, and other stereoregular polymers.

In polymerizing these compounds, a reaction between α-TiCl$_3$ and triethylaluminum produces a five coordinate titanium (III) complex arranged octahedrally. The catalyst surface has four Cl anions, an ethyl group, and a vacant catalytic site ($\square$) with the Ti(III) ion in the center of the octahedron. A polymerized ligand, such as ethylene, occupies the vacant site:

The next step is the cis insertion of the ethyl group, leaving a vacant site. In another step, ethylene occupies the vacant site. This process continues until the propagating chain terminates:

$$
\overset{C_2H_5}{\underset{CH_2}{\overset{|}{\underset{|}{-Ti}}}} \quad \longrightarrow \quad -\;Ti\;-\;CH_2CH_2C_2H_5 \quad \overset{H_2C=CH_2}{\underset{C_2H_4}{\longrightarrow}} \quad -Ti\;-\;CH_2CH_2C_2H_5
$$

When propylene is polymerized with free radicals or some ionic initiators, a mixture of three stereo-forms results (Figure 11-1).[10] These forms are

Atactic—the methyl groups are randomly distributed.
Isotactic—all methyl groups appear on one side of the polymer chain.
Syndiotactic—the methyl groups alternate regularly from one side to the other.

The isotactic form of propylene has better physical and mechanical properties than the three tactic form mixture (obtained from free radical polymerization). Isotactic polypropylene, in which all of the stereo cen-

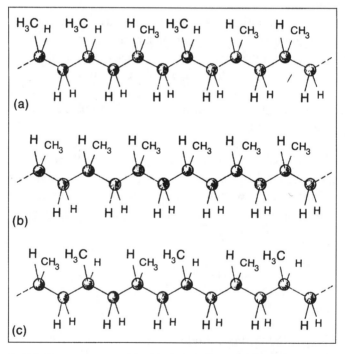

Figure 11-1. Propylene can undergo polymerization in three different ways to form atactic (a), isotactic (b), or syndiotactic polypropylene (c).[10]

ters of the polymer are the same, is a crystalline thermoplastic. By contrast, atactic polypropylene, in which the stereo centers are arranged randomly, is an amorphous gum elastomer. Polypropylene consisting of blocks of atactic and isotactic stereo sequences is rubbery.[11] Polymerizing propylene with Ziegler-Natta catalyst produces mainly isotactic polypropylene. The Cosse-Arlman model explains the formation of the stereoregular type by describing the crystalline structure of $\alpha TiCl_3$ as a hexagonal close packing with anion vacancies.[12] This structure allows for cis insertion. However, due to the difference in the steric requirements, one of the vacant sites available for the ligand to link with the titanium catalyst which has a greater affinity for the propagating polymer than the other site. Accordingly, the growing polymer returns rapidly back to that site as shown here:

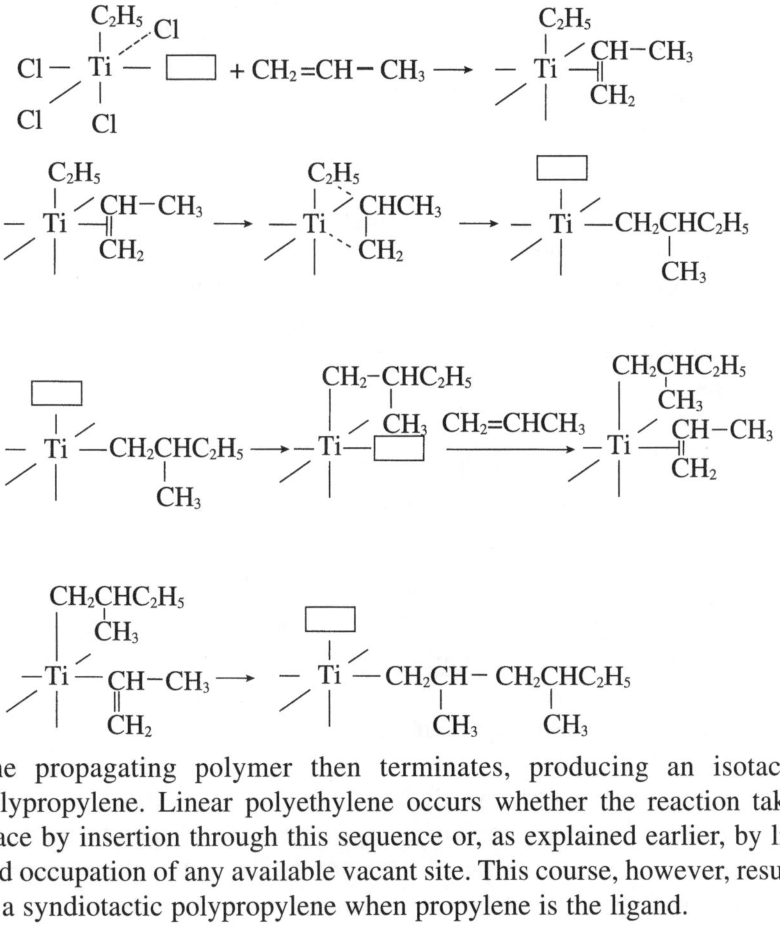

The propagating polymer then terminates, producing an isotactic polypropylene. Linear polyethylene occurs whether the reaction takes place by insertion through this sequence or, as explained earlier, by ligand occupation of any available vacant site. This course, however, results in a syndiotactic polypropylene when propylene is the ligand.

Adding hydrogen terminates the propagating polymer. The reaction between the polymer complex and the excess triethylaluminum also terminates the polymer. Treatment with alcohol or water releases the polymer:

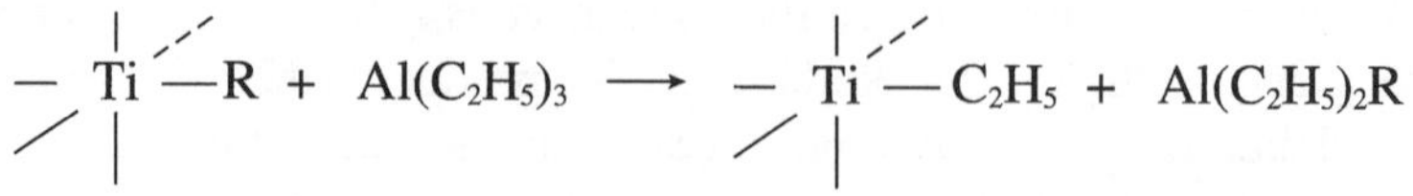

$$— \overset{|}{\underset{|}{Ti}}\!\!—R + Al(C_2H_5)_3 \longrightarrow — \overset{|}{\underset{|}{Ti}}\!\!—C_2H_5 + Al(C_2H_5)_2R$$

$$Al(C_2H_5)_2R + H_2O \longrightarrow RH + Al(C_2H_5)_2OH$$

R= growing polymer

A chain transfer reaction between the monomer and the growing polymer produces an unsaturated polymer. This occurs when the concentration of the monomer is high compared to the catalyst. Using ethylene as the monomer, the termination reaction has this representation:

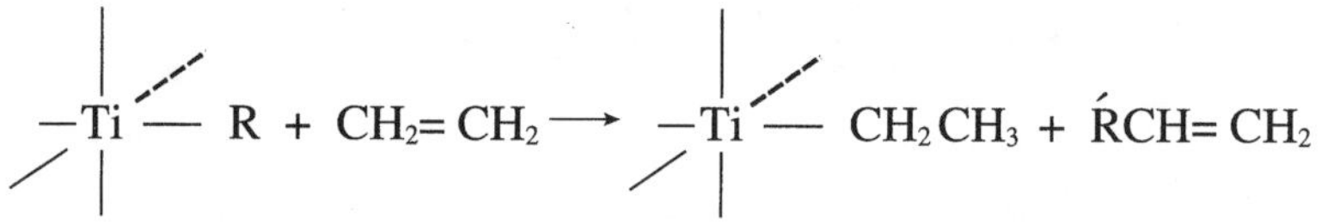

$$—\overset{|}{\underset{|}{Ti}}\!\!— R + CH_2=CH_2 \longrightarrow —\overset{|}{\underset{|}{Ti}}\!\!— CH_2CH_3 + \acute{R}CH=CH_2$$

A new generation coordination catalysts are metallocenes. The chiral form of metallocene produces isotactic polypropylene, whereas the achiral form produces atactic polypropylene. As the ligands rotate, the catalyst produces alternating blocks of isotactic and atactic polymer much like a miniature sewing machine which switches back and forth between two different kinds of stitches.[11]

CONDENSATION POLYMERIZATION
(Step-Reaction Polymerization)

Though less prevalent than addition polymerization, condensation polymerization produces important polymers such as polyesters, polyamides (nylons), polycarbonates, polyurethanes, and phenol-formaldehyde resins (Chapter 12).

In general, condensation polymerization refers to

1. A reaction between two different monomers. Each monomer possesses at least two similar functional groups that can react with the functional groups of the other monomer. For example, a reaction of a diacid and a dialcohol (diol) can produce polyesters:

$$\underset{\text{HOCR COH}}{\overset{O\quad O}{\overset{\|\quad\|}{}}} + \text{HOROH} \longrightarrow \underset{\text{HOCR COROH}}{\overset{O\quad O}{\overset{\|\quad\|}{}}} + H_2O$$

A similar reaction between a diamine and a diacid can also produce polyamides.

2. Reactions between one monomer species with two different functional groups. One functional group of one molecule reacts with the other functional group of the second molecule. For example, polymerization of an amino acid starts with condensation of two monomer molecules:

$$2\,H_2NRCOOH \longrightarrow \underset{\text{H}_2\text{NRCNHRCOOH}}{\overset{O}{\overset{\|}{}}} + H_2O$$

In these two examples, a small molecule (water) results from the condensation reaction.

Ring opening polymerization of lactams can also be considered a condensation reaction, although a small molecule is not eliminated. This type is noted later in this chapter under "Ring Opening Polymerization."

Condensation polymerization is also known as step-reaction polymerization because the reactions occur in steps. First, a dimer forms, then a trimer, next a tetramer, and so on until the polymer terminates. Although step polymerizations are generally slower than addition polymerizations, with long reaction times required for high conversions, the monomers disappear fast. The reaction medium contains only dimers, trimers, tetramers, and so on. For example, the dimer formed from the condensation of a diacid and a diol (reaction previously shown) has hydroxyl and carboxyl endings that can react with either a diacid or a diol to form a trimer:

$$\underset{\text{HOCR COR}'\text{OH}}{\overset{O\quad O}{\overset{\|\quad\|}{}}}
\begin{cases}
\xrightarrow{\ \underset{\text{HOCR COH}}{\overset{O\;\;O}{\overset{\|\;\;\|}{}}}\ } \underset{\text{HO-C-R-C-O-R}'\text{OC RCOH}}{\overset{O\;O\qquad\qquad O\;O}{\overset{\|\;\|\qquad\qquad\|\;\|}{}}} + H_2O \\[2ex]
\xrightarrow{\ \text{HOR}'\text{OH}\ } \underset{\text{HO-R}'\text{OCRCOR}'\text{OH}}{\overset{O\;O}{\overset{\|\;\|}{}}} + H_2O
\end{cases}$$

The compounds formed continue condensation as long as the species present have different endings. The polymer terminates by having one of the monomers in excess. This produces a polymer with similar endings. For example, a polyester formed with excess diol could be represented:

$$n \ HO\overset{O}{\overset{||}{C}} R\overset{O}{\overset{||}{C}}OH + m \ HOR'OH \longrightarrow HOR'O \, \overset{O \quad O}{\underset{}{\text{\Large(}} C R C O R O \text{\Large)}} H$$

where $m > n$ and $\overset{O \quad O}{\text{\Large(} CRCORO \text{\Large)}}$ is the repeating unit.

In these reactions, the monomers have two functional groups (whether one or two monomers are used), and a linear polymer results. With more than two functional groups present, crosslinking occurs and a thermosetting polymer results. Example of this type are polyurethanes and urea formaldehyde resins (Chapter 12).

Acid catalysts, such as metal oxides and sulfonic acids, generally catalyze condensation polymerizations. However, some condensation polymers form under alkaline conditions. For example, the reaction of formaldehyde with phenol under alkaline conditions produces methylolphenols, which further condense to a thermosetting polymer.

RING OPENING POLYMERIZATION

Ring opening polymerization produces a small number of synthetic commercial polymers. Probably the most important ring opening reaction is that of caprolactam for the production of nylon 6:

$$n \ \underset{}{\bigcirc}\!\!\text{\small C=O, NH} \longrightarrow \overset{O}{\underset{}{\text{\Large(} HN-(CH_2)_5-C \text{\Large)}_n}}$$

Although no small molecule gets eliminated, the reaction can be considered a condensation polymerization. Monomers suitable for polymerization by ring opening condensation normally possess two different functional groups within the ring. Examples of suitable monomers are lactams (such as caprolactam), which produce polyamides, and lactons, which produce polyesters.

Ring opening polymerization may also occur by an addition chain reaction. For example, a ring opening reaction polymerizes trioxane to a polyacetal in the presence of an acid catalyst. Formaldehyde also produces the same polymer:

$$\frac{n}{3} \ \underset{}{\bigcirc}\!\!\text{\small O, O, O} + H_2O \xrightarrow{\ H^+\ } HO \overset{}{\text{\Large(} CH_2O \text{\Large)}_n} H$$

Trioxane

Monomers used for ring opening polymerization (by addition) are cyclic compounds that open easily with the action of a catalyst during the reaction. Small strained rings are suitable for this type of reaction. For example, the action of a strong acid or a strong base could polymerize ethylene oxide to a high molecular-weight polymer.

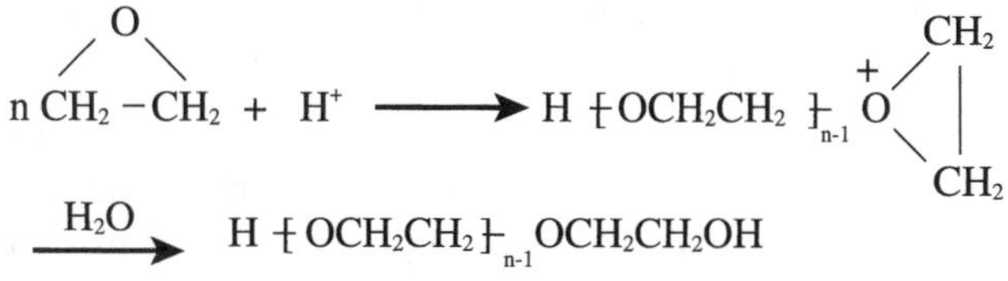

$$n \; CH_2-CH_2 \; + \; H^+ \longrightarrow H \; [OCH_2CH_2]_{n-1} \; O \begin{array}{c} CH_2 \\ | \\ CH_2 \end{array}$$

$$\xrightarrow{\;H_2O\;} \; H \; [OCH_2CH_2]_{n-1} \; OCH_2CH_2OH$$

These water soluble polymers are commercially known as carbowax.

The ring opening of cycloolefins is also possible with certain coordination catalysts. This simplified example shows cyclopentene undergoing a first-step formation of the dimer cyclodecadiene, and then incorporating additional cyclopentene monomer units to produce the solid, rubbery polypentamer:[13]

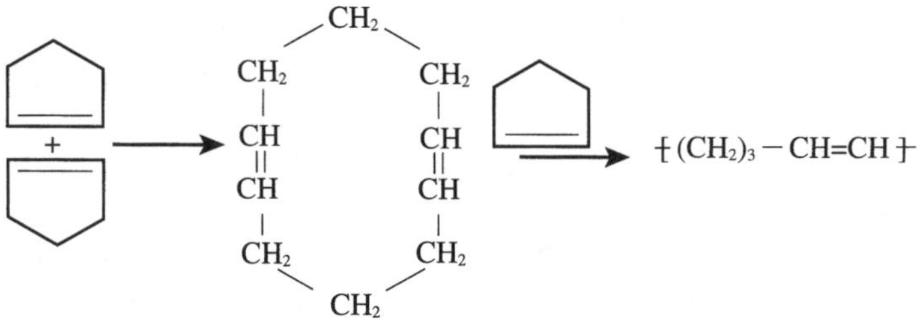

Cyclodecadiene   Polypentamer

Another example is the metathesis of cyclooctene, which produces polyoctenylene, an elastomor known as trans-polyoctenamer:[14]

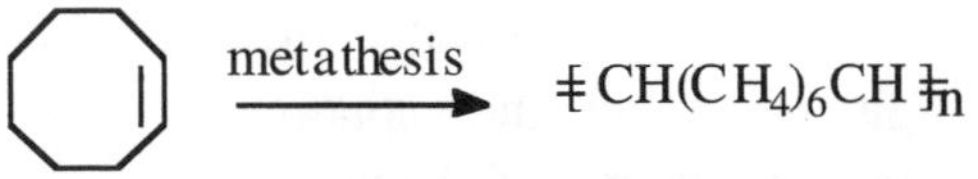

Chemische Werke Huls produces the polymer for use in blends with some conventional rubbers.[15] This metathetic reaction has become an important synthetic tool in the polymer field.[13,16] Catalyzed polymerization of cycloolefins has been reviewed by Tsonis.[17]

POLYMERIZATION TECHNIQUES

Polymerization reactions can occur in bulk (without solvent), in solution, in emulsion, in suspension, or in a gas-phase process. Interfacial polymerization is also used with reactive monomers, such as acid chlorides.

Polymers obtained by the *bulk technique* are usually pure due to the absence of a solvent. The purity of the final polymer depends on the purity of the monomers. Heat and viscosity are not easily controlled, as in other polymerization techniques, due to absence of a solvent, suspension, or emulsion medium. This can be overcome by carrying the reaction to low conversions and strong agitation. Outside cooling can also control the exothermic heat.

In *solution polymerization,* an organic solvent dissolves the monomer. Solvents should have low chain transfer activity to minimize chain transfer reactions that produce low-molecular-weight polymers. The presence of a solvent makes heat and viscosity control easier than in bulk polymerization. Removal of the solvent may not be necessary in certain applications such as coatings and adhesives.

Emulsion polymerization is widely used to produce polymers in the form of emulsions, such as paints and floor polishes. It also used to polymerize many water insoluble vinyl monomers, such as styrene and vinyl chloride. In emulsion polymerization, an agent emulsifies the monomers. Emulsifying agents should have a finite solubility. They are either ionic, as in the case of alkylbenzene sulfonates, or nonionic, like polyvinyl alcohol.

Water is extensively used to produce emulsion polymers with a sodium stearate emulsifier. The emulsion concentration should allow micelles of large surface areas to form. The micelles absorb the monomer molecules activated by an initiator (such as a sulfate ion radical $SO•_4^-$). X-ray and light scattering techniques show that the micelles start to increase in size by absorbing the macromolecules. For example, in the free radical polymerization of styrene, the micelles increased to 250 times their original size.

In *suspension polymerization,* the monomer gets dispersed in a liquid, such as water. Mechanical agitation keeps the monomer dispersed. Initiators should be soluble in the monomer. Stabilizers, such as talc or polyvinyl alcohol, prevent polymer chains from adhering to each other and keep the monomer dispersed in the liquid medium. The final polymer appears in a granular form.

Suspension polymerization produces polymers more pure than those from solution polymerization due to the absence of chain transfer reactions. As in a solution polymerization, the dispersing liquid helps control the reaction's heat.

Interfacial polymerization is mainly used in polycondensation reactions with very reactive monomers. One of the reactants, usually an acid chloride, dissolves in an organic solvent (such as benzene or toluene), and the second reactant, a diamine or a diacid, dissolves in water. This technique produces polycarbonates, polyesters, and polyamides. The reaction occurs at the interface between the two immiscible liquids, and the polymer is continuously removed from the interface.

PHYSICAL PROPERTIES OF POLYMERS

The properties of polymers determine whether they can be used as a plastic, a fiber, an elastomer, an adhesive, or a paint.

Important physical properties include the density, melt flow index, crystallinity, and average molecular weight. Mechanical properties of a polymer, such as modulus (the ratio of stress to strain), elasticity, and breaking strength, essentially follow from the physical properties.

The following sections describe some important properties of polymers.

CRYSTALLINITY

A polymer's tendency to have order and form crystallites derives from the regularity of the chains, presence (or absence) and arrangement of bulky groups, and the presence of secondary forces, such as hydrogen bonding. For example, isotactic polystyrene with phenyl groups arranged on one side of the polymer backbone is highly crystalline, while the atactic form (with a random arrangement of phenyl groups) is highly amorphous. Polyamides are also highly crystalline due to strong hydrogen bonding. High-density polyethylene exhibits no hydrogen bonding, but its linear structure makes it highly crystalline. Low-density polyethylene, on the other hand, has branches and a lower crystallinity. It does not pack as easily as the high-density polymer.

The mechanical and thermal behaviors depend partly on the degree of crystallinity. For example, highly disordered (dominantly amorphous) polymers make good elastomeric materials, while highly crystalline polymers, such as polyamides, have the rigidity needed for fibers. Crystallinity of polymers correlates with their melting points.

MELTING POINT

The freezing point of a pure liquid is the temperature at which the liquid's molecules lose transitional freedom and the solid's molecules become more ordered within a definite crystalline structure. Polymers, however, are non-homogeneous and do not have a definite crystallization temperature.

When a melted polymer cools, some polymer molecules line up and form crystalline regions within the melt. The rest of the polymer remains amorphous. The temperature at which these crystallites disappear when the crystalline polymer is gradually heated is called the crystalline melting temperature, T_m. Further cooling of the polymer below T_m changes the amorphous regions into a glass-like material. The temperature at which this change occurs is termed the glass transition temperature, T_g. Elastomeric materials usually have a low T_g (low crystallinity), while highly crystalline polymers, such as polyamides, have a relatively high T_g.

VISCOSITY

The viscosity of a substance measures its resistance to flow. The melt viscosity of a polymer increases as the molecular weight of the polymer rises. Polymers with high melt viscosities require higher temperatures for processing.

The melt flow index describes the viscosity of a solid plastic. It is the weight in grams of a polymer extruded through a defined orifice at a specified time. The melt viscosity and the melt flow index can measure the extent of polymerization. A polymer with a high melt flow index has a low melt viscosity, a lower molecular weight, and usually a lower impact tensile strength.

MOLECULAR WEIGHT

Polymerization usually produces macromolecules with varying chain lengths. As a result, polymers are described as polydisperse systems. Commercial polymers have molecular weights greater than 5,000 and contain macromolecules with variable molecular weights. The methods for determining the average molecular weights of polymers include measuring some colligative property, such as viscosity or sedimentation. Different methods do not correlate well, and determining the average molecular weight requires more than one method. Two methods normally determine the *number average* and the *weight average* molecular weights of the polymer.

Number Average Molecular Weight

The number average molecular weight (M_n) is related to the number of particles present in a sample. It is calculated by dividing the sum of the weights of all the species present (monomers, dimers, trimers, and so on) by the number of species present:

$$M_n = \frac{W}{\sum N_i} = \frac{\sum N_i M_i}{\sum N_i}$$

i = degree of polymerization (dimer, trimers, etc.)
N_i = number of each polymeric species
M_i = molecular weight of each polymer species
W = total weight of all polymer species.

M_n depends not on the molecular sizes of the particles but on the number of particles. Measuring colligative properties such as boiling point elevation, freezing point depression, and vapor pressure lowering can determine the number of particles in a sample.

Weight Average Molecular Weight

The weight average molecular weight (M_w) is the sum of the products of the weight of each species present and its molecular weight, divided by the sum of all the species' weights:

$$M_w = \frac{\sum W_i M_i}{W} = \frac{\sum W_i M_i}{\sum N_i M_i}$$

W_i = weight of each polymeric species
M_i = molecular weight of each polymeric species

Substituting $N_i M_i = W_i$, the weight average molecular weight can be defined as

$$M_w = \frac{\sum N_i M_i^2}{\sum N_i M_i}$$

Larger, heavier molecules contribute more to M_w than to M_n. Light scattering techniques and ultracentrifugation can determine M_w.

The following simple example illustrates the difference between M_n and M_w: Suppose a sample has six macro-molecules. Three of them have a molecular weight $= 1.0 \times 10^4$, two have a molecular weight $= 2.0 \times 10^4$, and one has a molecular weight $= 3.0 \times 10^4$:

$$M_n = \frac{(3.0 + 4.0 + 3.0)10^4}{6} = 1.7 \times 10^4$$

$$M_w = \frac{3(1.0 \times 10^4)^2 + 2(2.0 \times 10^4)^2 + 1(3.0 \times 10^4)^2}{3(1.0 \times 10^4) + 2(2.0 \times 10^4) + 1(3.0 \times 10^4)} = 2.0 \times 10^4$$

In monodispersed systems $M_n = M_w$.

The difference in the value between M_n and M_w. indicates the polydispersity of the polymer system. The closer M_n is to M_w, the narrower the molecular weight spread. Molecular weight distribution curves for polydispersed systems can be obtained by plotting the degree of polymerization i versus either the number fraction, N_i, or the weight fraction, W_i.

CLASSIFICATION OF POLYMERS

Synthetic polymers may be classified into addition or condensation polymers according to the type of reaction. A second classification depends on the monomer type and the linkage present in the repeating unit into polyolefins, polyesters, polyamides, etc. Other classifications depend on the polymerization technique (bulk, emulsion, suspension, etc.) or on the polymer's end use. The latter classifies polymers into three broad categories: plastics, elastomers, and synthetic fibers.

Plastics

Plastics are relatively tough substances with high molecular weight that can be molded with (or without) the application of heat. In general, plastics are subclassified into *thermoplastics,* polymers that can be resoftened by heat, and *thermosets,* which cannot be resoftened by heat.

Thermoplastics have moderate crystallinity. They can undergo large elongation, but this elongation is not as reversible as it is for elastomers. Examples of thermoplastics are polyethylene and polypropylene.

Thermosetting plastics are usually rigid due to high crosslinking between the polymer chains. Examples of this type are phenolfomaldehyde and polyurethanes. Crosslinking may also be promoted by using chemical agents such as sulfur or by heat treatment or irradiation with gamma rays, ultraviolet light, or energetic electrons. Recently, high energy ion beams were found to increase the hardness of the treated polymer drastically.[18]

Synthetic Rubber

Synthetic rubber (elastomers) are high molecular weight polymers with long flexible chains and weak intermolecular forces. They have low crystallinity (highly amorphous) in the unstressed state, segmental mobility, and high reversible elasticity. Elastomers are usually cross-linked to impart strength.

Synthetic Fibers

Synthetic fibers are long-chain polymers characterized by highly crystalline regions resulting mainly from secondary forces (e.g., hydrogen bonding). They have a much lower elasticity than plastics and elastomers. They also have high tensile strength, a light weight, and low moisture absorption.

REFERENCES

1. Fessenden, R., and Fessenden, J., *Organic Chemistry,* 4th Ed., Brooks/Cole Publishing Co., 1991, p. 926.
2. Hoffman, A. S. and Bacskai, R., Chapter 6 in *Copolymerization,* G. E. Ham, (ed.), Wiley-Interscience, New York, 1964.
3. Rodriguez, F., *Principles of Polymer Systems,* 3rd Ed., Hemisphere Publishing Corp., New York, 1989, p. 108.
4. Wiseman, P., *Petrochemicals,* Ellis Horwood Ltd., England, 1986, p. 45.
5. Seymor, R. and Corraher C. E., Jr., *Polymer Chemistry,* 2nd Ed., Dekker, New York, 1988, p. 284.
6. Kutz, I. and Berber, A., *J. Polymer Science,* Vol. 42, 1960, p. 299.
7. Stevens, M. P., *Polymer Chemistry,* Addison Wesley Publishing Co., London, 1975, p. 156.
8. Huheey, J. E., Chapter 11 in *Inorganic Chemistry,* 3rd Ed., Harper and Row Publishers, Inc., New York, 1983.
9. Allison, M. and Bennet, A., "Novel, Highly Active Iron and Cobalt Catalysts for Olefin Polymerization," *CHEMTECH,* July, 1999, pp. 24–28.

10. Watt, G. W., Hatch, L. F., and Lagowski, J. J., *Chemistry,* New York, W. W. Norton & Co., 1964, p. 449.
11. Baum, R. "Elastomeric Polypropylene Oscillating Catalyst Controls Microstructure," *Chemical and Engineering News,* Jan., 16, 1995, pp. 6–7.
12. Arlman, E. and Cossee, P. J., *Catal.* Vol. 3, 1964, p. 99.
13. Wagner, P. H., "Olefin Metathesis: Applications for the Nineties," *Chemistry and Industry,* 4 May 1992, pp. 330–333.
14. Parshall, G. W. and Nugent, W. A., "Functional Chemicals via Homogeneous Catalysis," *CHEMTECH,* Vol. 18, No. 5, May 1988, pp. 314–320.
15. Banks, R. L., in *"Applied Industrial Catalysis,"* Leach, B. E. (ed.) Academic, New York, 1984, pp. 234–235.
16. Platzer, N., *CHEMTECH,* Vol. 9, No. 1, 1979, pp. 16–20.
17. Tsonis, C. P., "Catalyzed Polymerization of Cycloolefins," *Journal of Applied Polymer Science,* Vol. 26, 1981, pp. 3525–3536.
18. Dagani, R. "Superhard-Surfaced Polymers Made by High-Energy Ion Irradiation," *Chemical and Engineering News,* Jan. 9, 1995, pp. 24–26.

CHAPTER TWELVE

Synthetic Petroleum-Based Polymers

INTRODUCTION

The synthetic polymer industry represents the major end use of many petrochemical monomers such as ethylene, styrene, and vinyl chloride. Synthetic polymers have greatly affected our lifestyle. Many articles that were previously made from naturally occurring materials such as wood, cotton, wool, iron, aluminum, and glass are being replaced or partially substituted by synthetic polymers. Clothes made from polyester, nylon, and acrylic fibers or their blends with natural fibers currently dominate the apparel market. Plastics are replacing many articles previously made of iron, wood, porcelain, or paper in thousands of diversified applications. Polymerization could now be tailored to synthesize materials stronger than steel.[1] For example, polyethylene fibers with a molecular weight of one million can be treated to be 10 times stronger than steel! However, its melting point is 148°C. A recently announced thermotropic liquid crystal polymer based on p-hydroxybenzoic acid, terephthalic acid, and p, p'-biphenol has a high melting point of 420°C and does not decompose up to 560°C. This polymer has an initial stress of 3.4×10^6 kg/mm^2, even after 6,000 hours of testing.[2]

The polymer field is versatile and fast growing, and many new polymers are continually being produced or improved. The basic chemistry principles involved in polymer synthesis have not changed much since the beginning of polymer production. Major changes in the last 70 years have occurred in the catalyst field and in process development. These improvements have a great impact on the economy. In the elastomer field, for example, improvements influenced the automobile industry and also related fields such as mechanical goods and wire and cable insulation.

This chapter discusses synthetic polymers based primarily on monomers produced from petroleum chemicals. The first section covers the synthesis of thermoplastics and engineering resins. The second part reviews thermosetting plastics and their uses. The third part discusses the chemistry of synthetic rubbers, including a brief review on thermoplastic elastomers, which are generally not used for tire production but to make other rubber products. The last section addresses synthetic fibers.

THERMOPLASTICS AND ENGINEERING RESINS

Thermoplastics are important polymeric materials that have replaced or substituted many naturally-derived products such as paper, wood, and steel. Plastics possess certain favorable properties such as light weight, corrosion resistance, toughness, and ease of handling. They are also less expensive. The major use of the plastics is in the packaging field. Many other uses include construction, electrical and mechanical goods, and insulation. One growing market that evolved fairly recently is engineering thermoplastics. This field includes polymers with special properties such as high thermal stability, toughness, and chemical and weather resistance. Nylons, polycarbonates, polyether sulfones, and polyacetals are examples of this group.

Another important and growing market for plastics is the automotive field. Many automobile parts are now made of plastics. Among the most used polymers are polystyrene polymers and copolymers, polypropylene, polycarbonates, and polyvinyl chloride. These materials reduce the cost and the weight of the cars. As a result, gasoline consumption is also reduced.

Most big-volume thermoplastics are produced by addition polymerization. Other thermoplastics are synthesized by condensation. Table 12-1 shows the major thermoplastics.[3]

POLYETHYLENE

Polyethylene is the most extensively used thermoplastic. The ever-increasing demand for polyethylene is partly due to the availability of the monomer from abundant raw materials (associated gas, LPG, naphtha). Other factors are its relatively low cost, ease of processing the polymer, resistance to chemicals, and its flexibility. World production of all polyethylene grades, approximately 100 billion pounds in 1997, is predicted

Table 12-1
Major thermoplastic polymers

Name	Family	Formula	Melting temp.	Density
Low-density polyethylene	Polyolefin	—CH₂—	110 (T_m)	0.910
High-density polyethylene	Polyolefin	—CH₂—	120 (T_m)	0.950
Polypropylene	Polyolefin	$\begin{array}{c}CH_3\\|\\—CH_2-CH—\end{array}$	175 (T_m)	0.902
Polyvinyl chloride	Vinyl	$\begin{array}{c}Cl\\|\\—CH_2-CH—\end{array}$	100 (T_g)	1.35
Polyvinyl acetate	Vinyl	$\begin{array}{c}O\\\|\|\\O-CCH_3\\\|\\—CH_2-CH—\end{array}$	—	—
Polystyrene	Styrenic	—CH₂-CH— (with phenyl group)	100 (T_g)	1.05
Acrylonitrilebutadiene styrene	Styrenic	—	—	—
Acrylonitrile styrene	Styrenic	—	—	—
Polymethylmeth- acrylate	Acrylic	—	—	—
Polyhexamethylene- adipate	Polyamide	—NH-(CH₂)₆-NH-Ç-(CH₂)₄-Ç— (with two C=O)	265 (T_m)	1.14
Polycaprolactam	Polyamide	—NH-(CH₂)₅-Ç— (with C=O)	225 (T_m)	1.14
Polyethylene- terephthalate	Polyester	—Ç—(benzene)—Ç-OCH₂ CH₂O— (with two C=O)	270 (T_m)	—
Polybutylene- terephthalate	Polyester	—Ç—(benzene)—Ç-O-(CH₂)₃-CH₂-O— (with two C=O)	250 (T_m)	1.3
Polycarbonates	Polyester	—O-R'-O-Ç-O-R-O-Ç— (with two C=O)	190 (T_g)	1.2
Polyacetals	Polyethers	—O-CH₂-O-R—	181 (T_m)	—

to reach 300 billion pounds in 2015, the largest increase for linear low density polyethylene.[4]

High-pressure polymerization of ethylene was introduced in the 1930s. The discovery of a new titanium catalyst by Karl Ziegler in 1953 revolutionized the production of linear unbranched polyethylene at lower pressures. The two most widely used grades of polyethylene are low-density polyethylene (LDPE) and high-density polyethylene (HDPE). Currently, a new LDPE grade has been introduced. It is a linear, low-density grade (LLDPE) produced like the high-density polymer at low pressures.

Polymerizing ethylene is a highly exothermic reaction. For each gram of ethylene consumed, approximately 3.5 KJ (850 cal) are released:[5]

$$nCH_2 =CH_2 \rightarrow \ +CH2–CH_2+_n \qquad \Delta H = -92KJ/mol$$

When ethylene is polymerized, the reactor temperature should be well controlled to avoid the endothermic decomposition of ethylene to carbon, methane, and hydrogen:

$$CH_2=CH_2 \rightarrow 2C + 2H_2$$

$$CH_2=CH_2 \rightarrow C + CH_4$$

Low-Density Polyethylene

Low-density polyethylene (LDPE) is produced under high pressure in the presence of a free radical initiator. As with many free radical chain addition polymerizations, the polymer is highly branched. It has a lower crystallinity compared to HDPE due to its lower capability of packing.

Polymerizing ethylene can occur either in a tubular or in a stirred autoclave reactor. In the stirred autoclave, the heat of the reaction is absorbed by the cold ethylene feed. Stirring keeps a uniform temperature throughout the reaction vessel and prevents agglomeration of the polymer.

In the tubular reactor, a large amount of reaction heat is removed through the tube walls.

Reaction conditions for the free radical polymerization of ethylene are 100–200°C and 100–135 atmospheres. Ethylene conversion is kept to a low level (10–25%) to control the heat and the viscosity. However, overall conversion with recycle is over 95%.

The polymerization rate can be accelerated by increasing the temperature, the initiator concentration, and the pressure. Degree of branching and molecular weight distribution depend on temperature and pressure. A higher density polymer with a narrower molecular weight distribution could be obtained by increasing the pressure and lowering the temperature.

The crystallinity of the polymer could be varied to some extent by changing the reaction conditions and by adding comonomers such as vinyl acetate or ethyl acrylate. The copolymers have lower crystallinity but better flexibility, and the resulting polymer has higher impact strength.[6]

High-Density Polyethylene

High-density polyethylene (HDPE) is produced by a low-pressure process in a fluid-bed reactor. Catalysts used for HDPE are either of the Zieglar-type (a complex of $Al(C_2H_5)_3$ and α-$TiCl_4$) or silica-alumina impregnated with a metal oxide such as chromium oxide or molybdenum oxide.

Reaction conditions are generally mild, but they differ from one process to another. In the newer Unipol process (Figure 12-1) used to produce both HDPE and LLDPE, the reaction occurs in the gas phase.[7] Ethylene and the comonomers (propene, 1-butene, etc.) are fed to the reactor containing a fluidized bed of growing polymer particles. Operation temperature and pressure are approximately 100°C and 20 atmospheres. A single-stage centrifugal compressor circulates unreacted ethylene. The circulated gas fluidizes the bed and removes some of the exothermic reaction heat. The product from the reactor is mixed with additives and then pelletized. New modifications for gas-phase processes have been reviewed by Sinclair.[8]

The polymerization of ethylene can also occur in a liquid-phase system where a hydrocarbon diluent is added. This requires a hydrocarbon recovery system.

High-density polyethylene is characterized by a higher crystallinity and higher melting temperature than LDPE due to the absence of branching.

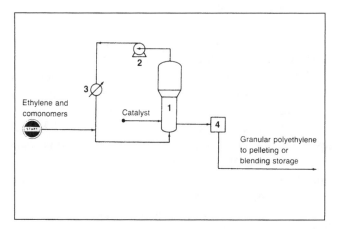

Figure 12-1. The Union Carbide Unipol process for producing HDPE[7]: (1) reactor, (2) single-stage centrifugal compressor, (3) heat exchanger, (4) discharge tank.

Some branching could be incorporated in the backbone of the polymer by adding variable amounts of comonomers such as hexene. These comonomers modify the properties of HDPE for specific applications.

Linear Low-Density Polyethylene

Linear low-density polyethylene (LLDPE) is produced in the gas phase under low pressure. Catalysts used are either Ziegler type or new generation metallocenes. The Union Carbide process used to produce HDPE could be used to produce the two polymer grades. Terminal olefins (C_4–C_6) are the usual comonomers to effect branching.

Developments in the gas-fluidized-bed polymerization reduced investments for high pressure processes used for LDPE. The new technology lowers capital and operating costs and reduces reactor purge/waste streams. Specific designed nozzles are located within the fluidized bed to disperse the hydrocarbons within the bed. The liquid injected through the nozzles quickly evaporates, hence removing the heat of polymerization. These processes can produce a wide range of polymers with different melt flow indices (MFI of <0.01 to >100) and densities of 890–970 Kg/m^3. Types of reactors and catalysts used for HDPE and LLDPE have been reviewed by Chinh and Power.[9]

LLDPE has properties between HDPE and LDPE. It has fewer branches, higher density, and higher crystallinity than LDPE.

Properties and Uses of Polyethylenes

Polyethylene is an inexpensive thermoplastic that can be molded into almost any shape, extruded into fiber or filament, and blown or precipitated into film or foil. Polyethylene products include packaging (largest market), bottles, irrigation pipes, film, sheets, and insulation materials.

Currently, high density polyethylene is the largest-volume thermoplastic. The 1997 U.S. production of HDPE was 12.5 billion pounds. LDPE was 7.7 billion pounds and LLDPE was 6.9 billion pounds.[10]

Because LDPE is flexible and transparent, it is mainly used to produce film and sheets. Films are usually produced by extrusion. Calendering is mainly used for sheeting and to a lesser extent for film production.

HDPE is important for producing bottles and hollow objects by blow molding. Approximately 64% of all plastic bottles are made from HDPE.[11] Injection molding is used to produce solid objects. Another important market for HDPE is irrigation pipes. Pipes made from HDPE are flexible, tough, and corrosion resistant. They could be used to carry abrasive materials such as gypsum. Table 12-2 shows the important properties of polyethylenes.

POLYPROPYLENE

Polypropylene (PP) is a major thermoplastic polymer. Although polypropylene did not take its position among the large volume polymers until fairly recently, it is currently the third largest thermoplastic after PVC. The delay in polypropylene development may be attributed to technical reasons related to its polymerization. Polypropylene produced by free radical initiation is mainly the atactic form. Due to its low crystallinity, it is not suitable for thermoplastic or fiber use. The turning point in polypropylene production was the development of a Ziegler-type catalyst by Natta to produce the stereoregular form (isotactic).

Catalysts developed in the titanium-aluminum alkyl family are highly reactive and stereoselective. Very small amounts of the catalyst are needed to achieve polymerization (one gram catalyst/300,000 grams polymer). Consequently, the catalyst entrained in the polymer is very small, and the catalyst removal step is eliminated in many new processes.[12] Amoco has introduced a new gas-phase process called "absolute gasphase" in which polymerization of olefins (ethylene, propylene) occurs in the total absence of inert solvents such as liquefied propylene in the reactor. Titanium residues resulting from the catalyst are less than 1 ppm, and aluminum residues are less than those from previous catalysts used in this application.[13]

Table 12-2
Important properties of polyethylenes

Polymer	Melting point range °C	Density g/cm^3	Degree of crystallinity %	Stiffness modules psi × 10^3
Branched, Low density	107–121	0.92	60–65	25–30
Medium density	—	0.935	75	60–65
Linear, High density				
Ziegler type	125–132	0.95	85	90–110
Phillips type	—	0.96	91	130–150

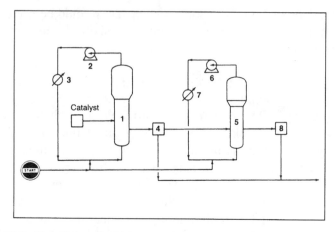

Figure 12-2. The Union Carbide gas-phase process for producing polypropylene[14]: (1) reactor, (2) centrifugal compressor, (3) heat exchanger, (4) product discharge tank (unreacted gas separated from product), (5) impact reactor, (6) compressor, (7) heat exchanger, (8) discharge tank (copolymer separated from reacted gas).

Polypropylene could be produced in a liquid or in a gas-phase process. Until 1980, the vertically stirred bed process of BASF was the only large-scale commercial gas phase process.[8] In the Union Carbide/Shell gas phase process (Figure 12-2), a wide range of polypropylenes are made in a fluidized bed gas phase reactor.[14] Melt index, atactic level, and molecular weight distribution are controlled by selecting the proper catalyst, adjusting operating conditions, and adding molecular weight control agents. This process is a modification of the polyethylene process (discussed before), but a second reactor is added. Homopolymers and random copolymers are produced in the first reactor, which operates at approximately 70°C and 35 atmospheres. Impact copolymers are produced in the second reactor (impact reactor) after transferring the polypropylene resin from the first reactor. Gaseous propylene and ethylene are fed to the impact reactor to produce the polymers' rubber phase. Operation of the impact reactor is similar to the initial one, but the second operates at lower pressure (approximately 17 atmospheres). The granular product is finally pelletized.

Random copolymers made by copolymerizing equal amounts of ethylene and propylene are highly amorphous, and they have rubbery properties.

An example of the liquid-phase polymerization is the Spheripol process (Figure 12-3), which uses a tubular reactor.[7] Copolymerization

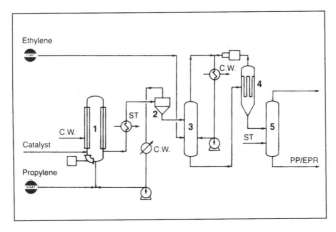

Figure 12-3. The Himont Inc. Spheripol process for producing polypropylene in a liquid-phase[7]: (1) tubular reactor, (2,4) two-stage flash pressure system (to separate unreacted monomer for recycle), (3) heterophasic copolymerization gas-phase reactor, (5) stripper.

occurs in a second gas phase reactor. Unreacted monomer is flashed in a two-stage pressure system and is recycled back to the reactor. Polymer yields of 30,000 or more Kg/Kg of supported catalyst are attainable, and catalyst residue removal from the polymer is not required. The product polymer has an isotactic index of 90–99%.

New generation metallocene catalysts can polymerize propylene in two different ways. Rigid chiral metallocene produce isotactic polypropylene whereas the achiral forms of the catalysts produce atactic polypropylene. The polymer microstructure is a function of the reaction conditions and the catalyst design.[15] Recent work has shown that the rate of ligand rotation in some unbridged metallocenes can be controlled so that the metallocene oscillates between two stereochemical states. One isomer produces isotactic polypropylene and the other produces the atactic polymer. As a result, alternating blocks of rigid isotactic and flexible atactic polypropylene grow within the same polymer chain.[16]

Properties and Uses of Polypropylene

The properties of commercial polypropylene vary widely according to the percentage of crystalline isotactic polymer and the degree of polymerization. Polypropylenes with a 99% isotactic index are currently produced.

Articles made from polypropylene have good electrical and chemical resistance and low water absorption. Its other useful characteristics are its light weight (lowest thermoplastic polymer density), high abrasion resistance, dimensional stability, high impact strength, and no toxicity. Table 12-3 shows the properties of polypropylene.

Polypropylene can be extruded into sheets and thermoformed by solid-phase pressure forming into thin-walled containers. Due to its light weight and toughness, polypropylene and its copolymers are extensively used in automobile parts.

Improvements in melt spinning techniques and film filament processes have made polypropylene accessible for fiber production. Low-cost fibers made from polypropylene are replacing those made from sisal and jute.

World demand for polypropylene is expected to be 30 billion pounds by 2002. This is the strongest growth forecast for any of the major thermoplastics (5.9%). Many of the resins new applications particularly in packaging come at the expense of PS and PVC, the two resins that have been the subject of regulatory restrictions related to solid waste issues and potential toxicity.[17]

Polyvinyl Chloride $+CH_2-CH+$
$$\underset{Cl}{|}$$

Polyvinyl chloride (PVC) is one of the most widely used thermoplastics. It can be extruded into sheets and film and blow molded into bottles. It is used in many common items such as garden hoses, shower curtains, irrigation pipes, and paint formulations.

PVC can be prepolymerized in bulk to approximately 7–8% conversion. It is then transferred to an autoclave where the particles are polymerized to a solid powder. Most vinyl chloride, however, is polymerized

Table 12-3
Properties of Polypropylene

Density, g/cm^3	0.90–0.91
Fill temperature, max. °C	130
Tensile strength, psi	3,200–5,000
Water absorption, 24 hr., %	0.01
Elongation, %	3–700
Melting point, T_m °C	176
Thermal expansion, 10^{-5} in./in. °C	5.8–10
Specific volume, cm^3/lb	30.4–30.8

in suspension reactors made of stainless steel or glass-lined. The peroxide used to initiate the reaction is dispersed in about twice its weight of water containing 0.01–1% of a stabilizer such as polyvinyl alcohol.[18]

In the European Vinyls Corp. process (Figure 12-4), vinyl chloride monomer (VCM) is dispersed in water and then charged with the additives to the reactor.[19] It is a stirred jacketed type ranging in size between 20–105m^3. The temperature is maintained between 53–70°C to obtain a polymer of a particular molecular weight. The heat of the reaction is controlled by cooling water in the jacket and by additional reflux condensers for large reactors. Conversion could be controlled between 85–95% as required by the polymer grade. At the end of the reaction, the PVC and water slurry are channelled to a blowdown vessel, from which part of unreacted monomer is recovered. The rest of the VCM is stripped, and the slurry is centrifuged to separate the polymer from both water and the initiator.

Polyvinyl chloride can also be produced in emulsion. Water is used as the emulsion medium. The particle size of the polymer is controlled using the proper conditions and emulsifier. Polymers produced by free radical initiators are highly branched with low crystallinity.

Vinyl chloride can be copolymerized with many other monomers to improve its properties. Examples of monomers used commercially are vinyl acetate, propylene, ethylene, and vinylidine chloride. The copolymer with ethylene or propylene ($T_g = 80°C$), which is rigid, is used for

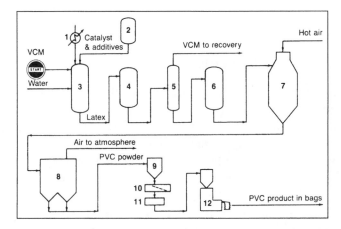

Figure 12-4. The European Vinyls Corp. process for producing polyvinyl chloride using suspension polymerization[19]: (1) reactor, (2) blow-down vessels (to separate unreacted monomer), (3) stripping column, (4) reacted monomer recovery, (5) slurry centrifuge, (6) slurry drier.

blow molding objects. Copolymers with 6–20% vinyl acetate (T_g = 50–80°C) are used for coatings.

Properties and Uses of Polyvinyl Chloride

Two types of the homopolymer are available, the flexible and the rigid. Both types have excellent chemical and abrasion resistance. The flexible types are produced with high porosity to permit plasticizer absorption. Articles made from the rigid type are hard and cannot be stretched more than 40% of their original length. An important property of PVC is that it is self-extinguishing due to presence of the chlorine atom.

Flexible PVC grades account for approximately 50% of PVC production. They go into such items as tablecloths, shower curtains, furniture, automobile upholstery, and wire and cable insulation.

Many additives are used with PVC polymers such as plasticizers, antioxidants, and impact modifiers. Heat stabilizers, which are particularly important with PVC resins, extend the useful life of the finished product. Plastic additives have been reviewed by Ainsworth.[20]

Rigid PVC is used in many items such as pipes, fittings roofing, automobile parts, siding, and bottles.

The 1997 U.S. production of PVC and its copolymers was approximately 14 billion pounds.

POLYSTYRENE

Polystyrene (PS) is the fourth big-volume thermoplastic. Styrene can be polymerized alone or copolymerized with other monomers. It can be polymerized by free radical initiators or using coordination catalysts. Recent work using group 4 metallocene combined with methylaluminoxane produce stereoregular polymer. When homogeneous titanium catalyst is used, the polymer was predominantly syndiotactic. The heterogeneous titanium catalyst gave predominantly the isotactic.[21] Copolymers with butadiene in a ratio of approximately 1:3 produces SBR, the most important synthetic rubber.

Copolymers of styrene-acrylonitrile (SAN) have higher tensile strength than styrene homopolymers. A copolymer of acrylonitrile, butadiene, and styrene (ABS) is an engineering plastic due to its better mechanical properties (discussed later in this chapter). Polystyrene is produced either by free radical initiators or by use of coordination catalysts. Bulk, suspension, and emulsion techniques are used with free radical initiators, and the polymer is atactic.

In a typical batch suspension process (Figure 12-5), styrene is suspended in water by agitation and use of a stabilizer.[14] The polymer forms beads. The bead/water slurry is separated by centrifugation, dried, and blended with additives.

Properties and Uses of Styrene Polymers

Polystyrene homopolymer produced by free radical initiators is highly amorphous (T_g = 100°C). The general purpose rubber (SBR), a block copolymer with 75% butadiene, is produced by anionic polymerization.

The most important use of polystyrene is in packaging. Molded polystyrene is used in items such as automobile interior parts, furniture, and home appliances. Packaging uses plus specialized food uses such as containers for carryout food are growth areas. Expanded polystyrene foams, which are produced by polymerizing styrene with a volatile solvent such as pentane, have low densities. They are used extensively in insulation and flotation (life jackets).

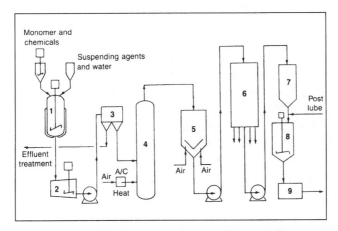

Figure 12-5. The Lummus Crest Inc. process for producing polystyrene[14]: (1) reactor, (2) holding tank (Polystyrene beads and water), (3) centrifuge, (4) pneumatic drier, (5) conditioning tank, (6) screening of beads, (7,8) lubrication and blending, (9) shipping product.

SAN (T_g = 105°C) is stiffer and has better chemical and heat resistance than the homopolymer. However, it is not as clear as polystyrene, and it is used in articles that do not require optical clarity, such as appliances and houseware materials.

ABS has a specific gravity of 1.03 to 1.06 and a tensile strength in the range of 6 to 7.5×10^3 psi. These polymers are tough plastics with outstanding mechanical properties. A wide variety of ABS modifications are available with heat resistance comparable to or better than polysulfones and polycarbonates (noted later in this section). Another outstanding property of ABS is its ability to be alloyed with other thermoplastics for improved properties. For example, ABS is alloyed with rigid PVC for a product with better flame resistance.

Among the major applications of ABS are extruded pipes and pipe fittings, appliances such as refrigerator door liners, and in molded automobile bodies.

World polystyrene production in 1997 was approximately 10 million tons. The demand is forecasted to reach 13 million tons by 2002.[22] The 1997 U.S. production of polystyrene polymers and copolymers was approximately 6.6 billion pounds. ABS, SAN, and other styrene copolymers were approximately 3 billion pounds for the same year.

NYLON RESINS

Nylon resins are important engineering thermoplastics. Nylons are produced by a condensation reaction of amino acids, a diacid and a diammine, or by ring opening lactams such as caprolactam. The polymers, however, are more important for producing synthetic fibers (discussed later in this chapter).

Important properties of nylons are toughness, abrasion and wear resistance, chemical resistance, and ease of processing. Reinforced nylons have higher tensile and impact strengths and lower expansion coefficients than metals. They are replacing metals in many of their applications. Objects made from nylons vary from extruded films used for pharmaceutical packaging to bearings and bushings, to cable and wire insulation.

THERMOPLASTIC POLYESTERS

Thermoplastic polyesters are among the large-volume engineering thermoplastics produced by condensation polymerization of terephthalic acid with ethylene glycol or 1,4-butanediol. These materials are used to produce film for magnetic tapes due to their abrasion and chemical resistance, low water absorption, and low gas permeability. Polyethylene terephthalate (PET) is also used to make plastic bottles (approximately 25% of plastic bottles are made from PET). Similar to nylons, the most important use of PET is for producing synthetic fibers (discussed later).

Polybutylene terephthalate (PBT) is another thermoplastic polyester produced by the condensation reaction of terephthalic acid and 1,4-butanediol:

$$
\text{HOC} \overset{O}{\underset{}{\|}} \longbigcirc \text{COH} \overset{O}{\underset{}{\|}} \quad + \quad \text{OH(CH}_2)_4\text{OH}
$$

$$
\longrightarrow \left[\text{C} \overset{O}{\underset{}{\|}} \longbigcirc \text{CO(CH}_2)_4 \overset{O}{\underset{}{\|}} - \text{O} \right] + 2\,\text{H}_2\text{O}
$$

The polymer is either produced in a bulk or a solution process. It is among the fastest growing engineering thermoplastics, and leads the market of reinforced plastics with an annual growth rate of 7.3%.[23]

The 1997 U.S. production of thermoplastic polyesters was approximately 4.3 billion pounds.

POLYCARBONATES

Polycarbonates (PC) are another group of condensation thermoplastics used mainly for special engineering purposes. These polymers are considered polyesters of carbonic acid. They are produced by the condensation of the sodium salt of bisphenol A with phosgene in the presence of an organic solvent. Sodium chloride is precipitated, and the solvent is removed by distillation:

$$
\text{Na O} \longbigcirc \overset{CH_3}{\underset{CH_3}{\overset{|}{\underset{|}{C}}}} \longbigcirc \text{O Na} \; + \; \text{Cl} - \overset{O}{\underset{}{\overset{\|}{C}}} - \text{Cl}
$$

$$
\longrightarrow \left[\text{O} \longbigcirc \overset{CH_3}{\underset{CH_3}{\overset{|}{\underset{|}{C}}}} \longbigcirc \text{O} - \overset{O}{\underset{}{\overset{\|}{C}}} \right] + 2\,\text{NaCl}
$$

Another method for producing polycarbonates is by an exchange reaction between bisphenol A or a similar bisphenol with diphenyl carbonate:

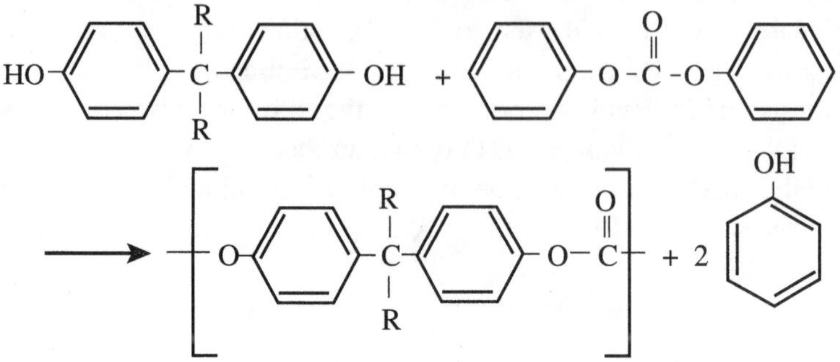

Diphenol carbonate is produced by the reaction of phosgene and phenol. A new approach to diphenol carbonate and non-phosgene route is by the reaction of CO and methyl nitrite using Pd/alumina. Dimethyl carbonate is formed which is further reacted with phenol in presence of tetraphenox titanium catalyst. Decarbonylation in the liquid phase yields diphenyl carbonate.

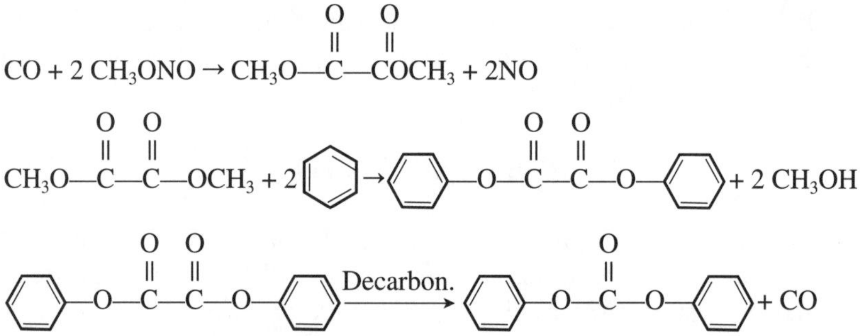

However, the reaction is equilibrium constained and requires a complicated processing scheme.[24]

Properties and Uses of Polycarbonates

Polycarbonates, known for their toughness in molded parts, typify the class of polymers known as engineering thermoplastics. These materials, designed to replace metals and glass in applications demanding strength and temperature resistance, offer advantages of light weight, low cost, and ease of fabrication.[25]

Materials made from polycarbonates are transparent, strong, and heat- and break-resistant. However, these materials are subject to stress cracking and can be attacked by weak alkalies and acids. Table 12-4 compares the properties of polycarbonates with other thermoplastic resins.[25]

Polycarbonates are used in a variety of articles such as laboratory safety shields, street light globes, and safety helmets. The maximum usage temperature for polycarbonate objects is 125°C.

POLYETHER SULFONES

Polyether sulfones (PES) are another class of engineering thermoplastics generally used for objects that require continuous use of temperatures around 200°C. They can also be used at low temperatures with no change in their physical properties.

Table 12-4
Properties of polycarbonates compared
with some thermoplastics[25]

Resin	Melting or glass transition temperature (°C)	tensile strength (MPa)	compressive strength (MPa)	flexural strength (MPa)	Izod impact, 1/8 in. (J/m)
PPO, impact modified	100–110	117–127	124–162	179–200	43–53
PC	149	65	86	93	850
PC, 30% glass	149	131	124	158	106
PC-ABS	149	48–50	76	89–94	560
Nylon 6/6, impact modified	240–260	48–55			160–210
Nylon 6/6, 33% glass	265	151–193	202	282	117–138
PBT	232–267	56	59–100	82–115	43–53
PBT, 30% glass	232–267	117–131	124–162	179–200	69–85
Acetal, homopolymer	181		124	96	69–122
ABS, impact modified	100–110	33–43	31–55	55–76	347–400
PPO, impact modified	135	48–55	69	56–76	320–370
PPO, 30% reinforced	100–110	117–127	123	138–158	90–112

Polyether sulfones can be prepared by the reaction of the sodium or potassium salt of bisphenol A and 4,4-dichlorodiphenyl sulfone. Bisphenol A acts as a nucleophile in the presence of the deactivated aromatic ring of the dichlorophenylsulfone. The reaction may also be catalyzed with Friedel-Crafts catalysts; the dichlorophenyl sulfone acts as an electrophile:

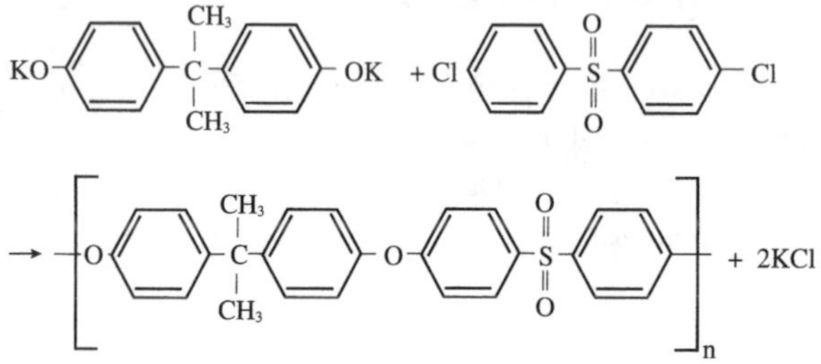

Polyether sulfones could also be prepared using one monomer:

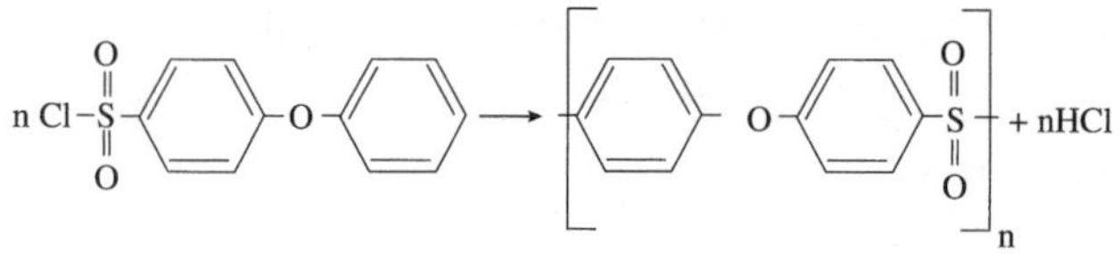

Properties and Uses of Aromatic Polyether Sulfones

In general, properties of polyether sulfones are similar to those of polycarbonates, but they can be used at higher temperatures. Figure 12-6 shows the maximum use temperature for several thermoplastics.[26]

Aromatic polyether sulfones can be extruded into thin films and foil and injection molded into various objects that need high-temperature stability.

POLY(PHENYLENE) OXIDE

Polyphenylene oxide (PPO) is produced by the condensation of 2,6-dimethylphenol. The reaction occurs by passing oxygen in the phenol solution in presence of Cu_2Cl_2 and pyridine:

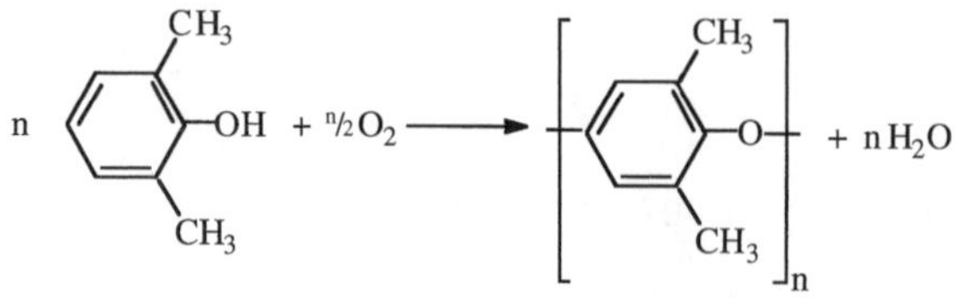

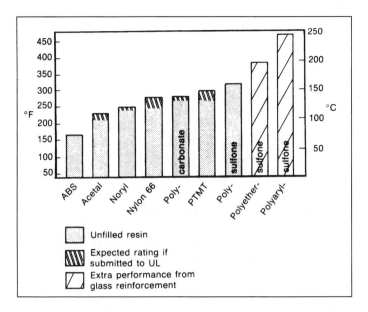

Figure 12-6. Maximum continuous use temperature of some engineering thermoplastics.[26]

PPO is an engineering thermoplastic with excellent properties. To improve its mechanical properties and dimensional stability, PPO can be blended with polystyrene and glass fiber. Articles made from PPO could be used up to 330°C: it is mainly used in items that require higher temperatures such as laboratory equipment, valves, and fittings.

POLYACETALS

Polyacetals are among the aliphatic polyether family and are produced by the polymerization of formaldehyde. They are termed polyacetals to distinguish them from polyethers produced by polymerizing ethylene oxide, which has two -CH_2- groups between the ether group. The polymerization reaction occurs in the presence of a Lewis acid and a small amount of water at room temperature. It could also be catalyzed with amines:

$$\text{n H–}\overset{\overset{\displaystyle O}{\|}}{C}\text{–H} + H_2O \rightarrow HO(CH_2O)_{n-1}CH_2OH$$

Polyacetals are highly crystalline polymers. The number of repeating units ranges from 500 to 3,000. They are characterized by high impact resistance, strength, and a low friction coefficient.

Articles made from polyacetals vary from door handles to gears and bushings, carburetor parts to aerosol containers. The major use of poly-acetals is for molded grades.

THERMOSETTING PLASTICS

This group includes many plastics produced by condensation polymer-ization. Among the important thermosets are the polyurethanes, epoxy resins, phenolic resins, and urea and melamine formaldehyde resins.

POLYURETHANES

Polyurethanes are produced by the condensation reaction of a polyol and a diisocyanate:

$$\text{OCN-R-NCO} + \text{HO-R'-OH} \longrightarrow \left[\begin{matrix} O & H & & H & O \\ \parallel & \mid & & \mid & \parallel \\ C-N-R-N-C-OR'-O \end{matrix} \right]$$

No by-product is formed from this reaction. Toluene diisocyanate (Chapter 10) is a widely used monomer. Diols and triols produced from the reaction of glycerol and ethylene oxide or propylene oxide are suit-able for producing polyurethanes.

Polyurethane polymers are either rigid or flexible, depending on the type of the polyol used. For example, triols derived from glycerol and propylene oxide are used for producing block slab foams. These polyols have moderate reactivity because the hydroxy groups are predominantly secondary. More reactive polyols (used to produce molding polyurethane foams) are formed by the reaction of polyglycols with ethylene oxide to give the more reactive primary group:[27]

$$
\begin{array}{c}
\text{OH} \\
\mid \\
\text{CH}_2\text{O--(RO)}_x\text{--CH}_2\text{CHCH}_3 \\
\mid \qquad\qquad \text{OH} \qquad O \\
\mid \qquad\qquad \mid \qquad\quad / \backslash \\
\text{CHO--(RO)}_y\text{--CH}_2\text{CHCH}_3 + \text{CH}_2\text{--CH}_2 \longrightarrow \\
\mid \qquad\qquad \text{OH} \\
\mid \qquad\qquad \mid \\
\text{CH}_2\text{O--(RO)}_z\text{--CH}_2\text{CHCH}_3
\end{array}
$$

$$
\begin{array}{c}
\text{CH}_2\text{--O--(RO)}_{x+1}\text{--CH}_2\text{CH}_2\text{OH} \\
\mid \\
\text{CH--O--(RO)}_{y+1}\text{--CH}_2\text{CH}_2\text{OH} \\
\mid \\
\text{CH}_2\text{--O--(RO)}_{z+1}\text{--CH}_2\text{CH}_2\text{OH}
\end{array}
$$

$$
\begin{array}{c}
\text{CH}_3 \\
\mid \\
R = \text{--CH}_2\text{--CH--}
\end{array}
$$

Other polyhydric compounds with higher functionality than glycerol (three-OH) are commonly used. Examples are sorbitol (six-OH) and sucrose (eight-OH). Triethanolamine, with three OH groups, is also used.

Diisocyanates generally employed with polyols to produce polyurethanes are 2,4-and 2,6-toluene diisocyanates prepared from dinitrotoluenes (Chapter 10):

2,4-Toluene diisocyanate

2,6-Toluene diisocyanate

MDI

A different diisocyanate used in polyurethane synthesis is methylene diisocyanate (MDI), which is prepared from aniline and formaldehyde. The diamine product is reacted with phosgene to get MDI.

The physical properties of polyurethanes vary with the ratio of the polyol to the diisocyanate. For example, tensile strength can be adjusted within a range of 1,200–600 psi; elongation, between 150–800%.[28]

Improved polyurethane can be produced by copolymerization. Block copolymers of polyurethanes connected with segments of isobutylenes exhibit high-temperature properties, hydrolytic stability, and barrier characteristics. The hard segments of polyurethane block polymers consist of $\{RNHCOO\}_n$, where R usually contains an aromatic moiety.[29]

Properties and Uses of Polyurethanes

The major use of polyurethanes is to produce foam. The density as well as the mechanical strength of the rigid and the flexible types varies widely with polyol type and reaction conditions. For example, polyurethanes could have densities ranging between 1–6 lb/ft^3 for the flexible types and 1–50 lb/ft^3 for the rigid types. Polyurethane foams have good abrasion resistance, low thermal conductance and good load-bearing characteristics. However, they have moderate resistance to organic solvents and are attacked by strong acids. Flame retardancy of polyurethanes could be improved by using special additives, spraying a coating material such as magnesium oxychloride, or by grafting a halogen phosphorous moiety to the polyol. Trichlorobutylene oxide is sometimes copolymerized with ethylene and propylene oxides to produce the polyol.

Major markets for polyurethanes are furniture, transportation, and building and construction. Other uses include carpet underlay, textural laminates and coatings, footwear, packaging, toys, and fibers.

The largest use for rigid polyurethane is in construction and industrial insulation due to its high insulating property. Figure 12-7 compares the degree of insulation of some insulating materials.[28]

Molded urethanes are used in items such as bumpers, steering wheels, instrument panels, and body panels. Elastomers from polyurethanes are characterized by toughness and resistance to oils, oxidation, and abrasion. They are produced using short-chain polyols such as polytetramethylene glycol from 1,4-butanediol. Polyurethanes are also used to produce fibers. Spandex (trade name) is a copolymer of polyurethane (85%) and polyesters.

Polyurethane networks based on triisocyante and diisocyanate connected by segments consisting of polyisobutylene are rubbery and exhibit high temperature properties, hydrolyic stability, and barrier characteristics.[29]

EPOXY RESINS

Epoxy resins are produced by reacting epichlorohydrin and a diphenol. Bisphenol A is the diphenol generally used. The reaction, a ring

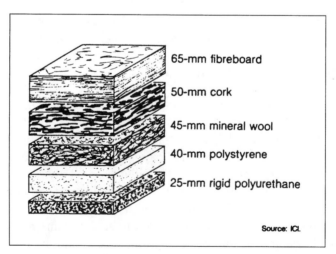

65-mm fibreboard

50-mm cork

45-mm mineral wool

40-mm polystyrene

25-mm rigid polyurethane

Source: ICI.

Figure 12-7. The comparative thickness for the same degree of insulation (dry conditions).[28]

opening polymerization of the epoxide ring, is catalyzed with strong bases such as sodium hydroxide. A nucleophilic attack of the phenoxy ion displaces a chloride ion and opens the ring:

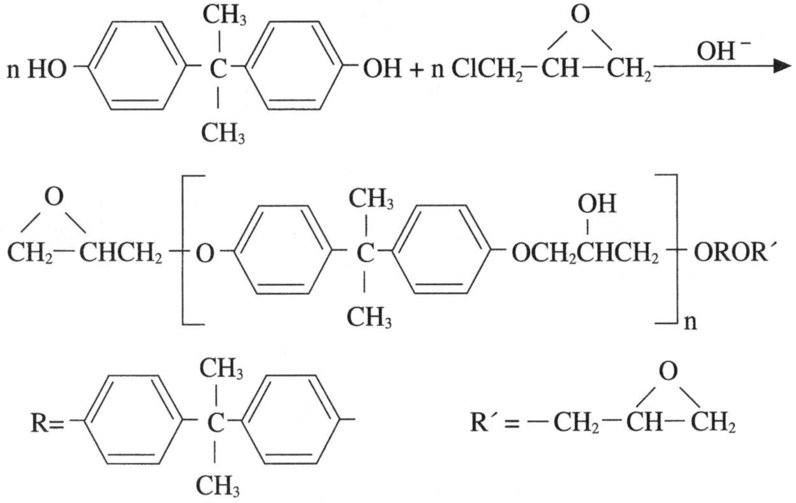

$$n \ HO-\underset{CH_3}{\overset{CH_3}{C}}-OH + n \ ClCH_2-CH-CH_2 \xrightarrow{OH^-}$$

$$CH_2-CHCH_2 \left[O-\underset{CH_3}{\overset{CH_3}{C}}-OCH_2CHCH_2 \right]_n OROR'$$

$$R=-\underset{CH_3}{\overset{CH_3}{C}}- \qquad R'=-CH_2-CH-CH_2$$

The linear polymer formed is cured by cross-linking either with an acid anhydride, which reacts with the -OH groups, or by an amine, which opens the terminal epoxide rings. Cresols and other bisphenols are also used for producing epoxy resins.

Properties and Uses of Epoxy Resins

Epoxy resins have a wide range of molecular weights ($\approx$1,000–10,000). Those with higher molecular weights, termed phenoxy, are hydrolyzed to transparent resins that do not have the epoxide groups. These could be used in molding purposes, or crosslinked by diisocyanates or by cyclic anhydrides.

Important properties of epoxy resins include their ability to adhere strongly to metal surfaces, their resistance to chemicals, and their high dimensional stability. They can also withstand temperatures up to 500°C.

Epoxy resins with improved stress cracking properties can be made by using toughening agents, such as carboxyl-terminated butadiene-acrylonitrile liquid polymers. The carboxyl group reacts with the terminal epoxy ring to form an ester. The ester, with its pendant hydroxyl groups, reacts with the remaining epoxide rings, then more crosslinking occurs by forming ether linkages. This material is tougher than epoxy resins and suitable for encapsulating electrical units.

Major uses of epoxy resins are coatings for appliance finishes, auto primers, adhesive, and in coatings for cans and drums. Interior coatings of drums used for chemicals and solvents manifests its chemical resistance.

In 1997, approximately 681 million pounds of unmodified epoxy resins were produced in the U.S.

UNSATURATED POLYESTERS

Unsaturated polyesters are a group of polymers and resins used in coatings or for castings with styrene. These polymers normally have maleic anhydride moiety or an unsaturated fatty acid to impart the required unsaturation. A typical example is the reaction between maleic anhydride and ethylene glycol:

$$
\begin{array}{c}
\text{(maleic anhydride)} + n\text{HOCH}_2-\text{CH}_2\text{OH} \\
\longrightarrow \left[\begin{array}{c} \overset{O}{\overset{\|}{C}} - \text{CH=CHC} \overset{O}{\overset{\|}{}} - \text{OCH}_2\text{CH}_2\text{O} \end{array} \right]_n + n\text{H}_2\text{O}
\end{array}
$$

Phthalic anhydride, a polyol, and an unsaturated fatty acid are usually copolymerized to unsaturated polyesters for coating purposes. Many other combinations in variable ratios are possible for preparing these resins. The 1998 U.S. production of polyesters was approximately 1.7 billion pounds.

PHENOL-FORMALDEHYDE RESINS

Phenol-formaldehyde resins are the oldest thermosetting polymers. They are produced by a condensation reaction between phenol and formaldehyde. Although many attempts were made to use the product and control the conditions for the acid-catalyzed reaction described by Bayer in 1872, there was no commercial production of the resin until the exhaustive work by Baekeland was published in 1909. In this paper, he describes the product as far superior to amber for pipe stem and similar articles, less flexible but more durable than celluloid, odorless, and fire-resistant.[30]

The reaction between phenol and formaldehyde is either base or acid catalyzed, and the polymers are termed resols (for the base catalyzed) and novalacs (for the acid catalyzed).

The first step in the base-catalyzed reaction is an attack by the phenoxide ion on the carbonyl carbon of formaldehyde, giving a mixture of ortho- and para-substituted mono-methylolphenol plus di- and trisubstituted methylol phenols:

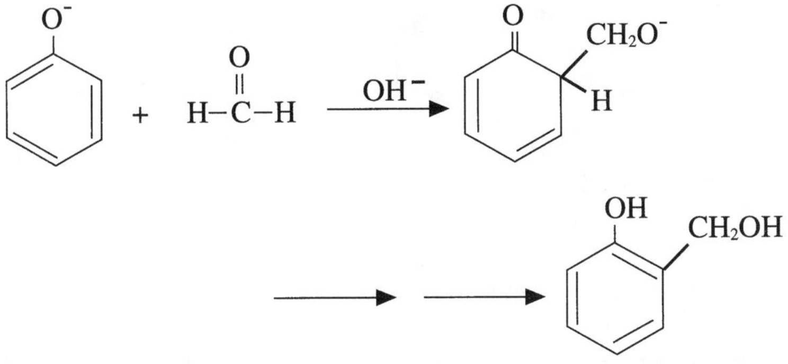

The second step is the condensation reaction between the methylolphenols with the elimination of water and the formation of the polymer. Crosslinking occurs by a reaction between the methylol groups and results in the formation of ether bridges. It occurs also by the reaction of the methylol groups and the aromatic ring, which forms methylene bridges. The formed polymer is a three-dimensional network thermoset:

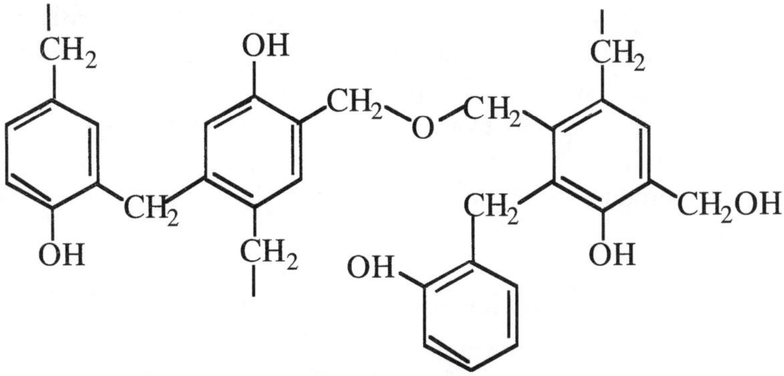

The acid-catalyzed reaction occurs by an electrophilic substitution where formaldehyde is the electrophile. Condensation between the methylol groups and the benzene rings results in the formation of methylene bridges. Usually, the ratio of formaldehyde to phenol is kept less than unity to produce a linear fusible polymer in the first stage. Crosslinking of the formed polymer can occur by adding more formaldehyde and a small amount of hexamethylene tetramine (hexamine,

$(CH_2)_6N_4)$. Hexamine decomposes in the presence of traces of moisture to formaldehyde and ammonia. This results in crosslinking and formation of a thermoset resin:

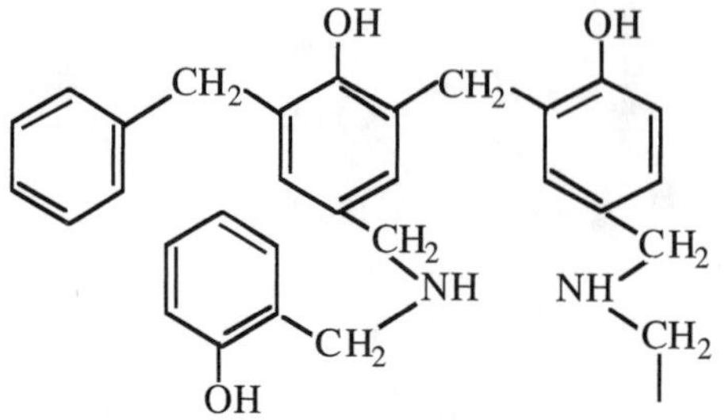

Properties and Uses of Phenolic Resins

Important properties of phenolic resins are their hardness, corrosion resistance, rigidity, and resistance to water hydrolysis. They are also less expensive than many other polymers.

Many additives are used with phenolic resins such as wood flour, oils, asbestos, and fiberglass. Fiberglass piping made with phenolic resins can operate at 150°C and pressure up to 150 psi.[31]

Molding applications dominate the market of phenolic resins. Articles produced by injection molding have outstanding heat resistance and dimensional stability. Compression-molded glass-filled phenolic disk brake pistons are replacing the steel ones in many automobiles because of their light weight and corrosion resistance.

Phenolics are also used in a variety of other applications such as adhesives, paints, laminates for building, automobile parts, and ion exchange resins. Global production of phenol-formaldehyde resins exceeded 5 billion pounds in 1997.

AMINO RESINS (Aminoplasts)

Amino resins are condensation thermosetting polymers of formaldehyde with either urea or melamine. Melamine is a condensation product of three urea molecules. It is also prepared from cyanamide at high pressures and temperatures:

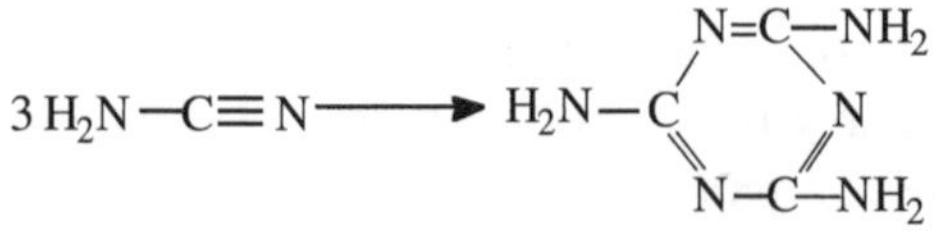

Melamine

Urea-Formaldehyde and Urea-Melamine Resins

The nucleophilic addition reaction of urea to formaldehyde produces mainly monomethylol urea and some dimethylol urea. When the mixture is heated in presence of an acid, condensation occurs, and water is released. This is accompanied by the formation of a crosslinked polymer:

$$2 \; O=C \big<^{NH_2}_{NH_2} \;+\; 3 \; H-\overset{\overset{O}{\|}}{C}-H \longrightarrow O=C \big<^{NHCH_2OH}_{NH_2} \;+\; O=C \big<^{NHCH_2OH}_{NHCH_2OH}$$

$$O=C \big<^{NHCH_2OH} \;+\; \overset{H_2N}{_{\diagup}}C=O \longrightarrow -\overset{\overset{O}{\|}}{C}-N\overset{H}{\underset{}{}}CH_2-\overset{H}{N}-\overset{\overset{O}{\|}}{C}- \;+\; H_2O$$

A similar reaction occurs between melamine and formaldehyde and produces methylolmelamines:

$$H_2N-C\overset{N=C-NH_2}{\underset{N-C-NH_2}{\diagdown\diagup}}N \;+\; 3\;H-\overset{\overset{O}{\|}}{C}-H \longrightarrow HOCH_2NH-C\overset{N=C-NHCH_2OH}{\underset{N-C-NHCH_2OH}{\diagdown\diagup}}N$$

A variety of methylols are possible due to the availability of six hydrogens in melamine. As with urea formaldehyde resins, polymerization occurs by a condensation reaction and the release of water.

Properties and Uses of Aminoplasts

Amino resins are characterized by being more clear and harder (tensile strength) than phenolics. However, their impact strength (breakability) and heat resistance are lower. Melamine resins have better heat and moisture resistance and better hardness than their urea analogs.

The most important use of amino resins is the production of adhesives for particleboard and hardwood plywood.

Compression and injection molding are used with amino resins to produce articles such as radio cabinets, buttons, and cover plates. Because melamine resins have lower water absorption and better chemical and heat resistance than urea resins, they are used to produce dinnerware and laminates used to cover furniture. Almost all molded objects use fillers such as cellulose, asbestos, glass, wood flour, glass fiber and paper. The 1997 U.S. production of amino resins was 2.6 billion pounds.

Polycyanurates

A new polymer type which emerged as important materials for circuit boards are polycyanurates. The simplest monomer is the dicyanate ester of bisphenol A. When polymerized, it forms three-dimensional, densly cross-linked structures through three-way cyanuric acid (2,4,6-triazinetriol):

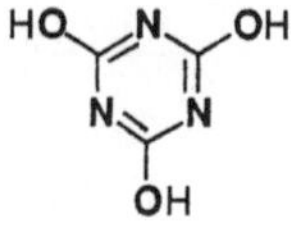

The cyanurate ring is formed by the trimerization of the cyanate ester.

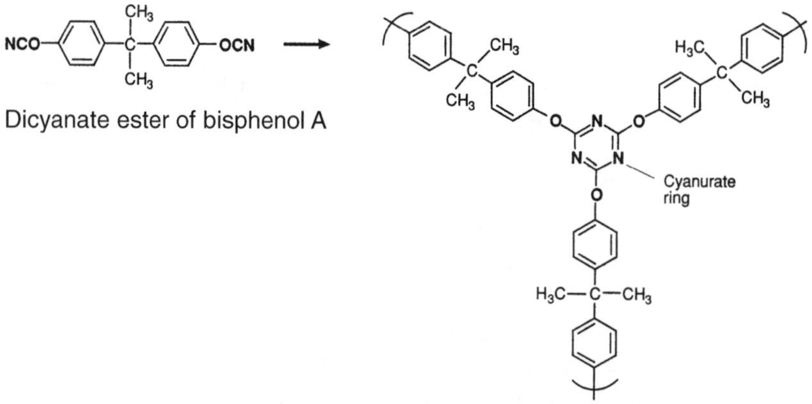

Dicyanate ester of bisphenol A

Other monomers, such as hexaflurobisphenol A and tetramethyl bisphenol F, are also used. These polymers are characterized by high glass transition temperatures ranging between 192 to >350°C.

The largest application of polycyanurates is in circuit boards. Their transparency to microwave and radar energy makes them useful for manufacturing the housing of radar antennas of military and reconnaissance planes. Their impact resistance makes them ideal for aircraft structures and engine pistons.[32]

SYNTHETIC RUBBER

Synthetic rubbers (elastomers) are long-chain polymers with special chemical and physical as well as mechanical properties. These materials have chemical stability, high abrasion resistance, strength, and good dimensional stability. Many of these properties are imparted to the original polymer through crosslinking agents and additives. Selected properties of some elastomers are shown in Table 12-5.[33] An important property of elastomeric materials is their ability to be stretched at least twice their original length and to return back to nearly their original length when released.

Natural rubber is an elastomer constituted of isoprene units. These units are linked in a cis-1,4-configuration that gives natural rubber the outstanding properties of high resilience and strength. Natural rubber occurs as a latex (water emulsion) and is obtained from *Hevea brasiliensis*, a tree that grows in Malaysia, Indonesia, and Brazil. Charles Goodyear (1839) was the first to discover that the latex could be vulcanized (crosslinked) by heating with sulfur or other agents. Vulcanization of rubber is a chemical reaction by which elastomer chains are linked together. The long chain molecules impart elasticity, and the crosslinks give load supporting strength.[34] Vulcanization of rubber has been reviewed by Hertz, Jr.[35] Synthetic rubbers include elastomers that could be crosslinked such as polybutadiene, polyisoprene, and ethylene-propylene-diene terepolymer. It also includes thermoplastic elastomers that are not crosslinked and are adapted for special purposes such as automobile bumpers and wire and cable coatings. These materials could be scraped and reused. However, they cannot replace all traditional rubber since they do not have the wide temperature performance range of thermoset rubber.[36]

The major use of rubber is for tire production. Non-tire consumption includes hoses, footwear, molded and extruded materials, and plasticizers.

Table 12-5
Selected properties of some elastomers[33]

	Durometer hardness range	Tensile strength at room temp, psi	Elongation at room temp, %	Temp. range of service °C	Weather resistance
Natural rubber	20–100	1,000–4,000	100–700	−55–80	Fair
Styrene-butadiene rubber (SBR)	40–100	1,000–3,500	100–700	−55–110	Fair
Polybutadiene	30–100	1,000–3,000	100–700	−60–100	Fair
Polyisoprene	20–100	1,000–4,000	100–750	−55–80	Fair
Polychloroprene	20–90	1,000–4,000	100–700	−55–100	Very good
Polyurethane	62–95 A 40–80 D	1,000–8,000	100–700	−70–120	Excellent
Polyisobutylene	30–100	1,000–3,000	100–700	−55–100	Very good

Worldwide use of synthetic rubber (not including thermoplastic elastomers) in 1997 was approximately 10.5 million metric tons.

Natural rubber use is currently about 6 million tons/year and is expected to grow at annual rate of 3.3%.

Thermoplastic elastomer consumption, approximately 0.8 million tons, is forecasted to reach over one million tons by the year 2000.

BUTADIENE POLYMERS AND COPOLYMERS

Butadiene could be polymerized using free radical initiators or ionic or coordination catalysts. When butadiene is polymerized in emulsion using a free radical initiator such as cumene hydroperoxide, a random polymer is obtained with three isomeric configurations, the 1,4-addition configuration dominating:

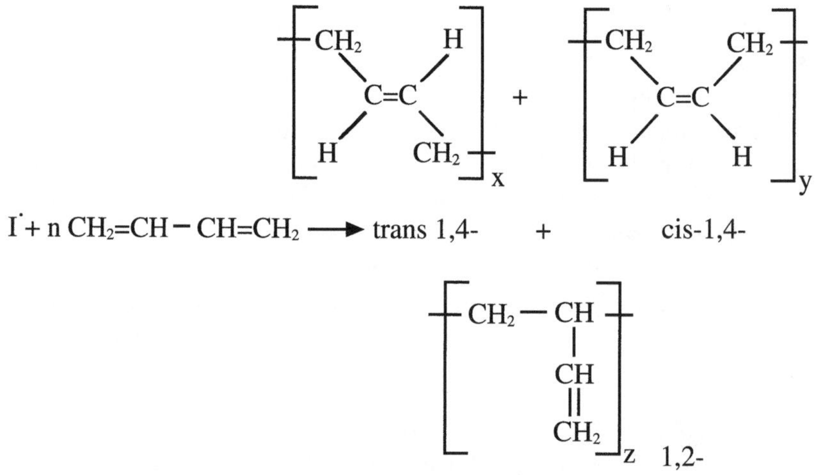

$$\left[\begin{array}{c} -CH_2 \qquad H \\ \diagdown \diagup \\ C=C \\ \diagup \diagdown \\ H \qquad CH_2- \end{array}\right]_x + \left[\begin{array}{c} -CH_2 \qquad CH_2- \\ \diagdown \diagup \\ C=C \\ \diagup \diagdown \\ H \qquad H \end{array}\right]_y$$

$$\overset{.}{I} + n\ CH_2{=}CH-CH{=}CH_2 \longrightarrow \text{trans 1,4-} \quad + \quad \text{cis-1,4-}$$

$$\left[\begin{array}{c} -CH_2-CH- \\ | \\ CH \\ \| \\ CH_2 \end{array}\right]_z \ \text{1,2-}$$

Polymerization of butadiene using anionic initiators (alkyllithium) in a nonpolar solvent produces a polymer with a high cis configuration.[37] A high cis-polybutadiene is also obtained when coordination catalysts are used.[38,39]

Properties and Uses of Polybutadiene

cis-1,4-Polybutadiene is characterized by high elasticity, low heat buildup, high abrasion resistance, and resistance to oxidation. However, it has a relatively low mechanical strength. This is improved by incorporating a cis, trans block copolymer or 1,2-(vinyl) block copolymer in the polybutadiene matrix.[40] Also, a small amount of natural rubber may be mixed with polybutadiene to improve its properties. trans 1,4-Polybutadiene is characterized by a higher glass transition temperature (T_g = $-14°C$) than the cis form ($T_g = -108°C$). The polymer has the toughness, resilience, and abrasion resistance of natural rubber ($T_g = -14°C$).

Styrene-Butadiene Rubber (SBR)

Styrene-butadiene rubber (SBR) is the most widely used synthetic rubber. It can be produced by the copolymerization of butadiene ($\approx 75\%$) and styrene ($\approx 25\%$) using free radical initiators. A random copolymer is obtained. The micro structure of the polymer is 60–68% trans, 14–19% cis, and 17–21% 1,2–. Wet methods are normally used to characterize polybutadiene polymers and copolymers. Solid state NMR provides a more convenient way to determine the polymer micro structure.[41]

Currently, more SBR is produced by copolymerizing the two monomers with anionic or coordination catalysts. The formed copolymer has better mechanical properties and a narrower molecular weight distribution. A random copolymer with ordered sequence can also be made in solution using butyllithium, provided that the two monomers are charged slowly.[42] Block copolymers of butadiene and styrene may be produced in solution using coordination or anionic catalysts. Butadiene polymerizes first until it is consumed, then styrene starts to polymerize. SBR produced by coordinaton catalysts has better tensile strength than that produced by free radical initiators.

The main use of SBR is for tire production. Other uses include footwear, coatings, carpet backing, and adhesives.

The 1997 U.S. production of SBR was approximately 940 million pounds.

NITRILE RUBBER (NBR)

Nitrile rubber is a copolymer of butadiene and acrylonitrile. It has the special property of being resistant to hydrocarbon liquids.

The copolymerization occurs in an aqueous emulsion. When free radicals are used, a random copolymer is obtained. Alternating copolymers are produced when a Zieglar-Natta catalyst is employed. Molecular weight can be controlled by adding modifiers and inhibitors. When the polymerization reaches approximately 65%, the reaction mixture is vacuum distilled in presence of steam to recover the monomer.

The ratio of acrylonitrile/butadiene could be adjusted to obtain a polymer with specific properties. Increasing the acrylonitrile ratio increases oil resistance of the rubber, but decreases its plasticizer compatibility.

NBR is produced in different grades depending on the end use of the polymer. Low acrylonitrile rubber is flexible at low temperatures and is generally used in gaskets, O-rings, and adhesives. The medium type is used in less flexible articles such as kitchen mats and shoe soles. High acrylonitrile polymers are more rigid and highly resistant to hydrocarbons and oils and are used in fuel tanks and hoses, hydraulic equipment, and gaskets. In 1997, the U.S. produced 86 million pounds of solid nitrile rubber.

POLYISOPRENE

Natural rubber is a stereoregular polymer composed of isoprene units attached in a cis configuration. This arrangement gives the rubber high resilience and strength.

Isoprene can be polymerized using free radical initiators, but a random polymer is obtained. As with butadiene, polymerization of isoprene can produce a mixture of isomers. However, because the isoprene molecule is asymmetrical, the addition can occur in 1,2-, 1,4- and 3,4- positions. Six tactic forms are possible from both 1,2- and 3,4- addition and two geometrical isomers from 1,4- addition (cis and trans):

cis-1,4-addition   trans-1,4-addition

1,2-addition   3,4-addition

Stereoregular polyisoprene is obtained when Zieglar-Natta catalysts or anionic initiators are used. The most important coordination catalyst is α-$TiCl_3$ cocatalyzed with aluminum alkyls. The polymerization rate and cis

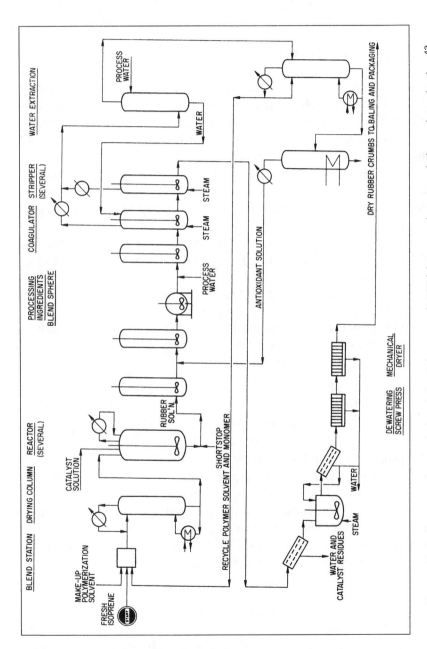

Figure 12-8. A process for producing 1,4-polyisoprene (>99%) by a continuous solution polymerization.[43]

content depends upon Al/Ti ratio, which should be greater than one. Lower ratios predominantly produce the trans structure. Figure 12-8 shows a process for producing cis-polyisoprene by a solution polymerization.[43]

Properties and Uses of Polyisoprene

Polyisoprene is a synthetic polymer (elastomer) that can be vulcanized by the addition of sulfur. cis-Polyisoprene has properties similar to that of natural rubber. It is characterized by high tensile strength and insensitivity to temperature changes, but it has low abrasion resistance. It is attacked by oxygen and hydrocarbons.

trans-Polyisoprene is similar to Gutta percha, which is produced from the leaves and bark of the sapotacea tree. It has different properties from the cis form and cannot be vulcanized. Few commercial uses are based on trans-polyisoprene.

Important uses of cis-polyisoprene include the production of tires, specialized mechanical products, conveyor belts, footwear, and insulation.

POLYCHLOROPRENE (Neoprene Rubber)

Polychloroprene is the oldest synthetic rubber. It is produced by the polymerization of 2-chloro-1,3-butadiene in a water emulsion with potassium sulfate as a catalyst:

$$n \ CH_2=\overset{\overset{\displaystyle Cl}{|}}{C}-CH=CH_2 \longrightarrow \left[CH_2-\overset{\overset{\displaystyle Cl}{|}}{C}=CH-CH_2\right]_n$$

The product is a random polymer that is vulcanized with sulfur or with metal oxides (zinc oxide, magnesium oxide etc.). Vulcanization with sulfur is very slow, and an accelerator is usually required.

Neoprene vulcanizates have a high tensile strength, excellent oil resistance (better than natural rubber), and heat resistance.

Neoprene rubber could be used for tire production, but it is expensive. Major uses include cable coatings, mechanical goods, gaskets, conveyor belts, and cables.

BUTYL RUBBER

Butyl rubber is a copolymer of isobutylene (97.5%) and isoprene (2.5%). The polymerization is carried out at low temperature (below

–95°C) using $AlCl_3$ cocatalyzed with a small amount of water. The cocatalyst furnishes the protons needed for the cationic polymerization:

$$AlCl_3 + H_2O \rightarrow H^+ \ (AlCl_3OH)^-$$

The product is a linear random copolymer that can be cured to a thermosetting polymer. This is made possible through the presence of some unsaturation from isoprene.

Butyl rubber vulcanizates have tensile strengths up to 2,000 psi, and are characterized by low permeability to air and a high resistance to many chemicals and to oxidation. These properties make it a suitable rubber for the production of tire inner tubes and inner liners of tubeless tires. The major use of butyl rubber is for inner tubes. Other uses include wire and cable insulation, steam hoses, mechanical goods, and adhesives. Chlorinated butyl is a low molecular weight polymer used as an adhesive and a sealant.

ETHYLENE-PROPYLENE RUBBER

Ethylene-propylene rubber (EPR) is a stereoregular copolymer of ethylene and propylene. Elastomers of this type do not possess the double bonds necessary for crosslinking. A third monomer, usually a monoconjugated diene, is used to provide the residual double bonds needed for cross linking. 1,4-Hexadiene and ethylidene norbornene are examples of these dienes. The main polymer chain is completely saturated while the unsaturated part is pending from the main chain. The product elastomer, termed ethylene-propylene terepolymer (EPT), can be crosslinked using sulfur. Crosslinking EPR is also possible without using a third component (a diene). This can be done with peroxides.

Important properties of vulcanized EPR and EPT include resistance to abrasion, oxidation, and heat and ozone; but they are susceptible to hydrocarbons.

The main use of ethylene-propylene rubber is to produce automotive parts such as gaskets, mechanical goods, wire, and cable coating. It may also be used to produce tires.

TRANSPOLYPENTAMER

Transpolypentamer (TPR) is produced by the ring cleavage of cyclopentene.[33,44] Cyclopentene is obtained from cracked naphtha or gas oil, which contain small amounts of cyclopentene, cyclopentadiene, and

dicyclopentadiene. Polymerization using organometallic catalysts produce a stereoregular product (trans 1,5-polypentamer):

$$n \, \bigtriangleup \longrightarrow \left[CH_2CH_2CH_2\overset{H}{\underset{H}{C}}{=}C \right]_n$$

Due to the presence of residual double bonds, the polymer could be
crosslinked with regular agents. TPR is a linear polymer with a high trans
configuration. It is highly amorphous at normal temperatures and has a
T_g of about 90°C and a density of 0.85.

THERMOPLASTIC ELASTOMERS

Thermoplastic elastomers (TPES), as the name indicates, are plastic
polymers with the physical properties of rubbers. They are soft, flexible,
and possess the resilience needed of rubbers. However, they are
processed like thermoplastics by extrusion and injection molding.

TPE's are more economical to produce than traditional thermoset
materials because fewer steps are required to manufacture them than to
manufacture and vulcanize thermoset rubber. An important property of
these polymers is that they are recyclable.

Thermoplastic elastomers are multiphase composites, in which the
phases are intimately depressed. In many cases, the phases are chemically bonded by block or graft copolymerization. At least one of the
phases consists of a material that is hard at room temperature.[45]

Currently, important TPE's include blends of semicrystalline thermoplastic polyolefins such as propylene copolymers, with ethylene-propylene terepolymer elastomer. Block copolymers of styrene with other
monomers such as butadiene, isoprene, and ethylene or ethylene/propy-
lene are the most widely used TPE's. Styrene-butadiene-styrene (SBS)
accounted for 70% of global styrene block copolymers (SBC). Currently,
global capacity of SBC is approximately 1.1 million tons. Polyurethane
thermoplastic elastomers are relatively more expensive then other TPE's.
However, they are noted for their flexibility, strength, toughness, and
abrasion and chemical resistance.[46] Blends of polyvinyl chloride with
elastomers such as butyl are widely used in Japan.[36]

Random block copolymers of polyesters (hard segments) and amorphous glycol soft segments, alloys of ethylene interpolymers, and chlorinated polyolefins are among the evolving thermoplastic elastomers.

Important properties of TPE's are their flexibility, softness, and resilience. However, compared to vulcanizable rubbers, they are inferior in resistance to deformation and solvents.

Important markets for TPE's include shoe soles, pressure sensitive adhesives, insulation, and recyclable bumpers.

SYNTHETIC FIBERS

Fibers are solid materials characterized by a high ratio of length to diameter. They can be manufactured from a natural origin such as silk, wool, and cotton, or derived from a natural fiber such as rayon. They may also be synthesized from certain monomers by polymerization (synthetic fibers). In general, polymers with high melting points, high crystallinity, and moderate thermal stability and tensile strengths are suitable for fiber production.

Man-made fibers include, in addition to synthetic fibers, those derived from cellulose (cotton, wood) but modified by chemical treatment such as rayon, cellophane, and cellulose acetate. These are sometimes termed "regenerated cellulose fibers." Rayon and cellophane have shorter chains than the original cellulose due to degradation by alkaline treatment. Cellulose acetates produced by reacting cellulose with acetic acid and modified or regenerated fibers are excluded from this book because they are derived from a plant source. Fibers produced by drawing metals or glass (SiO_2) such as glass wool are also excluded.

Major fiber-making polymers are those of polyesters, polyamides (nylons), polyacrylics, and polyolefins. Polyesters and polyamides are produced by step polymerization reactions, while polyacrylics and poly-olefins are synthesized by chain-addition polymerization.

POLYESTER FIBERS

Polyesters are the most important class of synthetic fibers. In general, polyesters are produced by an esterification reaction of a diol and a diacid. Carothers (1930) was the first to try to synthesize a polyester fiber by reacting an aliphatic diacid with a diol. The polymers were not suit-able because of their low melting points. However, he was successful in preparing the first synthetic fiber (nylon 66). In 1946, Whinfield and Dickson prepared the first polyester polymer by using terephthalic acid (an aromatic diacid) and ethylene glycol.

Polyesters can be produced by an esterification of a dicarboxylic acid and a diol, a transesterification of an ester of a dicarboxylic acid and a diol, or by the reaction between an acid dichloride and a diol.

The polymerization reaction could be generally represented by the esterification of a dicarboxylic acid and a diol as:

$$
nHO-\overset{\overset{\displaystyle O}{||}}{C}-R-\overset{\overset{\displaystyle O}{||}}{C}OH + nHOR'OH \longrightarrow \left[O-\overset{\overset{\displaystyle O}{||}}{C}-R-\overset{\overset{\displaystyle O}{||}}{C}-O-R'\right]_n + H_2O
$$

Less important methods are the self condensation of w-hydroxy acid and the ring opening of lactones and cyclic esters. In self condensation of w-hydroxy acids, cyclization might compete seriously with linear polymerization, especially when the hydroxyl group is in a position to give five or six membered lactones.

Polyethylene Terephthalate Production

Polyethylene terephthalate (PET) is produced by esterifying terephthalic acid (TPA) and ethylene glycol or, more commonly, by the transesterification of dimethyl terephthalate and ethylene glycol. This route is favored because the free acid is not soluble in many organic solvents. The reaction occurs in two stages (Figure 12-9).[47] Methanol is released in the first stage at approximately 200°C with the formation of bis(2-hydroxyethyl) terephthalate. In the second stage, polycondensation occurs, and excess ethylene glycol is driven away at approximately 280°C and at lower pressures (≈ 0.01 atm):

$CH_3O\overset{\overset{\displaystyle O}{||}}{C}$⟨⟩$\overset{\overset{\displaystyle O}{||}}{C}OCH_3$ + 2 $HOCH_2CH_2OH$ ⇌

$HOCH_2CH_2O\overset{\overset{\displaystyle O}{||}}{C}$⟨⟩$\overset{\overset{\displaystyle O}{||}}{C}OCH_2CH_2OH$ + 2 CH_3OH

(I)

(I) ⇌ $\left[\overset{\overset{\displaystyle O}{||}}{C}\text{⟨⟩}\overset{\overset{\displaystyle O}{||}}{C}OCH_2CH_2O\right]$ + $HO(CH_2)_2OH$

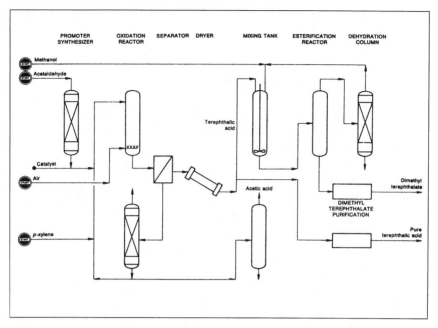

Figure 12-9. The Inventa AG Process for producing polyethylene-terephthalate.[47]

Using excess ethylene glycol is the usual practice because it drives the equilibrium to near completion and terminates the acid end groups. This results in a polymer with terminal -OH. When the free acid is used (esterification), the reaction is self catalyzed. However, an acid catalyst is used to compensate for the decrease in terephthalic acid as the esterification nears completion. In addition to the catalyst and terminator, other additives are used such as color improvers and dulling agents. For example, PET is delustred by the addition of titanium dioxide.

The molecular weight of the polymer is a function of the extent of polymerization and could be monitored through the melt viscosity. The final polymer may be directly extruded or transformed to chips, which are stored.

Batch polymerization is still used. However, most new processes use continuous polymerization and direct spinning.

An alternative route to PET is by the direct reaction of terephthalic acid and ethylene oxide. The product bis(2-hydroxyethyl)terephthalate reacts in a second step with TPA to form a dimer and ethylene glycol, which is released under reduced pressure at approximately 300°C.

This process differs from the direct esterification and the transesterification routes in that only ethylene glycol is released. In the former two routes, water or methanol are coproduced and the excess glycol released.

Properties and Uses of Polyesters

As mentioned earlier, polyethylene terephthalate is an important thermoplastic. However, most PET is consumed in the production of fibers.

Polyester fibers contain crystalline as well as noncrystalline regions. The degree of crystallinity and molecular orientation are important in determining the tensile strength of the fiber (between 18–22 denier) and its shrinkage. The degree of crystallinity and molecular orientation can be determined by X-ray diffraction techniques.[48]

Important properties of polyesters are the relatively high melting temperatures ($\approx 265°C$), high resistance to weather conditions and sunlight, and moderate tensile strength (Table 12-6).[49]

Melt spinning polyesters is preferred to solution spinning because of its lower cost. Due to the hydrophobic nature of the fiber, sulfonated terephthalic acid may be used as a comonomer to provide anionic sites for cationic dyes. Small amounts of aliphatic diacids such as adipic acid may also be used to increase the dyeability of the fibers by disturbing the fiber's crystallinity.

Polyester fibers can be blended with natural fibers such as cotton and wool. The products have better qualities and are used for men's and women's wear, pillow cases, and bedspreads. Fiberfill, made from polyesters, is used in mattresses, pillows, and sleeping bags. High-tenacity polymers for tire cord reinforcement are equivalent in strength to nylon tire cords and are superior because they do not "flat spot." V-belts and fire hoses made from industrial filaments are another market for polyesters.

POLYAMIDES (Nylon Fibers)

Polyamides are the second largest group of synthetic fibers after polyesters. However, they were the first synthetic fibers that appeared in the market in 1940. This was the result of the work of W. H. Carothers in USA who developed nylon 66. At about the same time nylon 6 was also developed in Germany by I. G. Farben. Both of these nylons still dominate the market for polyamides. However, due to patent restrictions and raw materials considerations, nylon 66 is most extensively produced in USA and nylon 6 is most extensively produced in Europe.

Table 12-6
Important properties of polyesters[49]

(Mechanical properties at 21°C, 65% relative humidity, using 60% minute strain rate)

| Property | Polyethyleneterephthalate PET | | | | Poly (cyclo-hexanedi-methylene terephtha-late) staple and tow |
| | Filament yarns | | Staple and tow | | |
	Regular tensile strength	High tensile strength	Regular tensile strength	High tensile strength	
Breaking strength (g/denier)*	2.8–5.6	6.0–9.5	2.2–6.0	5.8–6.0	2.5–3.0
Breaking elongation (%)	19–34	10–34	25–65	25–40	24–34
Initial modulus (g/denier)*	75–100	115–120	25–40	45–55	24–35
Elastic recovery (%r)	88–93 (at 5% elongation)	90 (at 5% elongation)	75–85 (at 5% elongation)	—	85–95 (at 2% elongation)
Moisture regain (%)**	0.4	0.4	0.4	0.4	0.34
Specific gravity	1.38	1.38	1.38	1.38	1.22
Melting temper- ature (°C)	265	265	265	265	290

*Grams per denier-grams of frorce per denier. Denier is linear density; the mass for 9,000 meters of fiber.
**The amout of moisture in the fiber at 21°C, 65% relative humidity.

Numbers that follow the word "nylon" denote the number of carbons present within a repeating unit and whether one or two monomers are being used in polymer formation. For nylons using a single monomer such as nylon 6 or nylon 12, the numbers 6 and 12 denote the number of carbons in caprolactam and laurolactam, respectively. For nylons using two monomers such as nylon 610, the first number, 6, indicates the number of carbons in the hexamethylene diamine and the other number, 10, is for the second monomer sebacic acid.

Polyamides are produced by the reaction between a dicarboxylic acid and a diamine (e.g., nylon 66), ring openings of a lactam, (e.g., nylon 6) or by the polymerization of w-amino acids (e.g., nylon 11). The production of some important nylons is discussed in the following sections.

Nylon 66 (Polyhexamethyleneadipate)

Nylon 66 is produced by the reaction of hexamethylenediamine and adipic acid (see Chapters 9 and 10 for the production of the two monomers). This produces hexamethylenediammonium adipate salt. The product is a dilute salt solution concentrated to approximately 60% and charged with acetic acid to a reactor where water is continuously removed. The presence of a small amount of acetic acid limits the degree of polymerization to the desired level:

$$nHOC(CH_2)_4COH + nH_2N(CH_2)_6\,NH_2 \longrightarrow [OC(CH_2)_4CO]_n^{2-}\;[H_3N(CH_2)_6NH_3]_n^{2+}$$

$$\longrightarrow [-C(CH_2)_4CNH(CH_2)_6\,NH\frac{}{}]_n + 2nH_2O$$

The temperature is then increased to 270–300°C and the pressure to approximately 16 atmospheres, which favors the formation of the polymer. The pressure is finally reduced to atmospheric to permit further water removal. After a total of three hours, nylon 66 is extruded under nitrogen pressure.

Nylon 6 (Polycaproamide)

Nylon 6 is produced by the polymerization of caprolactam. The monomer is first mixed with water, which opens the lactam ring and gives w-amino acid:

$$\text{Caprolactam} + H_2O \longrightarrow H_2N(CH_2)_5COOH$$

Caprolactam w-Amino acid

The formed amino acid reacts with itself or with caprolactam at approximately 250–280°C to form the polymer:

$$H_2N(CH_2)_5COOH + \text{Caprolactam} \longrightarrow \left[N-(CH_2)_5C \right]_n$$

Temperature control is important, especially for depolymerization, which is directly proportional to reaction temperature and water content. Figure 12-10 shows the Inventa-Fisher process.[50]

Nylon 12 (Polylaurylamide)

Nylon 12 is produced in a similar way to nylon 6 by the ring opening polymerization of laurolactam. The polymer has a lower water capacity than nylon 6 due to its higher hydrophobic properties. The polymeriza-

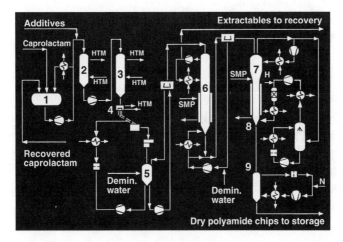

Figure 12-10. The Inventa-Fisher process for producing nylon 6 from caprolactam[50]: (1) Melting station, (2,3) polymerization reactors, (4) extruder, (5) intermediate vessel, (6) extraction column, (7,8) extraction columns, (9) cooling silo.

tion reaction is slower than for caprolactam. Higher temperatures are used to increase the rate of the reaction:

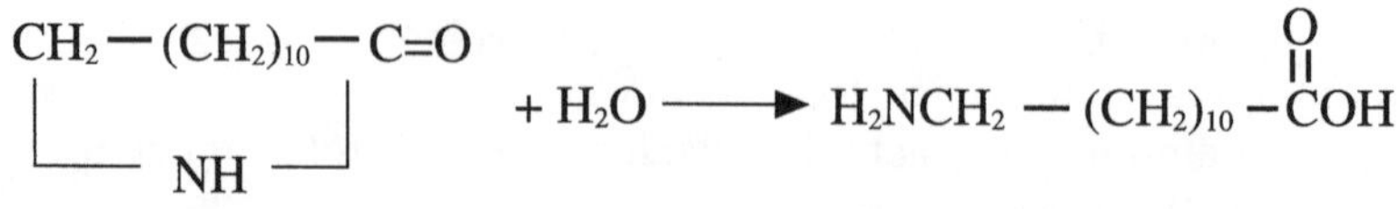

$$CH_2-(CH_2)_{10}-C{=}O \quad \underset{NH}{\big|_____\big|} \quad + H_2O \longrightarrow H_2NCH_2-(CH_2)_{10}-\overset{O}{\overset{\|}{C}}OH$$

$$n\,NH_2CH_2(CH_2)_{10}\overset{O}{\overset{\|}{C}}OH \longrightarrow \quad -\big[HN(CH_2)_{11}-\overset{O}{\overset{\|}{C}}\big]_n + nH_2O$$

The monomer (laurolactam) could be produced from 1,5,9-cyclododeca-triene, a trimer of butadiene (Chapter 9). The trimer is epoxidized with peracetic acid or acetaldehyde peracetate and then hydrogenated. The saturated epoxide is rearranged to the ketone with MgI_2 at 100°C.[51] This is then changed to the oxime and rearranged to laurolactam.

Nylon 4 (Polybutyramide)

Nylon 4 is produced by ring opening 2-pyrrolidone. Anionic polymerization is used to polymerize the lactam. Cocatalysts are used to increase the yield of the polymer. Carbon dioxide is reported to be an excellent polymerization activator.

Nylon 4 has a higher water absorption capacity than other nylons due to its lower hydrophobic property.

Nylon 11 (Polyundecanylamide)

Nylon 11 is produced by the condensation reaction of 11- aminounde-canoic acid. This is an example of the self condensation of an amino acid where only one monomer is used. The monomer is first suspended in water, then heated to melt the monomer and to start the reaction. Water is continuously removed to drive the equilibrium to the right. The polymer is finally withdrawn for storage:

$$n\,H_2N-CH_2(CH_2)_9\overset{O}{\overset{\|}{C}}OH \rightleftharpoons -\big[NH-(CH_2)_{10}\overset{O}{\overset{\|}{C}}\big]_n + nH_2O$$

Other Nylon Polymers

Many other nylons could be produced such as nylon nylon 5, nylon 7, nylon 610, and nylon 612. These have properties generally similar to those nylons described. Table 12-7 shows the monomers used to produce important nylons and their melting points.[52]

Table 12-7
Melting points of various nylons and the monomer formula[52]

Monomer	Nylon type	Monomer Formula	Approximate M.P. °C
Caprolactam	6	$CH_2-(CH_2)-_4C=0$ $\llcorner$—NH—$\lrcorner$	223
Adipic Acid	66	$HOOC-(CH_2)_4COOH$	265
Hexamethylenediamine		$H_2N-(CH_2)_6-NH_2$	
Sebacic Acid	610	$HOOC-(CH_2)_8COOH$	215
Hexamethylenediamine		$H_2N-(CH_2)_6-NH_2$	
11-Aminoundecanoic Acid	11	$H_2N-(CH_2)_{10}-COOH$	190
Laurolactam	12	$CH_2-(CH_2)_{10}-C=0$ $\llcorner$—NH—$\lrcorner$	119
Acrylamide	3	$CH_2=CHCONH_2$	320
2-pyrrolidone	4	$CH_2-(CH_2)_2-C=O$ $\llcorner$—NH—$\lrcorner$	265
Valerolactam	5	$CH_2-(CH_2)_3-C=0$ $\llcorner$—NH—$\lrcorner$	260
7-aminoheptanoic Acid or Ethyl 7-aminoheptanoate	7	$N_2H-(CH_2)_6-COOH$ $H_2H-(CH_2)_6-COOC_2H_5$	233
Capryllactam	8	$CH_2-(CH_2)_6-C=0$ $\llcorner$—NH—$\lrcorner$	200
9-Aminopelargonic acid	9	$H_2N-(CH_2)_8-COOH$	209
10-Aminodecanoic acid	10	$H_2N-(CH_2)_9-COOH$	188

Properties and Uses of Nylons

Nylons are generally characterized by relatively high melting points due to the presence of the amide linkage. They are also highly crystalline, and the degree of crystallinity depends upon factors such as the polymer structure and the distance between the amide linkages. An increase in polymer crystallinity increases its tensile strength, abrasion resistance, and modulus of elasticity.

Hydrogen bonding in polyamides is fairly strong and has a pronounced etfect on the physical properties of the polymer such as the crystallinity, melting point, and water absorption. For example, nylon 6, with six carbons, has a melting point of 223°C, while it is only 190°C for nylon 11. This reflects the higher hydrogen bonding in nylon 6 than in nylon 11.

Moisture absorption of nylons differs according to the distance between the amide groups. For example, nylon 4 has a higher moisture absorption than most other nylons, and it is approximately similar to that of cotton. This is a result of the higher hydrophilic character of nylon 4.

Nylons, however, are to some extent subject to deterioration by light. This has been explained on the basis of chain breaking and crosslinking. Nylons are liable to attack by mineral acids but are resistant to alkalies. They are difficult to ignite and are self-extinguishing.

In general, most nylons have remarkably similar properties, and the preference of using one nylon over the other is usually dictated by economic considerations except for specialized uses.

Nylons have a variety of uses ranging from tire cord to carpet to hosiery. The most important application is cord followed by apparel. Nylon staple and filaments are extensively used in the carpet industry. Nylon fiber is also used for a variety of other articles such as seat belts, monofilament finishes, and knitwear. Because of its high tenacity and elasticity, it is a valuable fiber for ropes, parachutes, and underwear.

The 1997 U.S. production of nylon fibers was approximately 2.9 billion pounds.

ACRYLIC AND MODACRYLIC FIBERS

Acrylic fibers are a major synthetic fiber class developed about the same time as polyesters. Modacrylic fibers are copolymers containing between 35–85% acrylonitrile. Acrylic fibers contain at least 85% acrylonitrile. Orlon is an acrylic fiber developed by DuPont in 1949; Dynel is a modacrylic fiber developed by Union Carbide in 1951.

Polyacrylics are produced by copolymerizing acrylonitrile with other monomers such as vinyl acetate, vinyl chloride, and acrylamide. Solution polymerization may be used where water is the solvent in the presence of a redox catalyst. Free radical or anionic initiators may also be used. The produced polymer is insoluble in water and precipitates. Precipitation polymerization, whether self nucleation or aggregate nucleation, has been reviewed by Juba.[53] The following equation is for an acrylonitrile polymer initiated by a free radical:

$$I^{\cdot} + n \ \ CH_2{=}CH{-}CN \longrightarrow \quad {-}\!\!\left[CH_2{-}\underset{\displaystyle \overset{\displaystyle |}{CN}}{CH}\right]_n$$

Copolymers of acrylonitrile are sensitive to heat, and melt spinning is not used. Solution spinning (wet or dry) is the preferred process where a polar solvent such as dimethyl formamide is used. In dry spinning the solvent is evaporated and recovered.

Dynel, a modacrylic fiber, is produced by copolymerizing vinyl chloride with acrylonitrile. Solution spinning is also used where the polymer is dissolved in a solvent such as acetone. After the solvent is evaporated, the fibers are washed and subjected to stretching, which extends the fiber 4–10 times of the original length.

Properties and Uses of Polyacrylics

Acrylic fibers are characterized by having properties similar to wool and have replaced wool in many markets such as blankets, carpets, and sweaters. Important properties of acrylics are resistance to solvents and sunlight, resistance to creasing, and quick drying.

Acrylic fiber breaking strength ranges between 22,000 and 39,000 psi and they have a water absorption of approximately 5%. Dynel, due to the presence of chlorine, is less flammable than many other synthetic fibers.

Major uses of acrylic fibers are woven and knitted clothing fabrics (for apparel), carpets, and upholstery.

CARBON FIBERS (Graphite Fibers)

Carbon fibers are special reinforcement types having a carbon content of 92–99 wt%. They are prepared by controlled pyrolysis of organic materials in fibrous forms at temperatures ranging from 1,000–3,000°C.

The commercial fibers are produced from rayon, polyacrylonitrile, and petroleum pitch. When acrylonitrile is heated in air at moderate temperatures ($\approx 220°C$), HCN is lost, and a ladder polymer is thought to be the intermediate:

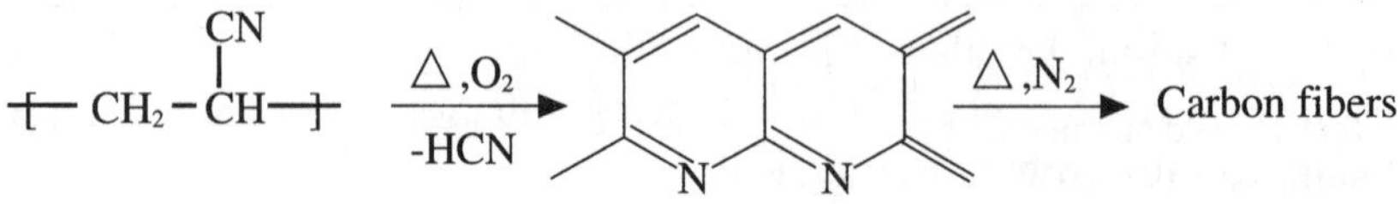

Further heating above 1700°C in the presence of nitrogen for a period of 24 hours produces carbon fiber. Carbon fibers are characterized by high strength, stiffness, low thermal expansion, and thermal and electrical conductivity, which makes them an attractive substitute for various metals and alloys.[54] These fibers have longitudinal tensile strengths and moduli ranging from 2.5–7.0 GPa and 230–590 GPa, respectively. A bending beam force detector was developed to measure longitudinal compressive strengths of polyacrylonitrile-based carbon fibers.[55]

Most carbon fiber composites are based mainly on thermosetting epoxy matrices.

Current U.S. production of carbon fibers is approximately ten million pounds/year.

POLYPROPYLENE FIBERS

Polypropylene fibers represent a small percent of the total polypropylene production. (Most polypropylene is used as a thermoplastic.) The fibers are usually manufactured from isotactic polypropylene.

Important characteristics of polypropylene are high abrasion resistance, strength, low static buildup, and resistance to chemicals. Crystallinity of fiber-grade polypropylene is moderate and when heated, it starts to soften at approximately 145°C and then melts at 170°C. The physical properties of fiber-grade polypropylene are given in Table 12-8. Melt spinning is normally used to produce the fibers.[56] The high MP of polypropylene is attributed to low entropy of fusion arising from stiffening of the chain.

Polypropylene fibers are used for face pile of needle felt, tufted carpets, upholstery fabrics, etc.

The total 1997 U.S. production of polyolefin fibers, including polypropylene fibers, was approximately 2.5 billion pounds.

Table 12-8
Physical properties of fiber-grade polypropylene[56]

Property	Fiber-grade homopolymer	Fiber-grade copolymer
Specific gravity at 23°C	0.905–0.910	0.895–0.905
Flow rate at 230°C, 2160 g load g/10 min	6	3
Tensile yield at 2 in./min psi	5000	4000
Stiffness in flexure 10^3 psi	190	150
Unnotched izod, impact at 0°F-ft-lb/in.	<10	>20
Melting point, dilatometer °C	172	170
Water adsorption, 24 hr. %	<0.01	<0.01
Environmental stress cracking % failure	none	none

REFERENCES

1. Wittcoff, H. A. "Polymers in Pursuit of Strength," *CHEMTECH,* Vol. 17, No. 3, 1987, pp. 156–166.
2. *Chemical Week,* No. 14, 1984, p. 13.
3. Hatch, L. F. and Matar, S., *From Hydrocarbons to Petrochemicals,* Gulf Publishing Co., Houston, 1981, p. 171.
4. Bennet, A., *CHEMTECH,* Vol. 29, No. 7, 1999, pp. 24–28.
5. El-Khadi, M. and David, O. F., Second Arab Conference on Petrochemicals, United Arab Emirates, Abu Dhabi, March 15–22, 1976.
6. Sittig, M., "Polyolefin Production Processes," *Chemical Technology Review* No. 79, New Jersey, Noyes Data Corp., 1976, p. 9.
7. "Petrochemical Handbook," *Hydrocarbon Processing,* Vol. 70, No. 3, 1991, p. 173.
8. Sinclair, K. B., "For Polyolefins: Estimate Gas Phase Production Costs," *Hydrocarbon Processing,* Vol. 64, No. 7, 1985, pp. 81–83.
9. Newton, D., Chinh, J. C., and Power, M., "Optimize Gas-phase Polyethylene," *Hydrocarbon Processing,* Vol. 77, 1998, pp. 85–91.
10. *Chemical and Engineering News,* June 29, 1998, p. 44
11. Sacks, W., "Packaging Containers," *CHEMTECH,* Vol. 18, No. 8, August 1988, pp. 480–483.
12. *Modern Plastics,* Vol . 52, No. 6, 1975, p. 6.
13. *Chemical and Engineering News,* March 30, 1992, p. 17.
14. "Petrochemical Handbook," *Hydrocarbon Processing,* Vol. 70, No. 3, 1991, p. 176.

15. *Chemical and Engineering News,* Jan. 16, 1995, pp. 6–7.
16. Golab, J. T., "Making Industrial Decisions with Combutational Chemistry," *CHEMTECH,* Vol. 28, No. 4, 1998, pp. 17–21.
17. *Hydrocarbon Processing,* Vol. 77, No. 11, 1998, p. 25
18. Rodriguez, F., *Principles of Polymer Systems,* 3rd Ed., Hemisphere Publishing Corp., New York, 1989, p. 466.
19. "Petrochemical Handbook," *Hydrocarbon Processing,* Vol. 70, No. 3, 1991, p. 180.
20. Ainsworth, S. J., "Plastics Additives," *Chemical and Engineering News,* August 31, 1992, pp. 34–39.
21. Kix, M. et al., *Polymer Bulletin,* Vol. 41, 1998, pp. 349–354.
22. *Hydrocarbon Processing,* Vol. 77, No. 9, 1998, p. 11
23. *Hydrocarbon Processing,* Vol. 70, No. 12, 1991, p. 29.
24. Piccolini, R. and Plotkin, J., "Patent Watch," *CHEMTECH,* Vol. 29, No. 3, 1999, p. 31.
25. Sikdar, S. K., "The World of Polycarbonates," *CHEMTECH,* Vol. 17, No. 2, 1987, pp. 112–117.
26. Leslie, V. J., Rose, J., Rudkin, G. O., and Fitzin, J. *CHEMTECH,* Vol. 5, No. 5, 1975, pp. 426–432.
27. *Guide to Plastics,* New York, McGraw Hill, Inc., 1976.
28. *Modern Plastics International,* Vol. 9, No. 4, 1979, p. 8–10.
29. Kennedy, J. P., "Polyurethanes Based on Polyisobutylenes," *CHEMTECH,* Vol. 16, No.11, 1986, pp. 694–697.
30. Baekeland, L. H., *The Journal of Industrial and Engineering Chemistry,* March 1909; *CHEMTECH,* Vol. 6, No. 11, 1979, pp. 40–53.
31. *Chemical Engineering,* Sept. 15, 1975, p. 106.
32. Stinson, S. "Polycyanurates Find Applications, Their Chemistry Remains Puzzling," *Chemical and Engineering News,* Sept. 12, 1994, pp. 30–31.
33. Hall, D. and Allen, E., *Chemistry,* Vol. 45, No. 6, 1972, pp. 6–12.
34. Coran, A. Y., *CHEMTECH,* Vol. 13, No. 2, 1983, p. 106.
35. Hertz, D. L., Jr., "Curing Rubber," *CHEMTECH,* Vol. 16, No. 7, 1986, pp. 444–447.
36. Reisch M. S., "Thermoplastic, Elastomers Bring New Vigor to Rubber Industry," *Chemical and Engineering News,* May 4, 1992, pp. 29–41.
37. Stevens, M. P., *Polymer Chemistry,* Addison Wesley Publishing Co. London, 1975, p. 156.
38. Natta, G. J., *J. Polymer Science,* Vol. 48, 1960, p. 219.

39. British Patent 848,065 to Phillips Petroleum Co., April 16, 1956.
40. Platzer, N., *CHEMTECH,* Vol. 9, No.1, 1979, pp. 16–20.
41. Jelinski, L. W., "NMR of Plastics," *CHEMTECH,* Vol. 16, No. 5, 1986, pp. 312–317.
42. Platzer, N. *CHEMTECH,* Vol. 7, No. 8, 1977, p. 637.
43. "Petrochemical Handbook," *Hydrocarbon Processing,* Vol. 54, No. 11, 1975, p. 194.
44. Dall'Asta, G. *Rubber Chemical Technology,* Vol. 47, 1974, p. 511.
45. Holden, G., *Condensed Encyclopedia of Polymer Science and Engineering*, John Wiley and Sons, 1990, pp. 296–297.
46. *Chemical Industries Newsletter,* Jan.–Mar., 1999, p. 12
47. "Petrochemical Handbook," *Hydrocarbon Processing,* Vol. 56, No. 11, 1977, p. 203.
48. Farrow, G. and Bagley, I., *Texas Research Journal,* Vol. 32, 1962, p. 587.
49. Brown, A. E. and Reinhart, K. A., *Science,* Vol. 173, No. 3994, 1971, p. 290.
50. "Petrochemical Handbook," *Hydrocarbon Processing,* Vol. 78, No. 3, 1999, p. 128.
51. Studiengesellschaft Kohle, German Patent, 1,075,610.
52. Hatch, L. F., *Studies on Petrochemicals,* New York, United Nations, 1966, pp. 511–522.
53. Juba, M. R., " A Review of Mechanistic Considerations and Process Design Parameters for Precipitation Polymerization," in *Polymerization Reactions and Processes,* ACS Symposium Series No. 104, Washington D.C., 1979, pp. 267–279.
54. Riggs, P. R., *Condensed Encyclopedia of Polymer Science and Engineering,* John Wiley and Sons, 1990, pp. 105–108.
55. Oya, N. and Johnson, J. "Direct Measurement of Longitudinal Compressive Strength in Carbon Fibers," *Carbon,* Vol. 37, No. 10, 1999, pp. 1539–1544.
56. Brownstein, E. E, International Seminar on Petrochemicals, October 25–30, 1975, Baghdad, Iraq.

APPENDIX ONE

Conversion Factors

To convert from	To	Multiply by
atmospheres	mm of mercury	760
atmospheres	pounds/sq. inch (psi)	14.696
barrels (oil)	gallons (U.S.)	42
bars	atmospheres	0.98692
≈	mm of mercury (0°C)	750.062
≈	pascal	1×10^5
Btu	calorie	252.15105
≈	joules	1.055×10^3
calories gram (mean/gram)	Btu (mean) pound	1.8
calories	Btu	3.9658×10^{-3}
≈	joules	4.1840
centimeters	angstrom	1×10^8
≈	feet	0.0328
≈	inches	0.3937
≈	meters	0.01
≈	microns	1×10^4
cubic feet (ft^3)	gallons (British)	6.2288
≈	gallons (US)	7.48052
≈	liters	28.317
cubic meters	barrels (US, liquid)	8.3865
≈	cubic feet (ft^3)	35.314445
≈	gallons (US)	264.173
≈	liters	999.973
feet	centimeters	30.48
gallons (U.S.)	cubic feet (ft^3)	0.1336805
≈	liters	3.78543
grams	ounces (avdp.)*	0.0352939
≈	ounces (troy)	0.0321507
grams/sq. centimeter	pounds/sq. foot	2.04817
inches	centimeters	2.540005
kilograms	pounds (avdp.)*	2.20462234
≈	pounds (troy)	2.6792285
liters	gallons (British)	0.219976
≈	gallons (U.S.)	0.2641776
meters	angstroms	1×10^{10}

To convert from	To	Multiply by
≈	inches	39.37
pounds/sq. inch pressure	kilopascal (kPa)	6.8948
pounds (avdp.)*	grams	453.59
≈	ounces (avdp.)*	16
pounds (avdp.)*	ounces (troy)	14.5833
pounds (troy)	ounces (troy)	12
pounds/sq. inch	grams/sq. centimeter	70.307
tons (metric)	kilograms	1000
≈	pounds (avdp.)*	2204.62
tons (short)	pounds	2000
watts (abs)	Btu (mean)/hour	3.41304

avdp. = avoirdupois

Temperature Conversion:

degree Celsius (°C) = (°F – 32) × 5/9
  degree Fahrenheit (°F)   = °C × 9/5 + 32
  degree Kelvin (°K) = °C + 273
  degree Rankine (°R)   = °F + 460

Selected Properties of Hydrogen, Important C_1-C_{10} Paraffins, Methylcyclopentane and Cyclohexane*

Hydrocarbon	Sp.Gr. 20/4°C	Boiling Point °C	Freezing Point °C	Heat of Combustion** K. Cal./mol
		Properties		
Hydrogen	0.08988 g/l[‡] (0.0694)[‡‡]	−252.8	−259.3	68.315
Methane	0.466[−164]	−164	−182	212.79
Ethane	0.572[−100/4] (1.049)[‡‡]	−88.6	−183.3	372.81
Propane	0.5853[−45/4] (1.562)[‡‡]	−42.1	−189.7	530.57
2-Methylpropane (isobutane)	0.5631	−11.1	−159.8	683.4
n-Butane	0.5788	−0.5	−138.4	—
2,2-Dimethylpropane (neopenane)	0.591	9.5	−16.5	—
2-Methylbutane (isopentane)	0.6201	27.8	−843.5	
n-Pentane	0.6262	36.1	−130	845.16
2,2-Dimethylbutane (neohexane)	0.6485	49.7	−99.9	—
2,3-Dimethylbutane	0.6616	58	−128.5	—
2-Methylpentane	0.6532	60.3	−153.7	—
3-Methylpentane	0.6645	63.3		
n-Hexane	0.6603	69	−95	—
n-Heptane	0.6837	98.4	−90.6	1149.9 (liquid)
n-Octane	0.7026	125.7	−56.8	1302.7 (liquid)
n-Nonane	0.7176	150.8	−51	—
n-Decane	0.7300	174.1	−29.7	1610.2 (liquid)

Properties

Methylcyclopentane	0.7486	71.8	−142.4	937.9 (liquid)
Cyclohexane	0.7785	80.7	6.5	936.87 (liquid)

*Handbook of Chemistry and Physics, *70th Ed. CRC Press, Boca Raton, Florida 1989.*

**Heat of combustion is the heat liberated or absorbed when one gram mole of the substance is completely oxidized to liquid water and CO_2 gas at one atmosphere and 20°C or 25°C. (C_1-C_5 hydrocarbons and cyclohexane at 25°C, others at 20°C). The gross heating value in Btu/ft^3 could be calculated as follows:*

Using ethane as an example:

$$CH_3CH_3(g) + 7/2O_2(g) \rightarrow 2CO_2(g) + 3H_2O\ (l)\ \Delta H = -372.81\ Kcal/mol$$

Volume of one mole gas at 25°C, one atm. = 24.45 l (Ideal gas at STP = 22.4 l)

$$Gross\ heating\ value = 372.81\frac{Kcal}{mol.} \times \frac{one\ mol.\ gas}{24.45\ l} \times \frac{3.9658\ btu}{Kcal}$$

$$\times \frac{28.317\ l}{ft^3} = 1722\ Btu/ft^3$$

‡*Density of gas*

‡‡*Sp.gr to air = 1,* Condensed Chemical Dictionary, *10th Ed., revised by Gessner, G. Hawley, Van Norstrand Reinhold Co., New York, 1981.*

Index

About the Authors

Sami Matar, Ph.D., is a retired professor of chemistry at King Fahd University of Petroleum and Minerals, Dharan, Saudi Arabia. He received a B.Sc. from the University of Cairo and a Ph.D. in chemistry from the University of Texas, Austin. Dr. Matar has served as associate member of the board of the Egyptian Petroleum Institute and general manager of the chemical and research laboratories of Suez Oil Processing Co. The author and contributor to many articles and books, Dr. Matar is also a member of the American Chemical Society and Society of Petroleum Engineers.

The late **Lewis F. Hatch, Ph.D.,** was well known and widely respected for his contributions to the fields of chemistry and petrochemical processing. He received his Ph.D. in chemistry from Purdue University and was the author of numerous books and technical publications.